HISTORY

OF THE

INDUCTIVE SCIENCES.

VOLUME I.

HISTORY

OF THE

INDUCTIVE SCIENCES,

FROM

THE EARLIEST TO THE PRESENT TIME.

BY WILLIAM WHEWELL, D.D.,

MASTER OF TRINITY COLLEGE, CAMBRIDGE.

THE THIRD EDITION, WITH ADDITIONS.

IN TWO VOLUMES.

VOLUME I.

NEW YORK:

D. APPLETON AND COMPANY,

346 & 348 BROADWAY.

M DCCC LVIII.

TO SIR JOHN FREDERICK WILLIAM HERSCHEL,

K. G. H.

My dear Herschel,

It is with no common pleasure that I take up my pen to dedicate these volumes to you. They are the result of trains of thought which have often been the subject of our conversation, and of which the origin goes back to the period of our early companionship at the University. And if I had ever wavered in my purpose of combining such reflections and researches into a whole, I should have derived a renewed impulse and increased animation from your delightful Discourse on a kindred subject. For I could not have read it without finding this portion of philosophy invested with a fresh charm; and though I might be well aware that I could not aspire to that large share of popularity which your work so justly gained, I should still have reflected, that something was due to the subject itself, and should have hoped that my own aim was so far similar to yours, that the present work might have a chance of exciting an interest in some of your readers. That it will interest you, I do not at all hesitate to believe.

If you were now in England I should stop here: but when a friend is removed for years to a far distant land, we seem to acquire a right to speak openly of his good qualities. I cannot, therefore, prevail upon myself to lay down my pen without alluding to the affectionate admiration of your moral and social, as well as intellectual excellencies, which springs up in the hearts of your friends, whenever you are thought of. They are much delighted to look upon the halo of deserved fame which plays round your head; but still more, to recollect, as one of them said, that your head is far from being the best part about you.

May your sojourn in the southern hemisphere be as happy and successful as its object is noble and worthy of you; and may your return home be speedy and prosperous, as soon as your purpose is attained.

Ever, my dear Herschel, yours,

W. WHEWELL.

March 22, 1837.

P. S. So I wrote nearly ten years ago, when you were at the Cape of Good Hope, employed in your great task of making a complete standard survey of the nebulæ and double stars visible to man. Now that you are, as I trust, in a few weeks about to put the crowning stone upon your edifice by the publication of your "Observations in the Southern Hemisphere," I cannot refrain from congratulating you upon having had your life ennobled by the conception and happy execution of so great a design, and once more offering you my wishes that you may long enjoy the glory you have so well won.

W. W.

TRINITY COLLEGE, Nov. 22, 1846.

PREFACE

TO THE THIRD EDITION.

IN the Prefaces to the previous Editions of this work, several remarks were made which it is not necessary now to repeat to the same extent. That a History of the Sciences, executed as this is, has some value in the eyes of the Public, is sufficiently proved by the circulation which it has obtained. I am still able to say that I have seen no objection urged against the plan of the work, and scarcely any against the details. The attempt to throw the history of each science into EPOCHS at which some great and cardinal discovery was made, and to arrange the subordinate events of each history as belonging to the PRELUDES and the SEQUELS of such Epochs, appears to be assented to, as conveniently and fairly exhibiting the progress of scientific truth. Such a view being assumed, as it was a constant light and guide to the writer in his task, so will it also, I think, make the view of the reader far more clear and comprehensive than it could otherwise be. With regard to the manner in which this plan has been carried into effect with reference to particular writers and their researches, as I have said, I have seen scarcely any objection made. I was aware, as I stated at the outset, of the difficulty and delicacy of the office which I had undertaken; but I had various considerations to encourage me to go through it; and I had a trust, which I have as yet seen nothing to disturb, that I should be able to speak impartially of the great scientific men of all ages, even of our own.

I have already said, in the Introduction, that the work aimed at being, not merely a narration of the facts in the history of Science, but a basis for the Philosophy of Science. It seemed to me that our study of the modes of discovering truth ought to be based upon a survey of the truths which have been discovered. This maxim, so stated, seems sufficiently self-evident; yet it has, even up to the present time, been very rarely acted on. Those who discourse concerning the nature of Truth and the mode of its discovery, still, commonly, make for themselves examples of truths, which for the most part are utterly frivolous and unsubstantial (as in most Treatises on Logic); or else they dig up, over and over, the narrow and special field of mathematical truth, which certainly cannot, of itself, exemplify the general mode by which man has attained to the vast body of certain truth which he now possesses.

Yet it must not be denied that the Ideas which form the basis of Mathematical Truth are concerned in the formation of Scientific Truth in general; and discussions concerning these Ideas are by no means necessarily barren of advantage. But it must be borne in mind that, besides these Ideas, there are also others, which no less lie at the root of Scientific Truth; and concerning which there have been, at various periods, discussions which have had an important bearing on the progress of Scientific Truth;—such as discussions concerning the nature and necessary attributes of Matter, of Force, of Atoms, of Mediums, of Kinds, of Organization. The controversies which have taken place concerning these have an important place in the history of Natural Science in its most extended sense. Yet it appeared convenient to carry on the history of Science, so far as it depends on Observation, in a line separate from these discussions concerning Ideas. The account of these discussions and the consequent controversies, therefore, though it be thoroughly historical, and, as appears to me, a very curious and interesting history, is reserved for the other work, the *Philosophy of the Inductive Sciences*. Such a history has, in truth, its natural place in the Philosophy of Science; for the Philosophy of Science at the present day must contain the result and summing up of all the truth which has been disentangled from error and confusion during these past controversies.

I have made a few Additions to the present Edition; partly, with a view of bringing up the history, at least of some of the Sciences, to the present time,—so far as those larger features of the History of Science are concerned, with which alone I have here to deal,—and partly also, especially in the First Volume, in order to rectify and enlarge some of the earlier portions of the history. Several works which have recently appeared suggested reconsideration of various points; and I hoped that my readers might be interested in the reflections so suggested.

I will add a few sentences from the Preface to the First Edition.

"As will easily be supposed, I have borrowed largely from other writers, both of the histories of special sciences and of philosophy in general.[1] I have done this without

[1] Among these, I may mention as works to which I have peculiar obligations, Tennemann's Geschichte der Philosophie; Degerando's Histoire Comparée des Systèmes de Philosophie; Montucla's Histoire des Mathématiques, with Delalande's continuation of it; Delambre's Astronomie Ancienne, Astronomie du Moyen Age, Astronomie Moderne, and Astronomie du Dix-huitième Siècle; Bailly's Histoire d'Astronomie Ancienne, and Histoire d'Astronomie Moderne; Voiron's Histoire

scruple, since the novelty of my work was intended to consist, not in its superiority as a collection of facts, but in the point of view in which the facts were placed. I have, however, in all cases, given references to my authorities, and there are very few instances in which I have not verified the references of previous historians, and studied the original authors. According to the plan which I have pursued, the history of each science forms a whole in itself, divided into distinct but connected members, by the *Epochs* of its successive advances. If I have satisfied the competent judges in each science by my selection of such epochs, the scheme of the work must be of permanent value, however imperfect may be the execution of any of its portions.

"With all these grounds of hope, it is still impossible not to see that such an undertaking is, in no small degree, arduous, and its event obscure. But all who venture upon such tasks must gather trust and encouragement from reflections like those by which their great forerunner prepared himself for his endeavors;—by recollecting that they are aiming to advance the best interests and privileges of man; and that they may expect all the best and wisest of men to join them in their aspirations and to aid them in their labors.

"'Concerning ourselves we speak not; but as touching the matter which we have in hand, this we ask;—that men deem it not to be the setting up of an Opinion, but the performing of a Work; and that they receive this as a certainty—that we are not laying the foundations of any sect or doctrine, but of the profit and dignity of mankind:—Fur-

---

d'Astronomie (published as a continuation of Bailly), Fischer's Geschichte der Physik, Gmelin's Geschichte der Chemie, Thomson's History of Chemistry, Sprengel's History of Medicine, his History of Botany, and in all branches of Natural History and Physiology, Cuvier's works; in their historical, as in all other portions, most admirable and instructive.

thermore, that being well disposed to what shall advantage themselves, and putting off factions and prejudices, they take common counsel with us, to the end that being by these our aids and appliances freed and defended from wanderings and impediments, they may lend their hands also to the labors which remain to be performed:—And yet, further, that they be of good hope; neither feign and imagine to themselves this our Reform as something of infinite dimension and beyond the grasp of mortal man, when, in truth, it is, of infinite error, the end and true limit; and is by no means unmindful of the condition of mortality and humanity, not confiding that such a thing can be carried to its perfect close in the space of one single day, but assigning it as a task to a succession of generations.'—BACON—INSTAURATIO MAGNA, *Præf. ad fin.*

" 'If there be any man who has it at heart, not merely to take his stand on what has already been discovered, but to profit by that, and to go on to something beyond;—not to conquer an adversary by disputing, but to conquer nature by working;—not to opine probably and prettily, but to know certainly and demonstrably;—let such, as being true sons of nature (if they will consent to do so), join themselves to us; so that, leaving the porch of nature which endless multitudes have so long trod, we may at last open a way to the inner courts. And that we may mark the two ways, that old one, and our new one, by familiar names, we have been wont to call the one the *Anticipation of the Mind*, the other, the *Interpretation of Nature*.'—INST. MAG. *Præf. ad Part.* ii.

CONTENTS

OF THE FIRST VOLUME.

BOOK I.

HISTORY OF THE GREEK SCHOOL PHILOSOPHY, WITH REFERENCE TO PHYSICAL SCIENCE.

BOOK II.

HISTORY OF THE PHYSICAL SCIENCES IN ANCIENT GREECE.

BOOK III.

HISTORY OF GREEK ASTRONOMY.

CHAPTER III.—INDUCTIVE EPOCH OF HIPPARCHUS.

CHAPTER IV.—SEQUEL TO THE INDUCTIVE EPOCH OF HIPPARCHUS.

BOOK IV.

HISTORY OF PHYSICAL SCIENCE IN THE MIDDLE AGES.

CHAPTER I.—ON THE INDISTINCTNESS OF IDEAS OF THE MIDDLE AGES.

BOOK V.

HISTORY OF FORMAL ASTRONOMY AFTER THE STATIONARY PERIOD.

CHAPTER III.—SEQUEL TO COPERNICUS. THE RECEPTION AND DEVELOPMENT OF THE COPERNICAN THEORY.

CHAPTER IV.—INDUCTIVE EPOCH OF KEPLER.

CHAPTER V.—SEQUEL TO THE EPOCH OF KEPLER. RECEPTION, VERIFICATION, AND EXTENSION OF THE ELLIPTICAL THEORY.

THE MECHANICAL SCIENCES.

BOOK VI.

HISTORY OF MECHANICS, INCLUDING FLUID MECHANICS.

CHAPTER I.—PRELUDE TO THE EPOCH OF GALILEO.

Chapter II.—Inductive Epoch of Galileo.—Discovery of the Laws of Motion in Simple Cases.

BOOK VII.

HISTORY OF PHYSICAL ASTRONOMY.

Chapter VI.—The Instruments and Aids of Astronomy during the Newtonian Period.

ADDITIONS TO THE THIRD EDITION.

Book I.—The Greek School Philosophy.

The Greek Schools.

Failure of the Greek Physical Philosophy.

Book II.—The Physical Sciences in Ancient Greece.

Book III.—The Greek Astronomy.

Earliest Stages of Astronomy.

Book IV.—Physical Science in the Middle Ages.

Progress in the Middle Ages.

Book V.—Formal Astronomy.

Prelude to Copernicus.

The Copernican Theory.

Sequel to Copernicus.

Book VI.—Mechanics.

Principles and Problems.

Book VII.—Physical Astronomy.

Prelude to Newton.

The Principia.

Verification and Completion of the Newtonian Theory.

Instruments.

INDEX OF PROPER NAMES.

The letters *a*, *b*, indicate vol. I., vol. II., respectively.

INDEX OF TECHNICAL TERMS.

A HISTORY OF THE INDUCTIVE SCIENCES.

INTRODUCTION.

"A just story of learning, containing the antiquities and originals of KNOWLEDGES, and their sects; their inventions, their diverse administrations and managings; their flourishings, their oppositions, decays, depressions, oblivions, removes; with the causes and occasions of them, and all other events concerning learning, throughout all ages of the world; I may truly affirm to be wanting.

"The use and end of which work I do not so much design for curiosity, or satisfaction of those that are the lovers of learning: but chiefly for a more serious and grave purpose; which is this, in few words—that it will make learned men more wise in the use and administration of learning."

BACON, *Advancement of Learning*, book ii.

INTRODUCTION.

IT is my purpose to write the History of some of the most important of the Physical Sciences, from the earliest to the most recent periods. I shall thus have to trace some of the most remarkable branches of human knowledge, from their first germ to their growth into a vast and varied assemblage of undisputed truths; from the acute, but fruitless, essays of the early Greek Philosophy, to the comprehensive systems, and demonstrated generalizations, which compose such sciences as the Mechanics, Astronomy, and Chemistry, of modern times.

The completeness of historical view which belongs to such a design, consists, not in accumulating all the details of the cultivation of each science, but in marking the larger features of its formation. The historian must endeavor to point out how each of the important advances was made, by which the sciences have reached their present position; and when and by whom each of the valuable truths was obtained, of which the aggregate now constitutes a costly treasure.

Such a task, if fitly executed, must have a well-founded interest for all those who look at the existing condition of human knowledge with complacency and admiration. The present generation finds itself the heir of a vast patrimony of science; and it must needs concern us to know the steps by which these possessions were acquired, and the documents by which they are secured to us and our heirs forever. Our species, from the time of its creation, has been travelling onwards in pursuit of truth; and now that we have reached a lofty and commanding position, with the broad light of day around us, it must be grateful to look back on the line of our vast progress;—to review the journey, begun in early twilight amid primeval wilds; for a long time continued with slow advance and obscure prospects; and gradually and in later days followed along more open and lightsome paths, in a wide and fertile region. The historian of science, from early periods to the present times, may hope for favor on the score of the mere subject of his narrative, and in virtue of the curiosity which the men of the present day may naturally feel respecting the events and persons of his story.

But such a survey may possess also an interest of another kind; it may be instructive as well as agreeable; it may bring before the reader the present form and extent, the future hopes and prospects of science, as well as its past progress. The eminence on which we stand may enable us to see the land of promise, as well as the wilderness through which we have passed. The examination of the steps by which our ancestors acquired our intellectual estate, may make us acquainted with our expectations as well as our possessions;—may not only remind us of what we have, but may teach us how to improve and increase our store. It will be universally expected that a History of Inductive Science should point out to us a philosophical distribution of the existing body of knowledge, and afford us some indication of the most promising mode of directing our future efforts to add to its extent and completeness.

To deduce such lessons from the past history of human knowledge, was the intention which originally gave rise to the present work. Nor is this portion of the design in any measure abandoned; but its execution, if it take place, must be attempted in a separate and future treatise, *On the Philosophy of the Inductive Sciences.* An essay of this kind may, I trust, from the progress already made in it, be laid before the public at no long interval after the present history.[1]

Though, therefore, many of the principles and maxims of such a work will disclose themselves with more or less of distinctness in the course of the history on which we are about to enter, the systematic and complete exposition of such principles must be reserved for this other treatise. My attempts and reflections have led me to the opinion, that justice cannot be done to the subject without such a division of it.

To this future work, then, I must refer the reader who is disposed to require, at the outset, a precise explanation of the terms which occur in my title. It is not possible, without entering into this philosophy, to explain adequately how science which is INDUCTIVE differs from that which is not so; or why some portions of *knowledge* may properly be selected from the general mass and termed SCIENCE. It will be sufficient at present to say, that the sciences of which we have

[1] The *Philosophy of the Inductive Sciences* was published shortly after the present work.

here to treat, are those which are commonly known as the *Physical Sciences;* and that by *Induction* is to be understood that process of collecting general truths from the examination of particular facts, by which such sciences have been formed.

There are, however, two or three remarks, of which the application will occur so frequently, and will tend so much to give us a clearer view of some of the subjects which occur in our history, that I will state them now in a brief and general manner.

Facts and Ideas.[2]—In the first place then, I remark, that, to the formation of science, two things are requisite;—Facts and Ideas; observation of Things without, and an inward effort of Thought; or, in other words, Sense and Reason. Neither of these elements, by itself, can constitute substantial general knowledge. The impressions of sense, unconnected by some rational and speculative principle, can only end in a practical acquaintance with individual objects; the operations of the rational faculties, on the other hand, if allowed to go on without a constant reference to external things, can lead only to empty abstraction and barren ingenuity. Real speculative knowledge demands the combination of the two ingredients;—right reason, and facts to reason upon. It has been well said, that true knowledge is the interpretation of nature; and therefore it requires both the interpreting mind, and nature for its subject; both the document, and the ingenuity to read it aright. Thus invention, acuteness, and connection of thought, are necessary on the one hand, for the progress of philosophical knowledge; and on the other hand, the precise and steady application of these faculties to facts well known and clearly conceived. It is easy to point out instances in which science has failed to advance, in consequence of the absence of one or other of these requisites; indeed, by far the greater part of the course of the world, the history of most times and most countries, exhibits a condition thus stationary with respect to knowledge. The facts, the impressions on the senses, on which the first successful attempts at physical knowledge proceeded, were as well known long before the time when they were thus turned to account, as at that period. The motions of the stars, and the effects of weight, were familiar to man before the rise of the Greek Astronomy and Mechanics: but the "diviner mind" was still absent; the act of thought had not been exerted, by which these facts were bound together under the form of laws and principles. And even at

---

[2] For the *Antithesis of Facts and Ideas*, see the *Philosophy*, book i. ch. 1, 2, 4, 5.

this day, the tribes of uncivilized and half-civilized man, over the whole face of the earth, have before their eyes a vast body of facts, of exactly the same nature as those with which Europe has built the stately fabric of her physical philosophy; but, in almost every other part of the earth, the process of the intellect by which these facts become science, is unknown. The scientific faculty does not work. The scattered stones are there, but the builder's hand is wanting. And again, we have no lack of proof that mere activity of thought is equally inefficient in producing real knowledge. Almost the whole of the career of the Greek schools of philosophy; of the schoolmen of Europe in the middle ages; of the Arabian and Indian philosophers; shows us that we may have extreme ingenuity and subtlety, invention and connection, demonstration and method; and yet that out of these germs, no physical science may be developed. We may obtain, by such means, Logic and Metaphysics, and even Geometry and Algebra; but out of such materials we shall never form Mechanics and Optics, Chemistry and Physiology. How impossible the formation of these sciences is without a constant and careful reference to observation and experiment;—how rapid and prosperous their progress may be when they draw from such sources the materials on which the mind of the philosopher employs itself;—the history of those branches of knowledge for the last three hundred years abundantly teaches us.

Accordingly, the existence of clear Ideas applied to distinct Facts will be discernible in the History of Science, whenever any marked advance takes place. And, in tracing the progress of the various provinces of knowledge which come under our survey, it will be important for us to see that, at all such epochs, such a combination has occurred; that whenever any material step in general knowledge has been made, —whenever any philosophical discovery arrests our attention,—some man or men come before us, who have possessed, in an eminent degree, a clearness of the ideas which belong to the subject in question, and who have applied such ideas in a vigorous and distinct manner to ascertained facts and exact observations. We shall never proceed through any considerable range of our narrative, without having occasion to remind the reader of this reflection.

Successive Steps in Science.[3]—But there is another remark which we must also make. Such sciences as we have here to do with are,

[3] Concerning *Successive Generalizations in Science*, see the *Philosophy*, book i. ch. 2, sect. 11.

commonly, not formed by a single act;—they are not completed by the discovery of one great principle. On the contrary, they consist in a long-continued advance; a series of changes; a repeated progress from one principle to another, different and often apparently contradictory. Now, it is important to remember that this contradiction is apparent only. The principles which constituted the triumph of the preceding stages of the science, may appear to be subverted and ejected by the later discoveries, but in fact they are (so far as they were true) taken up in the subsequent doctrines and included in them. They continue to be an essential part of the science. The earlier truths are not expelled but absorbed, not contradicted but extended; and the history of each science, which may thus appear like a succession of revolutions, is, in reality, a series of developments. In the intellectual, as in the material world,

> Omnia mutantur nil interit . . . . .
> Nec manet ut fuerat nec formas servat easdem,
> Sed tamen ipsa eadem est.
>
> All changes, naught is lost; the forms are changed,
> And that which has been is not what it was,
> Yet that which has been is.

Nothing which was done was useless or unessential, though it ceases to be conspicuous and primary.

Thus the final form of each science contains the substance of each of its preceding modifications; and all that was at any antecedent period discovered and established, ministers to the ultimate development of its proper branch of knowledge. Such previous doctrines may require to be made precise and definite, to have their superfluous and arbitrary portions expunged, to be expressed in new language, to be taken up into the body of science by various processes;—but they do not on such accounts cease to be true doctrines, or to form a portion of the essential constituents of our knowledge.

Terms record Discoveries.[4]—The modes in which the earlier truths of science are preserved in its later forms, are indeed various. From being asserted at first as strange discoveries, such truths come at last to be implied as almost self-evident axioms. They are recorded by some familiar maxim, or perhaps by some new word or phrase, which becomes part of the current language of the philosophical world; and thus asserts a principle, while it appears merely to indicate a transient

---

[4] Concerning *Technical Terms*, see *Philosophy*, book i. ch. 3.

notion;—preserves as well as expresses a truth;—and, like a medal of gold, is a treasure as well as a token. We shall frequently have to notice the manner in which great discoveries thus stamp their impress upon the terms of a science; and, like great political revolutions, are recorded by the change of the current coin which has accompanied them.

Generalization.—The great changes which thus take place in the history of science, the revolutions of the intellectual world, have, as a usual and leading character, this, that they are steps of *generalization;* transitions from particular truths to others of a wider extent, in which the former are included. This progress of knowledge, from individual facts to universal laws,—from particular propositions to general ones, —and from these to others still more general, with reference to which the former generalizations are particular,—is so far familiar to men's minds, that, without here entering into further explanation, its nature will be understood sufficiently to prepare the reader to recognize the exemplifications of such a process, which he will find at every step of our advance.

Inductive Epochs; Preludes; Sequels.—In our history, it is the *progress* of knowledge only which we have to attend to. This is the main action of our drama; and all the events which do not bear upon this, though they may relate to the cultivation and the cultivators of philosophy, are not a necessary part of our theme. Our narrative will therefore consist mainly of successive steps of generalization, such as have just been mentioned. But among these, we shall find some of eminent and decisive importance, which have more peculiarly influenced the fortunes of physical philosophy, and to which we may consider the rest as subordinate and auxiliary. These primary movements, when the Inductive process, by which science is formed, has been exercised in a more energetic and powerful manner, may be distinguished as the *Inductive Epochs* of scientific history; and they deserve our more express and pointed notice. They are, for the most part, marked by the great discoveries and the great philosophical names which all civilized nations have agreed in admiring. But, when we examine more clearly the history of such discoveries, we find that these epochs have not occurred suddenly and without preparation. They have been preceded by a period, which we may call their *Prelude*, during which the ideas and facts on which they turned were called into action; —were gradually evolved into clearness and connection, permanency and certainty; till at last the discovery which marks the epoch, seized and fixed forever the truth which had till then been obscurely and doubtfully discerned. And again, when this step has been made by the principal discoverers, there may generally be observed another period, which we may call the *Sequel* of the Epoch, during which the discovery has acquired a more perfect certainty and a more complete development among the leaders of the advance; has been diffused to the wider throng of the secondary cultivators of such knowledge, and traced into its distant consequences. This is a work, always of time and labor, often of difficulty and conflict. To distribute the History of science into such Epochs, with their Preludes and Sequels, if successfully attempted, must needs make the series and connections of its occurrences more distinct and intelligible. Such periods form resting-places, where we pause till the dust of the confused march is laid, and the prospect of the path is clear.

Inductive Charts.[5]—Since the advance of science consists in collecting by induction true general laws from particular facts, and in combining several such laws into one higher generalization, in which they still retain their truth; we might form a Chart, or Table, of the progress of each science, by setting down the particular facts which have thus been combined, so as to form general truths, and by marking the further union of these general truths into others more comprehensive. The Table of the progress of any science would thus resemble the Map of a River, in which the waters from separate sources unite and make rivulets, which again meet with rivulets from other fountains, and thus go on forming by their junction trunks of a higher and higher order. The representation of the state of a science in this form, would necessarily exhibit all the principal doctrines of the science; for each general truth contains the particular truths from which it was derived, and may be followed backwards till we have these before us in their separate state. And the last and most advanced generalization would have, in such a scheme, its proper place and the evidence of its validity. Hence such an *Inductive Table* of each science would afford a criterion of the correctness of our distribution of the inductive Epochs, by its coincidence with the views of the best judges, as to the substantial contents of the science in question. By forming, therefore, such Inductive Tables of the principal sciences of which I have here to speak, and by regulating by these tables, my views of the history of the sciences, I conceive that I have secured the distribution of my his-

---

[5] Inductive charts of the History of Astronomy and of Optics, such as are here referred to, are given in the *Philosophy*, book xi. ch. 6.

tory from material error; for no merely arbitrary division of the events could satisfy such conditions. But though I have constructed such charts to direct the course of the present history, I shall not insert them in the work, reserving them for the illustration of the philosophy of the subject; for to this they more properly belong, being a part of the *Logic of Induction*.

Stationary Periods.—By the lines of such maps the real advance of science is depicted, and nothing else. But there are several occurrences of other kinds, too interesting and too instructive to be altogether omitted. In order to understand the conditions of the progress of knowledge, we must attend, in some measure, to the failures as well as the successes by which such attempts have been attended. When we reflect during how small a portion of the whole history of human speculations, science has really been, in any marked degree, progressive, we must needs feel some curiosity to know what was doing in these *stationary* periods; what field could be found which admitted of so wide a deviation, or at least so protracted a wandering. It is highly necessary to our purpose, to describe the baffled enterprises as well as the achievements of human speculation.

Deduction.—During a great part of such stationary periods, we shall find that the process which we have spoken of as essential to the formation of real science, the conjunction of clear Ideas with distinct Facts, was interrupted; and, in such cases, men dealt with ideas alone. They employed themselves in reasoning from principles, and they arranged, and classified, and analyzed their ideas, so as to make their reasonings satisfy the requisitions of our rational faculties. This process of drawing conclusions from our principles, by rigorous and unimpeachable trains of demonstration, is termed *Deduction*. In its due place, it is a highly important part of every science; but it has no value when the fundamental principles, on which the whole of the demonstration rests, have not first been obtained by the induction of facts, so as to supply the materials of substantial truth. Without such materials, a series of demonstrations resembles physical science only as a shadow resembles a real object. To give a real significance to our propositions, Induction must provide what Deduction cannot supply. From a pictured hook we can hang only a pictured chain.

Distinction of common Notions and Scientific Ideas.[6]—When the

---

[6] Scientific Ideas depend upon certain *Fundamental Ideas*, which are enumerated in the *Philosophy*, book i. ch. 8.

notions with which men are conversant in the common course of practical life, which give meaning to their familiar language, and employment to their hourly thoughts, are compared with the Ideas on which exact science is founded, we find that the two classes of intellectual operations have much that is common and much that is different. Without here attempting fully to explain this relation (which, indeed, is one of the hardest problems of our philosophy), we may observe that they have this in common, that both are acquired by acts of the mind exercised in connecting external impressions, and may be employed in conducting a train of reasoning; or, speaking loosely (for we cannot here pursue the subject so as to arrive at philosophical exactness), we may say, that all notions and ideas are obtained by an *inductive*, and may be used in a *deductive* process. But scientific Ideas and common Notions differ in this, that the former are precise and stable, the latter vague and variable; the former are possessed with clear insight, and employed in a sense rigorously limited, and always identically the same; the latter have grown up in the mind from a thousand dim and diverse suggestions, and the obscurity and incongruity which belong to their origin hang about all their applications. Scientific Ideas can often be adequately exhibited for all the purposes of reasoning, by means of Definitions and Axioms; all attempts to reason by means of Definitions from common Notions, lead to empty forms or entire confusion.

Such common Notions are sufficient for the common practical conduct of human life: but man is not a practical creature merely; he has within him a *speculative* tendency, a pleasure in the contemplation of ideal relations, a love of knowledge *as* knowledge. It is this speculative tendency which brings to light the difference of common Notions and scientific Ideas, of which we have spoken. The mind analyzes such Notions, reasons upon them, combines and connects them; for it feels assured that intellectual things ought to be able to bear such handling. Even practical knowledge, we see clearly, is not possible without the use of the reason; and the speculative reason is only the reason satisfying itself of its own consistency. The speculative faculty cannot be controlled from acting. The mind cannot but claim a right to speculate concerning all its own acts and creations; yet, when it exercises this right upon its common practical notions, we find that it runs into barren abstractions and ever-recurring cycles of subtlety. Such Notions are like waters naturally stagnant; however much we urge and agitate them, they only revolve in stationary whirlpools. But the mind is capable of acquiring scientific Ideas, which are better fitted to undergo discussion and impulsion. When our speculations are duly fed from the springheads of Observation, and frequently drawn off into the region of Applied Science, we may have a living stream of consistent and progressive knowledge. That science may be both real as to its import, and logical as to its form, the examples of many existing sciences sufficiently prove.

School Philosophy.—So long, however, as attempts are made to form sciences, without such a verification and realization of their fundamental ideas, there is, in the natural series of speculation, no self-correcting principle. A philosophy constructed on notions obscure, vague, and unsubstantial, and held in spite of the want of correspondence between its doctrines and the actual train of physical events, may long subsist, and occupy men's minds. Such a philosophy must depend for its permanence upon the pleasure which men feel in tracing the operations of their own and other men's minds, and in reducing them to logical consistency and systematical arrangement.

In these cases the main subjects of attention are not external objects, but speculations previously delivered; the object is not to interpret nature, but man's mind. The opinions of the Masters are the facts which the Disciples endeavor to reduce to unity, or to follow into consequences. A series of speculators who pursue such a course, may properly be termed a *School*, and their philosophy a *School Philosophy;* whether their agreement in such a mode of seeking knowledge arise from personal communication and tradition, or be merely the result of a community of intellectual character and propensity. The two great periods of School Philosophy (it will be recollected that we are here directing our attention mainly to physical science) were that of the Greeks and that of the Middle Ages;—the period of the first waking of science, and that of its midday slumber.

What has been said thus briefly and imperfectly, would require great detail and much explanation, to give it its full significance and authority. But it seemed proper to state so much in this place, in order to render more intelligible and more instructive, at the first aspect, the view of the attempted or effected progress of science.

It is, perhaps, a disadvantage inevitably attending an undertaking like the present, that it must set out with statements so abstract; and must present them without their adequate development and proof. Such an Introduction, both in its character and its scale of execution, may be compared to the geographical sketch of a country, with which the historian of its fortunes often begins his narration. So much of Metaphysics is as necessary to us as such a portion of Geography is to the Historian of an Empire; and what has hitherto been said, is intended as a slight outline of the Geography of that Intellectual World, of which we have here to study the History.

The name which we have given to this History—A HISTORY OF THE INDUCTIVE SCIENCES—has the fault of seeming to exclude from the rank of Inductive Sciences those which are not included in the History; as Ethnology and Glossology, Political Economy, Psychology. This exclusion I by no means wish to imply; but I could find no other way of compendiously describing my subject, which was intended to comprehend those Sciences in which, by the observation of facts and the use of reason, systems of doctrine have been established which are universally received as truths among thoughtful men; and which may therefore be studied as examples of the manner in which truth is to be discovered. Perhaps a more exact description of the work would have been, *A History of the principal Sciences hitherto established by Induction.* I may add that I do not include in the phrase "Inductive Sciences," the branches of Pure Mathematics (Geometry, Arithmetic, Algebra, and the like), because, as I have elsewhere stated (*Phil. Ind. Sc.*, book ii. c. 1), these are not *Inductive* but *Deductive* Sciences. They do not infer true theories from observed facts, and more general from more limited laws: but they trace the conditions of all theory, the properties of space and number; and deduce results from ideas without the aid of experience. The History of these Sciences is briefly given in Chapters 13 and 14 of the Second Book of the *Philosophy* just referred to.

I may further add that the other work to which I refer, *the Philosophy of the Inductive Sciences*, is in a great measure historical, no less than the present *History.* That work contains the history of the Sciences so far as it depends on *Ideas;* the present work contains the history so far as it depends upon *Observation.* The two works resulted simultaneously from the same examination of the principal writers on science in all ages, and may serve to supplement each other.

BOOK I.

HISTORY

OF THE

GREEK SCHOOL PHILOSOPHY,

WITH REFERENCE TO

PHYSICAL SCIENCE.

Τίς γὰρ ἀρχὰ δέξατο ναυτιλίας;
Τίς δὲ κίνδυνος κρατεροῖς ἀδάμαν-
τος δῆσεν ἅλοις;
. . . . Ἐπεὶ δ' ἐμβόλου
Κρέμασαν ἀγκύρας ὕπερθεν
Χρυσέαν χείρεσσι λαβὼν φιάλαν
Ἄρχος ἐν πρύμνᾳ πατέρ Οὐρανιδᾶν
Ἐγχεικέραυνον Ζῆνα, καὶ ὠκυπόρους
Κυμάτων ῥίπας, ἀνέμων τ' ἐκάλει,
Νύκτας τε, καὶ πόντου κελεύθους,
Ἄματά τ' εὔφρονα, καὶ
Φιλίαν νόστοιο μοῖραν.

PINDAR. *Pyth.* iv. 124, 349.

Whence came their voyage? them what peril held
With adamantine rivets firmly bound?

* * * * * *

But soon as on the vessel's bow
The anchor was hung up,
Then took the Leader on the prow
In hands a golden cup,
And on great Father Jove did call,
And on the Winds and Waters all,
Swept by the hurrying blast;
And on the Nights, and Ocean Ways,
And on the fair auspicious Days,
And loved return at last.

BOOK I.

HISTORY OF THE GREEK SCHOOL PHILOSOPHY, WITH REFERENCE TO PHYSICAL SCIENCE.

CHAPTER I.

PRELUDE TO THE GREEK SCHOOL PHILOSOPHY.

Sect. 1.—*First Attempts of the Speculative Faculty in Physical Inquiries.*

AT an early period of history there appeared in men a propensity to pursue speculative inquiries concerning the various parts and properties of the material world. What they saw excited them to meditate, to conjecture, and to reason: they endeavored to account for natural events, to trace their causes, to reduce them to their principles. This habit of mind, or, at least that modification of it which we have here to consider, seems to have been first unfolded among the Greeks. And during that obscure introductory interval which elapsed while the speculative tendencies of men were as yet hardly disentangled from the practical, those who were most eminent in such inquiries were distinguished by the same term of praise which is applied to sagacity in matters of action, and were called *wise* men—σοφοί. But when it came to be clearly felt by such persons that their endeavors were suggested by the love of knowledge, a motive different from the motives which lead to the wisdom of active life, a name was adopted of a more appropriate, as well as of a more modest signification, and they were termed *philosophers*, or lovers of wisdom. This appellation is said[1] to have been first assumed by Pythagoras. Yet he, in Herodotus, instead of having this title, is called a powerful *sophist*—Ἑλλήνων οὐ τῷ ἀσθενεστάτῳ σοφιστῇ Πυθαγόρῃ;[2] the historian using this word, as it would seem, without intending to imply that misuse of reason which the term afterwards came to denote. The historians of literature

[1] Cic. Tusc. v. 3.

[2] Herod. iv. 95.

placed Pythagoras at the origin of the Italic School, one of the two main lines of succession of the early Greek philosophers: but the other, the Ionic School, which more peculiarly demands our attention, in consequence of its character and subsequent progress, is deduced from Thales, who preceded the age of *Philosophy*, and was one of the *sophi*, or "wise men of Greece."

The Ionic School was succeeded in Greece by several others; and the subjects which occupied the attention of these schools became very extensive. In fact, the first attempts were, to form systems which should explain the laws and causes of the material universe; and to these were soon added all the great questions which our moral condition and faculties suggest. The physical philosophy of these schools is especially deserving of our study, as exhibiting the character and fortunes of the most memorable attempt at universal knowledge which has ever been made. It is highly instructive to trace the principles of this undertaking; for the course pursued was certainly one of the most natural and tempting which can be imagined; the essay was made by a nation unequalled in fine mental endowments, at the period of its greatest activity and vigor; and yet it must be allowed (for, at least so far as physical science is concerned, none will contest this), to have been entirely unsuccessful. We cannot consider otherwise than as an utter failure, an endeavor to discover the causes of things, of which the most complete results are the Aristotelian physical treatises; and which, after reaching the point which these treatises mark, left the human mind to remain stationary, at any rate on all such subjects, for nearly two thousand years.

The early philosophers of Greece entered upon the work of physical speculation in a manner which showed the vigor and confidence of the questioning spirit, as yet untamed by labors and reverses. It was for later ages to learn that man must acquire, slowly and patiently, letter by letter, the alphabet in which nature writes her answers to such inquiries. The first students wished to divine, at a single glance, the whole import of her book. They endeavored to discover the origin and principle of the universe; according to Thales, *water* was the origin of all things, according to Anaximenes, *air;* and Heraclitus considered *fire* as the essential principle of the universe. It has been conjectured, with great plausibility, that this tendency to give to their Philosophy the form of a Cosmogony, was owing to the influence of the poetical Cosmogonies and Theogonies which had been produced and admired at a still earlier age. Indeed, such wide and ambitious doctrines as those which have been mentioned, were better suited to the dim magnificence of poetry, than to the purpose of a philosophy which was to bear the sharp scrutiny of reason. When we speak of the *principles* of things, the term, even now, is very ambiguous and indefinite in its import, but how much more was that the case in the first attempts to use such abstractions! The term which is commonly used in this sense (ἀρχή), signified at first *the beginning;* and in its early philosophical applications implied some obscure mixed reference to the mechanical, chemical, organic, and historical causes of the visible state of things, besides the theological views which at this period were only just beginning to be separated from the physical. Hence we are not to be surprised if the sources from which the opinions of this period appear to be derived are rather vague suggestions and casual analogies, than any reasons which will bear examination. Aristotle conjectures, with considerable probability, that the doctrine of Thales, according to which water was the universal element, resulted from the manifest importance of moisture in the support of animal and vegetable life.[3] But such precarious analyses of these obscure and loose dogmas of early antiquity are of small consequence to our object.

In more limited and more definite examples of inquiry concerning the causes of natural appearances, and in the attempts made to satisfy men's curiosity in such cases, we appear to discern a more genuine prelude to the true spirit of physical inquiry. One of the most remarkable instances of this kind is to be found in the speculations which Herodotus records, relative to the cause of the floods of the Nile. "Concerning the nature of this river," says the father of history,[4] "I was not able to learn any thing, either from the priests or from any one besides, though I questioned them very pressingly. For the Nile is flooded for a hundred days, beginning with the summer solstice; and after this time it diminishes, and is, during the whole winter, very small. And on this head I was not able to obtain any thing satisfactory from any one of the Egyptians, when I asked what is the power by which the Nile is in its nature the reverse of other rivers."

We may see, I think, in the historian's account, that the Grecian mind felt a craving to discover the reasons of things which other nations did not feel. The Egyptians, it appears, had no theory, and felt no want of a theory. Not so the Greeks; they had their reasons to render, though they were not such as satisfied Herodotus. "Some

---

[3] Metaph. i. 3.

[4] Herod. ii. 19.

of the Greeks," he says, "who wish to be considered great philosophers (Ἑλλήνων τινες επισήμοι-βουλόμενοι γενέσθαι σοφίην), have propounded three ways of accounting for these floods. Two of them," he adds, "I do not think worthy of record, except just so far as to mention them." But as these are some of the earliest Greek essays in physical philosophy, it will be worth while, even at this day, to preserve the brief notice he has given of them, and his own reasonings upon the same subject.

"One of these opinions holds that the Etesian winds [which blew from the north] are the cause of these floods, by preventing the Nile from flowing into the sea." Against this the historian reasons very simply and sensibly. "Very often when the Etesian winds do not blow, the Nile is flooded nevertheless. And moreover, if the Etesian winds were the cause, all other rivers, which have their course opposite to these winds, ought to undergo the same changes as the Nile; which the rivers of Syria and Libya so circumstanced do not."

"The next opinion is still more unscientific (ἀνεπιστημονεστέρη), and is, in truth, marvellous for its folly. This holds that the ocean flows all round the earth, and that the Nile comes out of the ocean, and by that means produces its effects." "Now," says the historian, "the man who talks about this ocean-river, goes into the region of fable, where it is not easy to demonstrate that he is wrong. I know of no such river. But I suppose that Homer and some of the earlier poets invented this fiction and introduced it into their poetry."

He then proceeds to a third account, which to a modern reasoner would appear not at all unphilosophical in itself, but which he, nevertheless, rejects in a manner no less decided than the others. "The third opinion, though much the most plausible, is still more wrong than the others; for it asserts an impossibility, namely, that the Nile proceeds from the melting of the snow. Now the Nile flows out of Libya, and through Ethiopia, which are very hot countries, and thus comes into Egypt, which is a colder region. How then can it proceed from snow?" He then offers several other reasons "to show," as he says, "to any one capable of reasoning on such subjects (ἀνδρί γε λογίζεσθαι τοιούτων πέρι οἵῳ τε ἐόντι), that the assertion cannot be true. The winds which blow from the southern regions are hot; the inhabitants are black; the swallows and kites (ἰκτῖνοι) stay in the country the whole year; the cranes fly the colds of Scythia, and seek their warm winter-quarters there; which would not be if it snowed ever so little." He adds another reason, founded apparently upon some limited empirical maxim of weather-wisdom taken from the climate of Greece. "Libya," he said, "has neither rain nor ice, and therefore no snow; *for*, in five days after a fall of snow there must be a fall of rain; so that if it snowed in those regions it must rain too." I need not observe that Herodotus was not aware of the difference between the climate of high mountains and plains in a torrid region; but it is impossible not to be struck both with the activity and the coherency of thought displayed by the Greek mind in this primitive physical inquiry.

But I must not omit the hypothesis which Herodotus himself proposes, after rejecting those which have been already given. It does not appear to me easy to catch his exact meaning, but the statement will still be curious. "If," he says, "one who has condemned opinions previously promulgated may put forward his own opinion concerning so obscure a matter, I will state why it seems to me that the Nile is flooded in summer." This opinion he propounds at first with an oracular brevity, which it is difficult to suppose that he did not intend to be impressive. "In winter the sun is carried by the seasons away from his former course, and goes to the upper parts of Libya. And *there, in short, is the whole account;* for that region to which this divinity (the sun) is nearest, must naturally be most scant of water, and the river-sources of that country must be dried up."

But the lively and garrulous Ionian immediately relaxes from this apparent reserve. "To explain the matter more at length," he proceeds, "it is thus. The sun when he traverses the upper parts of Libya, does what he commonly does in summer;—he *draws* the water to him (ἕλκει ἐπ' ἑωϋτὸν τὸ ὕδωρ), and having thus drawn it, he pushes it to the upper regions (of the air probably), and then the winds take it and disperse it till they dissolve in moisture. And thus the winds which blow from those countries, Libs and Notus, are the most moist of all winds. Now when the winter relaxes and the sun returns to the north, he still draws water from all the rivers, but they are increased by showers and rain torrents so that they are in flood till the summer comes; and then, the rain failing and the sun still drawing them, they become small. But the Nile, not being fed by rains, yet being drawn by the sun, is, alone of all rivers, much more scanty in the winter than in the summer. For in summer it is drawn like all other rivers, but in winter it alone has its supplies shut up. And in this way, I have been led to think the sun is the cause of the occurrence in question." We may remark that the historian here appears to ascribe the inequality of the Nile at different seasons to the influence of the sun upon its springs alone, the other cause of change, the rains, being here excluded; and that, on this supposition, the same relative effects would be produced whether the sun increase the sources in winter by melting the snows, or diminish them in summer by what he calls *drawing* them upwards.

This specimen of the early efforts of the Greeks in physical speculations, appears to me to speak strongly for the opinion that their philosophy on such subjects was the native growth of the Greek mind, and owed nothing to the supposed lore of Egypt and the East; an opinion which has been adopted with regard to the Greek Philosophy in general by the most competent judges on a full survey of the evidence.[5] Indeed, we have no evidence whatever that, at any period, the African or Asiatic nations (with the exception perhaps of the Indians) ever felt this importunate curiosity with regard to the definite application of the idea of cause and effect to visible phenomena; or drew so strong a line between a fabulous legend and a reason rendered; or attempted to ascend to a natural cause by classing together phenomena of the same kind. We may be well excused, therefore, for believing that they could not impart to the Greeks what they themselves did not possess; and so far as our survey goes, physical philosophy has its origin, apparently spontaneous and independent, in the active and acute intellect of Greece.

Sect. 2.—Primitive Mistake in Greek Physical Philosophy.

We now proceed to examine with what success the Greeks followed the track into which they had thus struck. And here we are obliged to confess that they very soon turned aside from the right road to truth, and deviated into a vast field of error, in which they and their successors have wandered almost to the present time. It is not necessary here to inquire why those faculties which appear to be bestowed upon us for the discovery of truth, were permitted by Providence to fail so signally in answering that purpose; whether, like the powers by which we seek our happiness, they involve a responsibility on our part, and may be defeated by rejecting the guidance of a higher faculty; or whether these endowments, though they did not immedi-

[5] Thirlwall, *Hist. Gr.*, ii. 130; and, as there quoted, Ritter, *Geschichte der Philosophie*, i. 159-173.

ately lead man to profound physical knowledge, answered some nobler and better purpose in his constitution and government. The fact undoubtedly was, that the physical philosophy of the Greeks soon became trifling and worthless; and it is proper to point out, as precisely as we can, in what the fundamental mistake consisted.

To explain this, we may in the first place return for a moment to Herodotus's account of the cause of the floods of the Nile.

The reader will probably have observed a remarkable phrase used by Herodotus, in his own explanation of these inundations. He says that the sun *draws*, or attracts, the water; a metaphorical term, obviously intended to denote some more general and abstract conception than that of the visible operation which the word primarily signifies. This abstract notion of "drawing" is, in the historian, as we see, very vague and loose; it might, with equal propriety, be explained to mean what we now understand by mechanical or by chemical attraction, or pressure, or evaporation. And in like manner, all the first attempts to comprehend the operations of nature, led to the introduction of abstract conceptions, often vague, indeed, but not, therefore, unmeaning; such as *motion* and *velocity*, *force* and *pressure*, *impetus* and *momentum* (ῥοπή). And the next step in philosophizing, necessarily was to endeavor to make these vague abstractions more clear and fixed, so that the logical faculty should be able to employ them securely and coherently. But there were two ways of making this attempt; the one, by examining the words only, and the thoughts which they call up; the other, by attending to the facts and things which bring these abstract terms into use. The latter, the method of *real* inquiry, was the way to success; but the Greeks followed the former, the *verbal* or *notional* course, and failed.

If Herodotus, when the notion of the sun's attracting the waters of rivers had entered into his mind, had gone on to instruct himself, by attention to facts, in what manner this notion could be made more definite, while it still remained applicable to all the knowledge which could be obtained, he would have made some progress towards a true solution of his problem. If, for instance, he had tried to ascertain whether this Attraction which the sun exerted upon the waters of rivers, depended on his influence at their fountains only, or was exerted over their whole course, and over waters which were not parts of rivers, he would have been led to reject his hypothesis; for he would have found, by observations sufficiently obvious, that the sun's Attraction, as shown in such cases, is a tendency to lessen all expanded and open collections of moisture, whether flowing from a spring or not; and it would then be seen that this influence, operating on the whole surface of the Nile, must diminish it as well as other rivers, in summer, and therefore could not be the cause of its overflow. He would thus have corrected his first loose conjecture by a real study of nature, and might, in the course of his meditations, have been led to available notions of Evaporation, or other natural actions. And, in like manner, in other cases, the rude attempts at explanation, which the first exercise of the speculative faculty produced, might have been gradually concentrated and refined, so as to fall in, both with the requisitions of reason and the testimony of sense.

But this was not the direction which the Greek speculators took. On the contrary; as soon as they had introduced into their philosophy any abstract and general conceptions, they proceeded to scrutinize these by the internal light of the mind alone, without any longer looking abroad into the world of sense. They took for granted that philosophy must result from the relations of those notions which are involved in the common use of language, and they proceeded to seek their philosophical doctrines by studying such notions. They ought to have reformed and fixed their usual conceptions by Observation; they only analyzed and expanded them by Reflection: they ought to have sought by trial, among the Notions which passed through their minds, some one which admitted of exact application to Facts; they selected arbitrarily, and, consequently, erroneously, the Notions according to which Facts should be assembled and arranged: they ought to have collected clear Fundamental Ideas from the world of things by *inductive* acts of thought; they only derived results by *Deduction* from one or other of their familiar Conceptions.[6]

When this false direction had been extensively adopted by the Greek philosophers, we may treat of it as the method of their *Schools*. Under that title we must give a further account of it.

---

[6] The course by which the Sciences were formed, and which is here referred to as that which the Greeks did *not* follow, is described in detail in the *Philosophy*, book xi., *Of the Construction of Science*.

CHAPTER II.

The Greek School Philosophy.

Sect. 1.—*The general Foundation of the Greek School Philosophy.*

THE physical philosophy of the Greek Schools was formed by looking at the material world through the medium of that common language which men employ to answer the common occasions of life; and by adopting, arbitrarily, as the grounds of comparison of facts, and of inference from them, notions more abstract and large than those with which men are practically familiar, but not less vague and obscure. Such a philosophy, however much it might be systematized, by classifying and analyzing the conceptions which it involves, could not overcome the vices of its fundamental principle. But before speaking of these defects, we must give some indications of its character.

The propensity to seek for principles in the common usages of language may be discerned at a very early period. Thus we have an example of it in a saying which is reported of Thales, the founder of Greek philosophy.[1] When he was asked, "What is the *greatest* thing?" he replied, "*Place;* for all other things are *in* the world, but the world is *in* it." In Aristotle we have the consummation of this mode of speculation. The usual point from which he starts in his inquiries is, that *we say* thus or thus in common language. Thus, when he has to discuss the question, whether there be, in any part of the universe, a Void, or space in which there is nothing, he inquires first in how many senses we say that one thing is *in* another. He enumerates many of these;[2] we say the part is in the whole, as the finger is *in* the hand; again we say, the species is in the genus, as man is included *in* animal; again, the government of Greece is *in* the king; and various other senses are described or exemplified, but of all these *the most proper* is when we say a thing is *in* a vessel, and generally, *in place.* He next examines what *place* is, and comes to this conclusion, that "if about a body there be another body including it, it is in place, and if not, not." A body *moves* when it changes its place; but

---

[1] Plut. *Conv. Sept. Sap.* Diog. Laert. i. 35.

[2] Physic. Ausc. iv. 3.

he adds, that if water be in a vessel, the vessel being at rest, the parts of the water may still move, for they are included by each other; so that while the whole does not change its place, the parts may change their places in a circular order. Proceeding then to the question of a *void*, he, as usual, examines the different senses in which the term is used, and adopts, as the most proper, *place without matter;* with no useful result, as we shall soon see.

Again,[3] in a question concerning mechanical action, he says, "When a man moves a stone by pushing it with a stick, *we say* both that the man moves the stone, and that the stick moves the stone, but the latter *more properly*."

Again, we find the Greek philosophers applying themselves to extract their dogmas from the most general and abstract notions which they could detect; for example,—from the conception of the Universe as One or as Many things. They tried to determine how far we may, or must, combine with these conceptions that of a whole, of parts, of number, of limits, of place, of beginning or end, of full or void, of rest or motion, of cause and effect, and the like. The analysis of such conceptions with such a view, occupies, for instance, almost the whole of Aristotle's *Treatise on the Heavens.*

The Dialogue of Plato, which is entitled *Parmenides*, appears at first as if its object were to show the futility of this method of philosophizing; for the philosopher whose name it bears, is represented as arguing with an Athenian named Aristotle,[4] and, by a process of metaphysical analysis, reducing him at least to this conclusion, "that whether *One* exist, or do not exist, it follows that both it and other things, with reference to themselves and to each other, all and in all respects, both are and are not, both appear and appear not." Yet the method of Plato, so far as concerns truths of that kind with which we are here concerned, was little more efficacious than that of his rival. It consists mainly, as may be seen in several of the dialogues, and especially in the *Timæus*, in the application of notions as loose as those of the Peripatetics; for example, the conceptions of the Good, the Beautiful, the Perfect; and these are rendered still more arbitrary, by assuming an acquaintance with the views of the Creator of the universe. The philosopher is thus led to maxims which agree with those

---

[3] Physic. Ausc. viii. 5.

[4] This Aristotle is not the Stagirite, who was forty-five years younger than Plato, but one of the "thirty tyrants," as they were called.

of the Aristotelians, that there can be no void, that things seek their own place, and the like.[5]

Another mode of reasoning, very widely applied in these attempts, was the doctrine of contrarieties, in which it was assumed, that adjectives or substantives which are in common language, or in some abstract mode of conception, opposed to each other, must point at some fundamental antithesis in nature, which it is important to study. Thus Aristotle[6] says, that the Pythagoreans, from the contrasts which number suggests, collected ten principles,—Limited and Unlimited, Odd and Even, One and Many, Right and Left, Male and Female, Rest and Motion, Straight and Curved, Light and Darkness, Good and Evil, Square and Oblong. We shall see hereafter, that Aristotle himself deduced the doctrine of Four Elements, and other dogmas, by oppositions of the same kind.

The physical speculator of the present day will learn without surprise, that such a mode of discussion as this, led to no truths of real or permanent value. The whole mass of the Greek philosophy, therefore, shrinks into an almost imperceptible compass, when viewed with reference to the progress of physical knowledge. Still the general character of this system, and its fortunes from the time of its founders to the overthrow of their authority, are not without their instruction, and, it may be hoped, not without their interest. I proceed, therefore, to give some account of these doctrines in their most fully developed and permanently received form, that in which they were presented by Aristotle.

Sect. 2.—The Aristotelian Physical Philosophy.

The principal physical treatises of Aristotle are, the eight Books of "Physical Lectures," the four Books "Of the Heavens," the two Books "Of Production and Destruction:" for the Book "Of the World" is now universally acknowledged to be spurious; and the "Meteorologics," though full of physical explanations of natural phenomena, does not exhibit the doctrines and reasonings of the school in so general a form; the same may be said of the "Mechanical Problems." The treatises on the various subjects of Natural History, "On Animals," "On the Parts of Animals," "On Plants," "On Physiognomonics," "On Colors," "On Sound," contain an extraordinary accumu-

---

[5] Timæus, p. 80.

[6] Metaph. l. 5.

lation of facts, and manifest a wonderful power of systematizing; but are not works which expound principles, and therefore do not require to be here considered.

The Physical Lectures are possibly the work concerning which a well-known anecdote is related by Simplicius, a Greek commentator of the sixth century, as well as by Plutarch. It is said, that Alexander the Great wrote to his former tutor to this effect; "You have not done well in publishing these lectures; for how shall we, your pupils, excel other men, if you make that public to all, which we learnt from you?" To this Aristotle is said to have replied: "My Lectures are published and not published; they will be intelligible to those who heard them, and to none besides." This may very easily be a story invented and circulated among those who found the work beyond their comprehension; and it cannot be denied, that to make out the meaning and reasoning of every part, would be a task very laborious and difficult, if not impossible. But we may follow the import of a large portion of the Physical Lectures with sufficient clearness to apprehend the character and principles of the reasoning; and this is what I shall endeavor to do.

The author's introductory statement of his view of the nature of philosophy falls in very closely with what has been said, that he takes his facts and generalizations as they are implied in the structure of language. "We must in all cases proceed," he says, "from what is known to what is unknown." This will not be denied; but we can hardly follow him in his inference. He adds, "We must proceed, therefore, from universal to particular. And something of this," he pursues, "may be seen in language; for names signify things in a general and indefinite manner, as *circle*, and by defining we unfold them into particulars." He illustrates this by saying, "thus children at first call all men *father*, and all women *mother*, but afterwards distinguish."

In accordance with this view, he endeavors to settle several of the great questions concerning the universe, which had been started among subtle and speculative men, by unfolding the meaning of the words and phrases which are applied to the most general notions of things and relations. We have already noticed this method. A few examples will illustrate it further:—Whether there was or was not a *void*, or place without matter, had already been debated among rival sects of philosophers. The antagonist arguments were briefly these:—There must be a void, because a body cannot move into a space except it is empty, and therefore without a void there could be no motion:—and, on the other hand, there is no void, for the interval between bodies are filled with air, and air is something. These opinions had even been supported by reference to experiment. On the one hand, Anaxagorus and his school had shown, that air, when confined, resisted compression, by squeezing a blown bladder, and pressing down an inverted vessel in the water; on the other hand, it was alleged that a vessel full of fine ashes held as much water as if the ashes were not there, which could only be explained by supposing void spaces among the ashes. Aristotle decides that there is no void, on such arguments as this:[7]—In a void there could be no difference of up and down; for as in nothing there are no differences, so there are none in a privation or negation; but a void is merely a privation or negation of matter; therefore, in a void, bodies could not move up and down, which it is in their nature to do. It is easily seen that such a mode of reasoning elevates the familiar forms of language and the intellectual connections of terms, to a supremacy over facts; making truth depend upon whether terms are or are not privative, and whether we say that bodies fall *naturally*. In such a philosophy every new result of observation would be compelled to conform to the usual combinations of phrases, as these had become associated by the modes of apprehension previously familiar.

It is not intended here to intimate that the common modes of apprehension, which are the basis of common language, are limited and casual. They imply, on the contrary, universal and necessary conditions of our perceptions and conceptions; thus all things are necessarily apprehended as existing in Time and Space, and as connected by relations of Cause and Effect; and so far as the Aristotelian philosophy reasons from these assumptions, it has a real foundation, though even in this case the conclusions are often insecure. We have an example of this reasoning in the eighth Book,[8] where he proves that there never was a time in which change and motion did not exist; "For if all things were at rest, the first motion must have been produced by some change in some of these things; that is, there must have been a change before the first change;" and again, "How can *before* and *after* apply when time is not? or how can time be when motion is not? If," he adds, "time is a numeration of motion, and if time be eternal, motion must be eternal." But he sometimes intro-

---

[7] Physic. Ausc. iv. 7, p. 215. [8] Ib. viii. 1, p. 258.

duces principles of a more arbitrary character; and besides the general relations of thought, takes for granted the inventions of previous speculators; such, for instance, as the then commonly received opinions concerning the frame of the world. From the assertion that motion is eternal, proved in the manner just stated, Aristotle proceeds by a curious train of reasoning, to identify this eternal motion with the diurnal motion of the heavens. "There must," he says, "be something which is the First Mover:"[9] this follows from the relation of causes and effects. Again, "Motion must go on constantly, and, therefore, must be either continuous or successive. Now what is continuous is more properly said to take place *constantly*, than what is successive. Also the continuous is better; but we always suppose that which is better to take place in nature, if it be possible. The motion of the First Mover will, therefore, be continuous, if such an eternal motion be possible." We here see the vague judgment of *better* and *worse* introduced, as that of *natural* and *unnatural* was before, into physical reasonings.

I proceed with Aristotle's argument.[10] "We have now, therefore, to show that there may be an infinite single, continuous motion, and that this is circular." This is, in fact, proved, as may readily be conceived, from the consideration that a body may go on perpetually revolving uniformly in a circle. And thus we have a demonstration, on the principles of this philosophy, that there is and must be a First Mover, revolving eternally with a uniform circular motion.

Though this kind of philosophy may appear too trifling to deserve being dwelt upon, it is important for our purpose so far as to exemplify it, that we may afterwards advance, confident that we have done it no injustice.

I will now pass from the doctrines relating to the motions of the heavens, to those which concern the material elements of the universe. And here it may be remarked that the tendency (of which we are here tracing the development) to extract speculative opinions from the relations of words, must be very natural to man; for the very widely accepted doctrine of the Four Elements which appears to be founded on the opposition of the adjectives *hot* and *cold*, *wet* and *dry*, is much older than Aristotle, and was probably one of the earliest of philosophical dogmas. The great master of this philosophy, however, puts the opinion in a more systematic manner than his predecessors.

---

[9] Physic. Ausc. viii. 6. p. 258. [10] Ib. viii. 8.

"We seek," he says,[11] "the principles of sensible things, that is, of tangible bodies. We must take, therefore, not all the contrarieties of quality, but those only which have reference to the touch. Thus black and white, sweet and bitter, do not differ as tangible qualities, and therefore must be rejected from our consideration.

"Now the contrarieties of quality which refer to the touch are these: hot, cold; dry, wet; heavy, light; hard, soft; unctuous, meagre; rough, smooth; dense, rare." He then proceeds to reject all but the four first of these, for various reasons; heavy and light, because they are not active and passive qualities; the others, because they are combinations of the four first, which therefore he infers to be the four elementary qualities.

"[12]Now in four things there are six combinations of two; but the combinations of two opposites, as hot and cold, must be rejected; we have, therefore, four elementary combinations, which agree with the four apparently elementary bodies. Fire is hot and dry; air is hot and wet (for steam is air); water is cold and wet, earth is cold and dry."

It may be remarked that this disposition to assume that some common elementary quality must exist in the cases in which we habitually apply a common adjective, as it began before the reign of the Aristotelian philosophy, so also survived its influence. Not to mention other cases, it would be difficult to free Bacon's *Inquisitio in naturam calidi*, "Examination of the nature of heat," from the charge of confounding together very different classes of phenomena under the cover of the word *hot.*

The correction of these opinions concerning the elementary composition of bodies belongs to an advanced period in the history of physical knowledge, even after the revival of its progress. But there are some of the Aristotelian doctrines which particularly deserve our attention, from the prominent share they had in the very first beginnings of that revival; I mean the doctrines concerning motion.

These are still founded upon the same mode of reasoning from adjectives; but in this case, the result follows, not only from the opposition of the words, but also from the distinction of their being *absolutely* or *relatively* true. "Former writers," says Aristotle, "have considered heavy and light *relatively* only, taking cases, where both things have weight, but one is lighter than the other; and they imagined that, in

[11] De Gen. et Corrupt. ii. 2. [12] Ib. iii. 3.

this way, they defined what was *absolutely* (ἁπλῶς) heavy and light." We now know that things which rise by their lightness do so only because they are pressed upwards by heavier surrounding bodies; and this assumption of absolute levity, which is evidently gratuitous, or rather merely nominal, entirely vitiated the whole of the succeeding reasoning. The inference was, that fire must be absolutely light, since it tends to take its place above the other three elements; earth absolutely heavy, since it tends to take its place below fire, air, and water. The philosopher argued also, with great acuteness, that air, which tends to take its place below fire and above water, must do so *by its nature*, and not in virtue of any combination of heavy and light elements. "For if air were composed of the parts which give fire its levity, joined with other parts which produce gravity, we might assume a quantity of air so large, that it should be lighter than a small quantity of fire, having more of the light parts." It thus follows that each of the four elements tends to its own place, fire being the highest, air the next, water the next, and earth the lowest.

The whole of this train of errors arises from fallacies which have a verbal origin;—from considering light as opposite to heavy; and from considering levity as a quality of a body, instead of regarding it as the effect of surrounding bodies.

It is worth while to notice that a difficulty which often embarrasses persons on their entrance upon physical speculations,—the difficulty of conceiving that up and down are different directions in different places,—had been completely got over by Aristotle and the Greek philosophers. They were steadily convinced of the roundness of the earth, and saw that this truth led to the conclusion that all heavy bodies tend in converging directions to the centre. And, they added, as the heavy tends to the centre, the light tends to the exterior, "for Exterior is opposite to Centre as heavy is to light."[13]

The tendencies of bodies downwards and upwards, their weight, their fall, their floating or sinking, were thus accounted for in a manner which, however unsound, satisfied the greater part of the speculative world till the time of Galileo and Stevinus, though Archimedes in the mean time published the true theory of floating bodies, which is very different from that above stated. Other parts of the doctrines of motion were delivered by the Stagirite in the same spirit and with the same success. The motion of a body which is thrown along the

---

[13] De Cœlo, iv. 4.

ground diminishes and finally ceases; the motion of a body which falls from a height goes on becoming quicker and quicker; this was accounted for on the usual principle of opposition, by saying that the former is a *violent*, the latter a *natural* motion. And the later writers of this school expressed the characters of such motions in verse. The rule of natural motion was[14]

> Principium tepeat, medium cum fine calebit.
> Cool at the first, it warm and warmer glows.

And of violent motion, the law was—

> Principium fervet, medium calet, ultima friget.
> Hot at the first, then barely warm, then cold.

It appears to have been considered by Aristotle a difficult problem to explain why a stone thrown from the hand continues to move for some time, and then stops. If the hand was the cause of the motion, how could the stone move at all when left to itself? if not, why does it ever stop? And he answers this difficulty by saying,[15] "that there is a motion communicated to the air, the successive parts of which urge the stone onwards; and that each part of this medium continues to act for some while after it has been acted on, and the motion ceases when it comes to a particle which cannot act after it has ceased to be acted on." It will be readily seen that the whole of this difficulty, concerning a body which moves forward and is retarded till it stops, arises from ascribing the retardation, not to the real cause, the surrounding resistances, but to the body itself.

One of the doctrines which was the subject of the warmest discussion between the defenders and opposers of Aristotle, at the revival of physical knowledge, was that in which he asserts,[16] "That body is heavier than another which in an equal bulk moves downward quicker." The opinion maintained by the Arisotelians at the time of Galileo was, that bodies fall quicker exactly in proportion to their weight. The master himself asserts this in express terms, and reasons upon it.[17] Yet in another passage he appears to distinguish between weight and actual motion downwards.[18] "In physics, we call bodies heavy and light from their *power* of motion; but these names are not applied to their actual operations (ἐνέργειαις) except any one thinks

---

[14] Alsted. Encyc. tom. i. p. 687. [15] Phys. Ausc. viii. 10. [16] De Cœlo, iv. 1, p. 308.
[17] Ib. iii. 2. [18] Ib. iv. 1, p. 307.

momentum (ῥοπή) to be a word of both applications. But heavy and light are, as it were, the *embers* or *sparks* of motion, and therefore proper to be treated of here."

The distinction just alluded to, between Power or Faculty of Action, and actual Operation or Energy, is one very frequently referred to by Aristotle; and though not by any means useless, may easily be so used as to lead to mere verbal refinements instead of substantial knowledge.

The Aristotelian distinction of Causes has not any very immediate bearing upon the parts of physics of which we have here mainly spoken; but it was so extensively accepted, and so long retained, that it may be proper to notice it.[19] "One kind of Cause is the matter of which any thing is made, as bronze of a statue, and silver of a vial; another is the form and pattern, as the Cause of an octave is the ratio of two to one; again, there is the Cause which is the origin of the production, as the father of the child; and again, there is the End, or that for the sake of which any thing is done, as health is the cause of walking." These four kinds of Cause, the *material*, the *formal*, the *efficient*, and the *final*, were long leading points in all speculative inquiries; and our familiar forms of speech still retain traces of the influence of this division.

It is my object here to present to the reader in an intelligible shape, the principles and mode of reasoning of the Aristotelian philosophy, not its results. If this were not the case, it would be easy to excite a smile by insulating some of the passages which are most remote from modern notions. I will only mention, as specimens, two such passages, both very remarkable.

In the beginning of the book "On the Heavens," he proves[20] the world to be *perfect*, by reasoning of the following kind: "The bodies of which the world is composed are solids, and therefore have three dimensions: now three is the most perfect number; it is the first of numbers, for of *one* we do not speak as a number; of *two* we say *both*; but *three* is the first number of which we say *all*; moreover, it has a beginning, a middle, and an end."

The reader will still perceive the verbal foundations of opinions thus supported.

"The simple elements must have simple motions, and thus fire and air have their natural motions upwards, and water and earth have

[19] Phys. ii. 3. [20] De Cœlo, i. 1.

their natural motions downwards; but besides these motions, there is motion in a circle, which is unnatural to these elements, but which is a more perfect motion than the other, because a circle is a perfect line, and a straight line is not; and there must be something to which this motion is natural. From this it is evident," he adds, with obvious animation, "that there is some essence of body different from those of the four elements, more divine than those, and superior to them. If things which move in a circle move contrary to nature, it is marvellous, or rather absurd, that this, the unnatural motion, should alone be continuous and eternal; for unnatural motions decay speedily. And so, from all this, we must collect, that besides the four elements which we have here and about us, there is another removed far off, and the more excellent in proportion as it is more distant from us." This fifth element was the "*quinta essentia*," of after writers, of which we have a trace in our modern literature, in the word *quintessence*.

Sect. 3.—Technical Forms of the Greek Schools.

We have hitherto considered only the principle of the Greek Physics; which was, as we have seen, to deduce its doctrines by an analysis of the notions which common language involves. But though the Grecian philosopher began by studying words in their common meanings, he soon found himself led to fix upon some special shades or applications of these meanings as the permanent and standard notion, which they were to express; that is, he made his language *technical*. The invention and establishment of technical terms is an important step in any philosophy, true or false; we must, therefore, say a few words on this process, as exemplified in the ancient systems.

1. *Technical Forms of the Aristotelian Philosophy.*—We have already had occasion to cite some of the distinctions introduced by Aristotle, which may be considered as technical; for instance, the classification of Causes as *material*, *formal*, *efficient*, and *final;* and the opposition of Qualities as *absolute* and *relative*. A few more of the most important examples may suffice. An analysis of objects into *Matter* and *Form*, when metaphorically extended from visible objects to things conceived in the most general manner, became an habitual hypothesis of the Aristotelian school. Indeed this metaphor is even yet one of the most significant of those which we can employ, to suggest one of the most comprehensive and fundamental antitheses with which philosophy has to do;—the opposition of sense and reason, of

impressions and laws. In this application, the German philosophers have, up to the present time, rested upon this distinction a great part of the weight of their systems; as when Kant says, that Space and Time are the *Forms of Sensation.* Even in our own language, we retain a trace of the influence of this Aristotelian notion, in the word *Information*, when used for that knowledge which may be conceived as moulding the mind into a definite shape, instead of leaving it a mere mass of unimpressed susceptibility.

Another favorite Aristotelian antithesis is that of *Power* and *Act* (δύναμις, ἐνέργεια). This distinction is made the basis of most of the physical philosophy of the school; being, however, generally introduced with a peculiar limitation. Thus, Light is defined to be "the Act of what is lucid, as being lucid. And if," it is added, "the lucid be so in power but not in act, we have darkness." The reason of the limitation, "as being lucid," is, that a lucid body may act in other ways; thus a torch may move as well as shine, but its moving is not its act *as being a lucid* body.

Aristotle appears to be well satisfied with this explanation, for he goes on to say, "Thus light is not Fire, nor any body whatever, or the emanation of any body (for that would be a kind of body), but it is the presence of something like Fire in the body; it is, however, impossible that two bodies should exist in the same place, so that it is not a body;" and this reasoning appears to leave him more satisfied with his doctrine, that Light is an *Energy* or *Act.*

But we have a more distinctly technical form given to this notion. Aristotle introduced a word formed by himself, to express the act which is thus opposed to inactive power: this is the celebrated word ἐντελέχεια. Thus the noted definition of Motion in the third book of the Physics,[21] is that it is "the *Entelechy*, or Act, of a movable body in respect of being movable;" and the definition of the Soul is[22] that it is "the *Entelechy* of a natural body which has life by reason of its power." This word has been variously translated by the followers of Aristotle, and some of them have declared it untranslatable. *Act* and *Action* are held to be inadequate substitutes; the *very act, ipse cursus actionis*, is employed by some; *primus actus* is employed by many, but another school use *primus actus* of a non-operating form. Budæus uses *efficacia.* Cicero[23] translates it "quasi quandam continuatam motionem, et perennem;" but this paraphrase, though it may

[21] Phys. iii. 1. [22] De Animâ, ii. 1. [23] Tusc. i. 10.

fall in with the description of the soul, which is the subject with which Cicero is concerned, does not appear to agree with the general applications of the term. Hermolaus Barbarus is said to have been so much oppressed with this difficulty of translation, that he consulted the evil spirit by night, entreating to be supplied with a more common and familiar substitute for this word: the mocking fiend, however, suggested only a word equally obscure, and the translator, discontented with this, invented for himself the word *perfectihabia*.

We need not here notice the endless apparatus of technicalities which was, in later days, introduced into the Aristotelian philosophy; but we may remark, that their long continuance and extensive use show us how powerful technical phraseology is, for the perpetuation either of truth or error. The Aristotelian terms, and the metaphysical views which they tend to preserve, are not yet extinct among us. In a very recent age of our literature it was thought a worthy employment by some of the greatest writers of the day, to attempt to expel this system of technicalities by ridicule.

"Crambe regretted extremely that *substantial forms*, a race of harmless beings, which had lasted for many years, and afforded a comfortable subsistence to many poor philosophers, should now be hunted down like so many wolves, without a possibility of retreat. He considered that it had gone much harder with them than with *essences*, which had retired from the schools into the apothecaries' shops, where some of them had been advanced to the degree of *quintessences*.[24]

We must now say a few words on the technical terms which others of the Greek philosophical sects introduced.

2. *Technical Forms of the Platonists.*—The other sects of the Greek philosophy, as well as the Aristotelians, invented and adopted technical terms, and thus gave fixity to their tenets and consistency to their traditionary systems; of these I will mention a few.

A technical expression of a contemporary school has acquired perhaps greater celebrity than any of the terms of Aristotle. I mean the *Ideas* of Plato. The account which Aristotle gives of the origin of these will serve to explain their nature.[25] "Plato," says he, "who, in his youth, was in habits of communication first with Cratylus and the Heraclitean opinions, which represent all the objects of sense as being in a perpetual flux, so that concerning these no science nor certain

---

[24] Martinus Scriblerus, cap. vii.

[25] Arist. Metaph. i. 6. The same account is repeated, and the subject discussed, Metaph. xii. 4.

knowledge can exist, entertained the same opinions at a later period also. When, afterwards, Socrates treated of moral subjects, and gave no attention to physics, but, in the subjects which he did discuss, arrived at universal truths, and before any man, turned his thoughts to definitions, Plato adopted similar doctrines on this subject also; and construed them in this way, that these truths and definitions must be applicable to something else, and not to sensible things: for it was impossible, he conceived, that there should be a general common definition of any sensible object, since such were always in a state of change. The things, then, which were the subjects of universal truths he called *Ideas;* and held that objects of sense had their names according to Ideas and after them; so that things participated in that Idea which had the same name as was applied to them."

In agreement with this, we find the opinions suggested in the *Parmenides* of Plato, the dialogue which is considered by many to contain the most decided exposition of the doctrine of Ideas. In this dialogue, Parmenides is made to say to Socrates, then a young man,[26] "O Socrates, philosophy has not yet claimed you for her own, as, in my judgment, she will claim you, and you will not dishonor her. As yet, like a young man as you are, you look to the opinions of men. But tell me this: it appears to you, as you say, that there are certain *Kinds* or *Ideas* (εἰδῆ) of which things partake and receive applications according to that of which they partake: thus those things which partake of *Likeness* are called *like;* those things which partake of *Greatness* are called *great;* those things which partake of *Beauty* and *Justice* are called *beautiful* and *just*." To this Socrates assents. And in another part of the dialogue he shows that these Ideas are not included in our common knowledge, from whence he infers that they are objects of the Divine mind.

In the Phædo the same opinion is maintained, and is summed up in this way, by a reporter of the last conversation of Socrates,[27] εἶναι τι ἕκαστον τῶν εἰδῶν, καὶ τούτων τ'ἄλλα μεταλαμβάνοντα αὐτῶν τούτων τὴν ἐπωνυμίαν ἴσχειν; "that each *Kind* has an existence, and that other things partake of these Kinds, and are called according to the Kind of which they partake."

The inference drawn from this view was, that in order to obtain true and certain knowledge, men must elevate themselves, as much as possible, to these Ideas of the qualities which they have to consider:

---

[26] Parmenid. p. 131. [27] Phædo, p. 102.

and as things were thus called after the Ideas, the Ideas had a priority and pre-eminence assigned them. The *Idea* of Good, Beautiful, and Wise was the "First Good," the "First Beautiful," the "First Wise." This dignity and distinction were ultimately carried to a large extent. Those Ideas were described as eternal and self-subsisting, forming an "Intelligible World," full of the models or archetypes of created things. But it is not to our purpose here to consider the Platonic Ideas in their theological bearings. In physics they were applied in the same form as in morals. The *primum calidum*, *primum frigidum* were those Ideas of fundamental Principles by participation of which, all things were hot or cold.

This school did not much employ itself in the development of its principles as applied to physical inquiries: but we are not without examples of such speculations. Plutarch's Treatise Περὶ τοῦ Πρώτου Ψυχροῦ, "On the First Cold," may be cited as one. It is in reality a discussion of a question which has been agitated in modern times also; —whether cold be a positive quality or a mere privation. "Is there, O Favorinus," he begins, "a First Power and Essence of the Cold, as Fire is of the Hot; by a certain presence and participation of which all other things are cold: or is rather coldness a privation of heat, as darkness is of light, and rest of motion?"

3. *Technical Forms of the Pythagoreans.*—The *Numbers* of the Pythagoreans, when propounded as the explanation of physical phenomena, as they were, are still more obscure than the Ideas of the Platonists. There were, indeed, considerable resemblances in the way in which these two kinds of notions were spoken of. Plato called his Ideas *unities*, *monads;* and as, according to him, Ideas, so, according to the Pythagoreans, Numbers, were the causes of things being what they are.[28] But there was this difference, that things shared the nature of the Platonic Ideas "by participation," while they shared the nature of Pythagorean Numbers "by imitation." Moreover, the Pythagoreans followed their notion out into much greater development than any other school, investing particular numbers with extraordinary attributes, and applying them by very strange and forced analogies. Thus the number Four, to which they gave the name of *Tetractys*, was held to be the most perfect number, and was conceived to correspond to the human soul, in some way which appears to be very imperfectly understood by the commentators of this philosophy.

---

[28] Arist. Metaph. i. 6.

It has been observed by a distinguished modern scholar,[29] that the place which Pythagoras ascribed to his numbers is intelligible only by supposing that he confounded, first a numerical unit with a geometrical point, and then this with a material atom. But this criticism appears to place systems of physical philosophy under requisitions too severe. If all the essential properties and attributes of things were fully represented by the relations of number, the philosophy which supplied such an explanation of the universe, might well be excused from explaining also that existence of objects which is distinct from the existence of all their qualities and properties. The Pythagorean love of numerical speculations might have been combined with the doctrine of atoms, and the combination might have led to results well worth notice. But so far as we are aware, no such combination was attempted in the ancient schools of philosophy; and perhaps we of the present day are only just beginning to perceive, through the disclosures of chemistry and crystallography, the importance of such a line of inquiry.

4. *Technical Forms of the Atomists and Others.*—The atomic doctrine, of which we have just spoken, was one of the most definite of the physical doctrines of the ancients, and was applied with most perseverance and knowledge to the explanation of phenomena. Though, therefore, it led to no success of any consequence in ancient times, it served to transmit, through a long series of ages, a habit of really physical inquiry; and, on this account, has been thought worthy of an historical disquisition by Bacon.[30]

The technical term, *Atom*, marks sufficiently the nature of the opinion. According to this theory, the world consists of a collection of simple particles, of one kind of matter, and of indivisible smallness (as the name indicates), and by the various configurations and motions of these particles, all kinds of matter and all material phenomena are produced.

To this, the Atomic Doctrine of Leucippus and Democritus, was opposed the *Homoiomeria* of Anaxagoras; that is, the opinion that material things consist of particles which are homogeneous in each kind of body, but various in different kinds: thus for example, since by food the flesh and blood and bones of man increase, the author of this doctrine held that there are in food particles of flesh, and blood,

[29] Thirlwall's *Hist. Gr.* ii. 142.

[30] Parmenidis et Telesii et præcipue Democriti Philosophia, &c., Works, vol. ix. 317.

and bone. As the former tenet points to the corpuscular theories of modern times, so the latter may be considered as a dim glimpse of the idea of chemical analysis. The Stoics also, who were, especially at a later period, inclined to materialist views, had their technical modes of speaking on such subjects. They asserted that matter contained in itself tendencies or dispositions to certain forms, which dispositions they called λόγοι σερματικοί, *seminal proportions*, or *seminal reasons*.

Whatever of sound view, or right direction, there might be in the notions which suggested these and other technical expressions, was, in all the schools of philosophy (so far as physics was concerned) quenched and overlaid by the predominance of trifling and barren speculations; and by the love of subtilizing and commenting upon the works of earlier writers, instead of attempting to interpret the book of nature. Hence these technical terms served to give fixity and permanence to the traditional dogmas of the sect, but led to no progress of knowledge.

The advances which were made in physical science proceeded, not from these schools of philosophy (if we except, perhaps, the obligations of the science of Harmonics to the Pythagoreans), but from reasoners who followed an independent path. The sequel of the ambitious hopes, the vast schemes, the confident undertakings of the philosophers of ancient Greece, was an entire failure in the physical knowledge of which it is our business to trace the history. Yet we are not, on that account, to think slightingly of these early speculators. They were men of extraordinary acuteness, invention, and range of thought; and, above all, they had the merit of first completely unfolding the speculative faculty—of starting in that keen and vigorous chase of knowledge out of which all the subsequent culture and improvement of man's intellectual stores have arisen. The sages of early Greece form the heroic age of science. Like the first navigators in their own mythology, they boldly ventured their untried bark in a distant and arduous voyage, urged on by the hopes of a supernatural success; and though they missed the imaginary golden prize which they sought, they unlocked the gates of distant regions, and opened the seas to the keels of the thousands of adventurers who, in succeeding times, sailed to and fro, to the indefinite increase of the mental treasures of mankind.

But inasmuch as their attempts, in one sense, and at first, failed, we must proceed to offer some account of this failure, and of its nature and causes.

CHAPTER III.

Failure of the Physical Philosophy of the Greek Schools.

Sect. 1.—*Result of the Greek School Philosophy.*

THE methods and forms of philosophizing which we have described as employed by the Greek Schools, failed altogether in their application to physics. No discovery of general laws, no explanation of special phenomena, rewarded the acuteness and boldness of these early students of nature. Astronomy, which made considerable progress during the existence of the sects of Greek philosophers, gained perhaps something by the authority with which Plato taught the supremacy and universality of mathematical rule and order; and the truths of Harmonics, which had probably given rise to the Pythagorean passion for numbers, were cultivated with much care by that school. But after these first impulses, the sciences owed nothing to the philosophical sects; and the vast and complex accumulations and apparatus of the Stagirite do not appear to have led to any theoretical physical truths.

This assertion hardly requires proof, since in the existing body of science there are no doctrines for which we are indebted to the Aristotelian School. Real truths, when once established, remain to the end of time a part of the mental treasure of man, and may be discerned through all the additions of later days. But we can point out no physical doctrine now received, of which we trace the anticipation in Aristotle, in the way in which we see the Copernican system anticipated by Aristarchus, the resolution of the heavenly appearances into circular motions suggested by Plato, and the numerical relations of musical intervals ascribed to Pythagoras. But it may be worth while to look at this matter more closely.

Among the works of Aristotle are thirty-eight chapters of "Problems," which may serve to exemplify the progress he had really made in the reduction of phenomena to laws and causes. Of these Problems, a large proportion are physiological, and these I here pass by, as not illustrative of the state of physical knowledge. But those which are properly physical are, for the most part, questions concerning such facts and difficulties as it is the peculiar business of theory to explain. Now it may be truly said, that in scarcely any one instance are the answers, which Aristotle gives to his questions, of any value. For the most part, indeed, he propounds his answer with a degree of hesitation or vacillation which of itself shows the absence of all scientific distinctness of thought; and the opinions so offered never appear to involve any settled or general principle.

We may take, as examples of this, the problems of the simplest kind, where the principles lay nearest at hand—the mechanical ones. "Why," he asks,[1] "do small forces move great weights by means of a lever, when they have thus to move the lever added to the weight? Is it," he suggests, "because a greater radius moves faster?" "Why does a small wedge split great weights?[2] Is it because the wedge is composed of two opposite levers?" "Why,[3] when a man rises from a chair, does he bend his leg and his body to acute angles with his thigh? Is it because a right angle is connected with equality and rest?" "Why[4] can a man throw a stone further with a sling than with his hand? Is it that when he throws with his hand he moves the stone from rest, but when he uses the sling he throws it already in motion?" "Why,[5] if a circle be thrown on the ground, does it first describe a straight line and then a spiral, as it falls? Is it that the air first presses equally on the two sides and supports it, and afterwards presses on one side more?" "Why[6] is it difficult to distinguish a musical note from the octave above? Is it that proportion stands in the place of equality?" It must be allowed that these are very vague and worthless surmises; for even if we were, as some commentators have done, to interpret some of them so as to agree with sound philosophy, we should still be unable to point out, in this author's works, any clear or permanent apprehension of the general principles which such an interpretation implies.

Thus the Aristotelian physics cannot be considered as otherwise than a complete failure. It collected no general laws from facts; and consequently, when it tried to explain facts, it had no principles which were of any avail.

The same may be said of the physical speculations of the other schools of philosophy. They arrived at no doctrines from which they could deduce, by sound reasoning, such facts as they saw; though they

---

[1] Mech. Prob. 4. [2] Ib. 18. [3] Ib. 31. [4] Ib. 13.
[5] Περὶ Ἄψυχα. 11. [6] Περὶ Ἁρμον. 14.

often venture so far to trust their principles as to infer from them propositions beyond the domain of sense. Thus, the principle that each element seeks *its own place*, led to the doctrine that, the place of fire being the highest, there is, above the air, a Sphere of Fire—of which doctrine the word *Empyrean*, used by our poets, still conveys a reminiscence. The Pythagorean tenet that ten is a perfect number,[7] led some persons to assume that the heavenly bodies are in number ten; and as nine only were known to them, they asserted that there was an *antichthon*, or *counter-earth*, on the other side of the sun, invisible to us. Their opinions respecting numerical ratios, led to various other speculations concerning the distances and positions of the heavenly bodies: and as they had, in other cases, found a connection between proportions of distance and musical notes, they assumed, on this suggestion, *the music of the spheres*.

Although we shall look in vain in the physical philosophy of the Greek Schools for any results more valuable than those just mentioned, we shall not be surprised to find, recollecting how much an admiration for classical antiquity has possessed the minds of men, that some writers estimate their claims much more highly than they are stated here. Among such writers we may notice Dutens, who, in 1766, published his "Origin of the Discoveries attributed to the Moderns; in which it is shown that our most celebrated Philosophers have received the greatest part of their knowledge from the Works of the Ancients." The thesis of this work is attempted to be proved, as we might expect, by very large interpretations of the general phrases used by the ancients. Thus, when Timæus, in Plato's dialogue, says of the Creator of the world,[8] "that he infused into it two powers, the origins of motions, both of that of the same thing and of that of different things;" Dutens[9] finds in this a clear indication of the projectile and attractive forces of modern science. And in some of the common declamation of the Pythagoreans and Platonists concerning the general prevalence of numerical relations in the universe, he discovers their acquaintance with the law of the inverse square of the distance by which gravitation is regulated, though he allows[10] that it required all the penetration of Newton and his followers to detect this law in the scanty fragments by which it is transmitted.

Argument of this kind is palpably insufficient to cover the failure of the Greek attempts at a general physical philosophy; or rather we

---

[7] Arist. Metaph. i. 5. [8] Tim. 96. [9] 3d ed. p. 83. [10] Ib. p. 88.

may say, that such arguments, since they are as good as can be brought in favor of such an opinion, show more clearly how entire the failure was. I proceed now to endeavor to point out its causes.

Sect. 2.—Cause of the Failure of the Greek Physical Philosophy.

THE cause of the failure of so many of the attempts of the Greeks to construct physical science is so important, that we must endeavor to bring it into view here; though the full development of such subjects belongs rather to the Philosophy of Induction. The subject must, at present, be treated very briefly.

I will first notice some errors which may naturally occur to the reader's mind, as possible causes of failure, but which, we shall be able to show, were not the real reasons in this case.

The cause of failure was *not the neglect of facts.* It is often said that the Greeks disregarded experience, and spun their philosophy out of their own thoughts alone; and this is supposed by many to be their essential error. It is, no doubt, true, that the disregard of experience is a phrase which may be so interpreted as to express almost any defect of philosophical method; since coincidence with experience is requisite to the truth of all theory. But if we fix a more precise sense on our terms, I conceive it may be shown that the Greek philosophy did, in its opinions, recognize the necessity and paramount value of observations; did, in its origin, proceed upon observed facts; and did employ itself to no small extent in classifying and arranging phenomena. We must endeavor to illustrate these assertions, because it is important to show that these steps alone do not necessarily lead to science.

1. The acknowledgment of experience as the main ground of physical knowledge is so generally understood to be a distinguishing feature of later times, that it may excite surprise to find that Aristotle, and other ancient philosophers, not only asserted in the most pointed manner that all our knowledge must begin from experience, but also stated in language much resembling the habitual phraseology of the most modern schools of philosophizing, that particular facts must be *collected;* that from these, general principles must be obtained by *induction;* and that these principles, when of the most general kind, are *axioms.* A few passages will show this.

"The way[11] must be the same," says Aristotle, in speaking of the rules of reasoning, "with respect to philosophy, as it is with respect to

[11] Anal. Prior. i. 30.

any art or science whatever; we must collect the facts, and the things to which the facts happen, in each subject, and provide as large a supply of these as possible." He then proceeds to say that "we are not to look at once at all this collected mass, but to consider small and definite portions" . . . "And thus it is the office of observation to supply principles in each subject; for instance, astronomical observation supplies the principles of astronomical science. For the phenomena being properly assumed, the astronomical demonstrations were from these discovered. And the same applies to every art and science. So that if we take the facts (τὰ ὑπάρχοντα) belonging to each subject, it is *our* task to mark out clearly the course of the demonstrations. For if *in our natural history* (κατὰ τὴν ἱστορίαν) we have omitted nothing of the facts and properties which belong to the subject, we shall learn what we can demonstrate and what we cannot."

These facts, τὰ ὑπάρχοντα, he, at other times, includes in the term *sensation*. Thus, he says,[12] "It is obvious that if any sensation is wanting, there must be also some knowledge wanting which we are thus prevented from having, since we arrive at knowledge either by induction or by demonstration. Demonstration proceeds from universal propositions, Induction from particulars. But we cannot have universal theoretical propositions except from induction; and we cannot make inductions without having sensation; for sensation has to do with particulars."

In another place,[13] after stating that principles must be prior to, and better known than conclusions, he distinguishes such principles into absolutely prior, and prior relative to us: "The prior principles, relative to us, are those which are nearer to the sensation; but the principles absolutely prior are those which are more remote from the sensation. The most general principles are the more remote, the more particular are nearer. The general principles which are necessary to knowledge are *axioms*."

We may add to these passages, that in which he gives an account of the way in which Leucippus was led to the doctrine of atoms. After describing the opinions of some earlier philosophers, he says,[14] "Thus, proceeding in violation of sensation, and disregarding it, because, as they held, they must follow reason, some came to the conclusion that the universe was one, and infinite, and at rest. As it appeared, however, that though this ought to be by reasoning, it

---

[12] Anal. Post. i. 18. [13] Ib. i. 2. [14] De Gen. et Cor. i. 8.

would go near to madness to hold such opinions in practice (for no one was ever so mad as to think fire and ice to be one), Leucippus, therefore, pursued a line of reasoning which was in accordance with sensation, and which was not irreconcilable with the production and decay, the motion and multitude of things." It is obvious that the school to which Leucippus belonged (the Eclectic) must have been, at least in its origin, strongly impressed with the necessity of bringing its theories into harmony with the observed course of nature.

2. Nor was this recognition of the fundamental value of experience a mere profession. The Greek philosophy did, in its beginning, proceed upon observation. Indeed it is obvious that the principles which it adopted were, in the first place, assumed in order to account for some classes of facts, however imperfectly they might answer their purpose. The principle of things seeking their own places, was invented in order to account for the falling and floating of bodies. Again, Aristotle says, that heat is that which brings together things of the same kind, cold is that which brings together things whether of the same or of different kinds: it is plain that in this instance he intended by his principle to explain some obvious facts, as the freezing of moist substances, and the separation of heterogeneous things by fusion; for, as he adds, if fire brings together things which are akin, it will separate those which are not akin. It would be easy to illustrate the remark further, but its truth is evident from the nature of the case; for no principles could be accepted for a moment, which were the result of an arbitrary caprice of the mind, and which were not in some measure plausible, and apparently confirmed by facts.

But the works of Aristotle show, in another way, how unjust it would be to accuse him of disregarding facts. Many large treatises of his consist almost entirely of collections of facts, as for instance, those "On Colors," "On Sounds," and the collection of Problems to which we have already referred; to say nothing of the numerous collection of facts bearing on natural history and physiology, which form a great portion of his works, and are even now treasuries of information. A moment's reflection will convince us that the physical sciences of our own times, for example, Mechanics and Hydrostatics, are founded almost entirely upon facts with which the ancients were as familiar as we are. The defect of their philosophy, therefore, wherever it may lie, consists neither in the speculative depreciation of the value of facts, nor in the practical neglect of their use.

3. Nor again, should we hit upon the truth, if we were to say that

Aristotle, and other ancient philosophers, did indeed collect facts; but that they took no steps in classifying and comparing them; and that thus they failed to obtain from them any general knowledge. For, in reality, the treatises of Aristotle which we have mentioned, are as remarkable for the power of classifying and systematizing which they exhibit, as for the industry shown in the accumulation. But it is not classification of facts merely which can lead us to knowledge, except we adopt that special arrangement, which, in each case, brings into view the principles of the subject. We may easily show how unprofitable an arbitrary or random classification is, however orderly and systematic it may be.

For instance, for a long period all unusual fiery appearances in the sky were classed together as *meteors.* Comets, shooting-stars, and globes of fire, and the aurora borealis in all its forms, were thus grouped together, and classifications of considerable extent and minuteness were proposed with reference to these objects. But this classification was of a mixed and arbitrary kind. Figure, color, motion, duration, were all combined as characters, and the imagination lent its aid, transforming these striking appearances into fiery swords and spears, bears and dragons, armies and chariots. The facts so classified were, notwithstanding, worthless; and would not have been one jot the less so, had they and their classes been ten times as numerous as they were. No rule or law that would stand the test of observation was or could be thus discovered. Such classifications have, therefore, long been neglected and forgotten. Even the ancient descriptions of these objects of curiosity are unintelligible, or unworthy of trust, because the spectators had no steady conception of the usual order of such phenomena. For, however much we may fear to be misled by preconceived opinions, the caprices of imagination distort our impressions far more than the anticipations of reason. In this case men had, indeed we may say with regard to many of these meteors, they still have, no science: not for want of facts, nor even for want of classification of facts; but because the classification was one in which no real principle was contained.

4. Since, as we have said before, two things are requisite to science, —Facts and Ideas; and since, as we have seen, Facts were not wanting in the physical speculations of the ancients, we are naturally led to ask, Were they then deficient in Ideas? Was there a want among them of mental activity, and logical connection of thought? But it is so obvious that the answer to this inquiry must be in the negative, that we need not dwell upon it. No one who knows any thing of the

history of the ancient Greek mind, can question, that in acuteness, in ingenuity, in the power of close and distinct reasoning, they have never been surpassed. The common opinion, which considers the defect of their philosophical character to reside rather in the exclusive activity of such qualities, than in the absence of them, is at least so far just.

5. We come back again, therefore, to the question, What was the radical and fatal defect in the physical speculations of the Greek philosophical schools?

To this I answer: The defect was, that though they had in their possession Facts and Ideas, *the Ideas were not distinct and appropriate to the Facts.*

The peculiar characteristics of scientific ideas, which I have endeavored to express by speaking of them as *distinct* and *appropriate to the facts*, must be more fully and formally set forth, when we come to the philosophy of the subject. In the mean time, the reader will probably have no difficulty in conceiving that, for each class of Facts, there is some special set of Ideas, by means of which the facts can be included in general scientific truths; and that these Ideas, which may thus be termed *appropriate*, must be possessed with entire distinctness and clearness, in order that they may be successfully applied. It was the want of Ideas having this reference to material phenomena, which rendered the ancient philosophers, with very few exceptions, helpless and unsuccessful speculators on physical subjects.

This must be illustrated by one or two examples. One of the facts which Aristotle endeavors to explain is this; that when the sun's light passes through a hole, whatever be the form of the hole, the bright image, if formed at any considerable distance from the hole, is round, instead of imitating the figure of the hole, as shadows resemble their objects in form. We shall easily perceive this appearance to be a necessary consequence of the circular figure of the sun, if we conceive light to be diffused from the luminary by means of straight *rays* proceeding from every point of the sun's disk and passing through every point within the boundary of the hole. By attending to the consequences of this mode of conception, it will be seen that each point of the hole will be the vertex of a double cone of rays which has the sun's disk for its base on one side and an image of the sun on the other; and the figure of the image of the hole will be determined by supposing a series of equal bright circles, images of the sun, to be placed along the boundary of an image equal to the hole itself. The figure of the image thus determined will partake of the form of the hole, and of the circular form of the sun's image: but these circular images become larger and larger as they are farther from the hole, while the central image of the hole remains always of the original size; and thus at a considerable distance from the hole, the trace of the hole's form is nearly obliterated, and the image is nearly a perfect circle. Instead of this distinct conception of a cone of rays which has the sun's disk for its basis, Aristotle has the following loose conjecture.[15] "Is it because light is emitted in a conical form; and of a cone, the base is a circle; so that on whatever the rays of the sun fall, they appear more circular?" And thus though he applies the notion of rays to this problem, he possesses this notion so *indistinctly* that his explanation is of no value. He does not introduce into his explanation the consideration of the sun's circular figure, and is thus prevented from giving a true account of this very simple optical phenomenon.

6. Again, to pass to a more extensive failure: why was it that Aristotle, knowing the property of the lever, and many other mechanical truths, was unable to form them into a science of mechanics, as Archimedes afterwards did?

The reason was, that, instead of considering rest and motion directly, and distinctly, with reference to the Idea of Cause, that is Force, he wandered in search of reasons among other ideas and notions, which could not be brought into steady connection with the facts;—the ideas of properties of circles, of proportions of velocities,—the notions of "strange" and "common," of "natural" and "unnatural." Thus, in the Proem to his Mechanical Problems, after stating some of the difficulties which he has to attack, he says, "Of all such cases, the circle contains the principle of the cause. And this is what might be looked for; for it is nothing absurd, if something *wonderful* is derived from something more wonderful still. Now the most wonderful thing is, that opposites should be combined; and the circle is constituted of such combinations of opposites. For it is constructed by a stationary point and a moving line, which are contrary to each other in nature; and hence we may the less be surprised at the resulting contrarieties. And in the first place, the circumference of the circle, though a line without breadth, has opposite qualities; for it is both *convex* and *concave*. In the next place, it has, at the same time, opposite motions, for it moves forward and backward at the same time. For the circumference, setting out from any point, comes to the same point again, so

---

[15] Problem. 15, *ὅσα μαθηματίκης*, &c.

that by a continuous progression, the last point becomes the first. So that, as was before stated, it is not surprising that the circle should be the principle of all wonderful properties."

Aristotle afterwards proceeds to explain more specially how he applies the properties of the circle in this case. "The reason," he says, in his fourth Problem, "why a force, acting at a greater distance from the fulcrum, moves a weight more easily, is, that it describes a greater circle." He had already asserted that when a body at the end of a lever is put in motion, it may be considered as having two motions; one in the direction of the tangent, and one in the direction of the radius; the former motion is, he says, *according to nature*, the latter, *contrary to nature*. Now in the smaller circle, the motion, contrary to nature, is more considerable than it is in the larger circle. "Therefore," he adds, "the mover or weight at the larger arm will be transferred further by the same force than the weight moved, which is at the extremity of the shorter arm."

These loose and inappropriate notions of "natural" and "unnatural" motions, were unfit to lead to any scientific truths; and, with the habits of thought which dictated these speculations a perception of the true grounds of mechanical properties was impossible.

7. Thus, in this instance, the error of Aristotle was the neglect of the Idea *appropriate* to the facts, namely, the Idea of Mechanical Cause, which is Force; and the substitution of vague or inapplicable notions involving only relations of space or emotions of wonder. The errors of those who failed similarly in other instances, were of the same kind. To detail or classify these would lead us too far into the philosophy of science; since we should have to enumerate the Ideas which are appropriate, and the various classes of Facts on which the different sciences are founded,—a task not to be now lightly undertaken. But it will be perceived, without further explanation, that it is necessary, in order to obtain from facts any general truth, that we should apply to them that appropriate Idea, by which permanent and definite relations are established among them.

In such Ideas the ancients were very poor, and the stunted and deformed growth of their physical science was the result of this penury. The Ideas of Space and Time, Number and Motion, they did indeed possess distinctly; and so far as these went, their science was tolerably healthy. They also caught a glimpse of the Idea of a Medium by which the qualities of bodies, as colors and sounds, are perceived. But the idea of Substance remained barren in their hands;

in speculating about elements and qualities, they went the wrong way, assuming that the properties of Compounds must *resemble* those of the Elements which determine them; and their loose notions of Contrariety never approached the form of those ideas of Polarity, which, in modern times, regulate many parts of physics and chemistry.

If this statement should seem to any one to be technical or arbitrary, we must refer, for the justification of it, to the Philosophy of Science, of which we hope hereafter to treat. But it will appear, even from what has been here said, that there are certain Ideas or Forms of mental apprehension, which may be applied to Facts in such a manner as to bring into view fundamental principles of science; while the same Facts, however arrayed or reasoned about, so long as these appropriate ideas are not employed, cannot give rise to any exact or substantial knowledge.

[2d Ed.] This account of the cause of failure in the physical speculations of the ancient Greek philosophers has been objected to as unsatisfactory. I will offer a few words in explanation of it.

The mode of accounting for the failure of the Greeks in physics is, in substance;—that the Greeks in their physical speculations fixed their attention upon the wrong aspects and relations of the phenomena; and that the aspects and relations in which phenomena are to be viewed in order to arrive at scientific truths may be arranged under certain heads, which I have termed *Ideas;* such as Space, Time, Number, Cause, Likeness. In every case, there is an Idea to which the phenomena may be referred, so as to bring into view the Laws by which they are governed; this Idea I term the *appropriate* Idea in such case; and in order that the reference of the phenomena to the Law may be clearly seen, the Idea must be *distinctly* possessed.

Thus the reason of Aristotle's failure in his attempts at Mechanical Science is, that he did not refer the facts to the appropriate Idea, namely Force, the Cause of Motion, but to relations of Space and the like; that is, he introduces *Geometrical* instead of *Mechanical* Ideas. It may be said that we learn little by being told that Aristotle's failure in this and the like cases arose from his referring to the wrong class of Ideas; or, as I have otherwise expressed it, fixing his attention upon the wrong aspects and relations of the facts; since, it may be said, this is only to state in other words that he *did* fail. But this criticism is, I think, ill-founded. The account which I have given is not only a statement that Aristotle, and others who took a like course, did fail; but also, that they failed in one certain point out of several which are enumerated. They did not fail because they neglected to observe facts; they did not fail because they omitted to class facts; they did not fail because they had not ideas to reason from; but they failed because they did not take the right ideas in each case. And so long as they were in the wrong in this point, no industry in collecting facts, or ingenuity in classing them and reasoning about them, could lead them to solid truth.

Nor is this account of the nature of their mistake without its instruction for us; although we are not to expect to derive from the study of their failure any technical rule which shall necessarily guide us to scientific discovery. For their failure teaches us that, in the formation of science, an Error in the Ideas is as fatal to the discovery of Truth as an Error in the Facts; and may as completely impede the progress of knowledge. I have in Books II. to X. of the *Philosophy*, shown historically how large a portion of the progress of Science consists in the establishment of Appropriate Ideas as the basis of each science. Of the two main processes by which science is constructed, as stated in Book XI. of that work, namely the *Explication of Conceptions* and the *Colligation of Facts*, the former must precede the latter. In Book XII. chap. 5, of the *Philosophy*, I have stated the maxim concerning appropriate Ideas in this form, that *the Idea and the Facts must be homogeneous.*

When I say that the failure of the Greeks in physical science arose from their not employing *appropriate* Ideas to connect the facts, I do not use the term "appropriate" in a loose popular sense; but I employ it as a somewhat technical term, to denote *the* appropriate Idea, out of that series of Ideas which have been made (as I have shown in the *Philosophy*) the foundation of sciences; namely, Space, Time, Number, Cause, Likeness, Substance, and the rest. It appears to me just to say that Aristotle's failure in his attempts to deal with problems of equilibrium, arose from his referring to circles, velocities, notions of natural and unnatural, and the like,—conceptions depending upon Ideas of Space, of Nature, &c.—which are not appropriate to these problems, and from his missing the Idea of Mechanical Force or Pressure, which is the appropriate Idea.

I give this, not as an account of *all* failures in attempts at science, but only as the account of such radical and fundamental failures as this of Aristotle; who, with a knowledge of the facts, failed to connect them into a really scientific view. If I had to compare rival theories of a more complex kind, I should not necessarily say that one involved an appropriate Idea and the other did not, though I might judge one to be true and the other to be false. For instance, in comparing the emissive and the undulatory theory of light, we see that both involve the same Idea;—the Idea of a Medium acting by certain mechanical properties. The question there is, What is the true view of the mechanism of the Medium?

It may be remarked, however, that the example of Aristotle's failure in physics, given in p. 87, namely, his attempted explanation of the round image of a square hole, is a specimen rather of *indistinct* than of inappropriate ideas.

The geometrical explanation of this phenomenon, which I have there inserted, was given by Maurolycus, and before him, by Leonardo da Vinci.

We shall, in the next Book, see the influence of the appropriate general Ideas, in the formation of various sciences. It need only be observed, before we proceed, that, in order to do full justice to the physical knowledge of the Greek Schools of philosophy, it is not necessary to study their course after the time of their founders. Their fortunes, in respect of such acquisitions as we are now considering, were not progressive. The later chiefs of the Schools followed the earlier masters; and though they varied much, they added little. The Romans adopted the philosophy of their Greek subjects; but they were always, and, indeed, acknowledged themselves to be, inferior to their teachers. They were as arbitrary and loose in their ideas as the Greeks, without possessing their invention, acuteness, and spirit of system.

In addition to the vagueness which was combined with the more elevated trains of philosophical speculation among the Greeks, the Romans introduced into their treatises a kind of declamatory rhetoric, which arose probably from their forensic and political habits, and which still further obscured the waning gleams of truth. Yet we may also trace in the Roman philosophers to whom this charge mostly applies (Lucretius, Pliny, Seneca), the national vigor and ambition. There is something Roman in the public spirit and anticipation of universal empire which they display, as citizens of the intellectual republic. Though they speak sadly or slightingly of the achievements of their own generation, they betray a more abiding and vivid belief in the dignity and destined advance of human knowledge as a whole, than is obvious among the Greeks.

We must, however, turn back, in order to describe steps of more definite value to the progress of science than those which we have hitherto noticed.

BOOK II.

HISTORY

OF THE

PHYSICAL SCIENCES

IN

ANCIENT GREECE.

Ναρθηκοπλήρωτον δὲ θηρῶμαι πυρὸς
Πηγὴν κλοπαίαν, ἣ διδάσκαλος τέχνης
Πάσης βροτοῖς πεφῆνε καὶ μέγας πόρος.
Prom. Vinct. 109.

I brought to earth the spark of heavenly fire,
Concealed at first, and small, but spreading soon
Among the sons of men, and burning on,
Teacher of art and use, and fount of power.

INTRODUCTION.

IN order to the acquisition of any such exact and real knowledge of nature as that which we properly call Physical Science, it is requisite, as has already been said, that men should possess Ideas both distinct and appropriate, and should apply them to ascertained Facts. They are thus led to propositions of a general character, which are obtained by Induction, as will elsewhere be more fully explained. We proceed now to trace the formation of Sciences among the Greeks by such processes. The provinces of knowledge which thus demand our attention are, Astronomy, Mechanics and Hydrostatics, Optics and Harmonics ; of which I must relate, first, the earliest stages, and next, the subsequent progress.

Of these portions of human knowledge, Astronomy is, beyond doubt or comparison, much the most ancient and the most remarkable ; and probably existed, in somewhat of a scientific form, in Chaldea and Egypt, and other countries, before the period of the intellectual activity of the Greeks. But I will give a brief account of some of the other Sciences before I proceed to Astronomy, for two reasons ; first, because the origin of Astronomy is lost in the obscurity of a remote antiquity ; and therefore we cannot exemplify the conditions of the first rise of science so well in that subject as we can in others which assumed their scientific form at known periods ; and next, in order that I may not have to interrupt, after I have once begun it, the history of the only progressive Science which the ancient world produced.

It has been objected to the arrangement here employed that it is not symmetrical ; and that Astronomy, as being one of the Physical Sciences, ought to have occupied a chapter in this Second Book, instead of having a whole Book to itself (Book III). I do not pretend that the arrangement is symmetrical, and have employed it only on the ground of convenience. The importance and extent of the history of Astronomy are such that this science could not, with a view to our purposes, be made co-ordinate with Mechanics or Optics.

CHAPTER I.

Earliest Stages of Mechanics and Hydrostatics.

Sect. 1.—*Mechanics.*

ASTRONOMY is a science so ancient that we can hardly ascend to a period when it did not exist; Mechanics, on the other hand, is a science which did not begin to be till after the time of Aristotle; for Archimedes must be looked upon as the author of the first sound knowledge on this subject. What is still more curious, and shows remarkably how little the continued progress of science follows inevitably from the nature of man, this department of knowledge, after the right road had been fairly entered upon, remained absolutely stationary for nearly two thousand years; no single step was made, in addition to the propositions established by Archimedes, till the time of Galileo and Stevinus. This extraordinary halt will be a subject of attention hereafter; at present we must consider the original advance.

The great step made by Archimedes in Mechanics was the establishing, upon true grounds, the general proposition concerning a straight lever, loaded with two heavy bodies, and resting upon a fulcrum. The proposition is, that two bodies so circumstanced will balance each other, when the distance of the smaller body from the fulcrum is greater than the distance of the other, in exactly the same proportion in which the weight of the body is less.

This proposition is proved by Archimedes in a work which is still extant, and the proof holds its place in our treatises to this day, as the simplest which can be given. The demonstration is made to rest on assumptions which amount in effect to such Definitions and Axioms as these: That those bodies are of equal weight which balance each other at equal arms of a straight lever; and that in every heavy body there is a definite point called a *Centre of Gravity*, in which point we may suppose the weight of the body collected.

The principle, which is really the foundation of the validity of the demonstration thus given, and which is the condition of all experimental knowledge on the subject, is this: that when two equal weights are supported on a lever, they act on the fulcrum of the lever with the same effect as if they were both together supported immediately at that point. Or more generally, we may state the principle to be this: that the pressure by which a heavy body is supported continues the same, however we alter the form or position of the body, so long as the magnitude and material continue the same.

The experimental truth of this principle is a matter of obvious and universal experience. The weight of a basket of stones is not altered by shaking the stones into new positions. We cannot make the direct burden of a stone less by altering its position in our hands; and if we try the effect on a balance or a machine of any kind, we shall see still more clearly and exactly that the altered position of one weight, or the altered arrangement of several, produces no change in their effect, so long as their point of support remains unchanged.

This general fact is obvious, when we possess in our minds the ideas which are requisite to apprehend it clearly. But when we are so prepared, the truth appears to be manifest, even independent of experience, and is seen to be a rule to which experience must conform. What, then, is the leading idea which thus enables us to reason effectively upon mechanical subjects? By attention to the course of such reasonings, we perceive that it is the idea of *Pressure;* Pressure being conceived as a measurable effect of heavy bodies at rest, distinguishable from all other effects, such as motion, change of figure, and the like. It is not here necessary to attempt to trace the history of this idea in our minds; but it is certain that such an idea may be distinctly formed, and that upon it the whole science of statics may be built. *Pressure*, *load*, *weight*, are names by which this idea is denoted when the effect tends directly downwards; but we may have pressure without motion, or *dead pull*, in other cases, as at the critical instant when two nicely-matched wrestlers are balanced by the exertion of the utmost strength of each.

Pressure in any direction may thus exist without any motion whatever. But the causes which produce such pressure are capable of producing motion, and are generally seen producing motion, as in the above instance of the wrestlers, or in a pair of scales employed in weighing; and thus men come to consider pressure as the exception, and motion as the rule: or perhaps they image to themselves the motion which *might* or *would* take place; for instance, the motion which the arms of a lever *would* have if they *did* move. They turn away from the case really before them, which is that of bodies at rest, and balancing each other, and pass to another case, which is arbitrarily assumed to represent the first. Now this arbitrary and capricious evasion of the question we consider as opposed to the introduction of the distinct and proper idea of Pressure, by means of which the true principles of this subject can be apprehended.

We have already seen that Aristotle was in the number of those who thus evaded the difficulties of the problem of the lever, and consequently lost the reward of success. He failed, as has before been stated, in consequence of his seeking his principles in notions, either vague and loose, as the distinction of natural and unnatural motions, or else inappropriate, as the circle which the weight *would* describe, the velocity which it *would* have if it moved; circumstances which are not part of the fact under consideration. The influence of such modes of speculation was the main hindrance to the prosecution of the true Archimedean form of the science of Mechanics.

The mechanical doctrine of Equilibrium, is *Statics*. It is to be distinguished from the mechanical doctrine of Motion, which is termed *Dynamics*, and which was not successfully treated till the time of Galileo.

Sect. 2.—Hydrostatics.

Archimedes not only laid the foundations of the Statics of solid bodies, but also solved the principal problem of *Hydrostatics*, or the Statics of Fluids; namely, the conditions of the floating of bodies. This is the more remarkable, since not only did the principles which Archimedes established on this subject remain unpursued till the revival of science in modern times, but, when they were again put forward, the main proposition was so far from obvious that it was termed, and is to this day called, the *hydrostatic paradox*. The true doctrine of Hydrostatics, however, assuming the Idea of Pressure, which it involves, in common with the Mechanics of solid bodies, requires also a distinct Idea of a Fluid, as a body of which the parts are perfectly movable among each other by the slightest partial pressure, and in which all pressure exerted on one part is transferred to all other parts. From this idea of Fluidity, necessarily follows that multiplication of pressure which constitutes the hydrostatic paradox; and the notion being seen to be verified in nature, the consequences were also realized as facts. This notion of Fluidity is expressed in the postulate which stands at the head of Archimedes' "Treatise on Floating Bodies." And from this principle are deduced the solutions, not only of the simple problems of the science, but of some problems of considerable complexity.

The difficulty of holding fast this Idea of Fluidity so as to trace its consequences with infallible strictness of demonstration, may be judged of from the circumstance that, even at the present day, men of great talents, not unfamiliar with the subject, sometimes admit into their reasonings an oversight or fallacy with regard to this very point. The importance of the Idea when clearly apprehended and securely held, may be judged of from this, that the whole science of Hydrostatics in its most modern form is only the development of the Idea. And what kind of attempts at science would be made by persons destitute of this Idea, we may see in the speculations of Aristotle concerning light and heavy bodies, which we have already quoted; where, by considering light and heavy as opposite qualities, residing in things themselves, and by an inability to apprehend the effect of surrounding fluids in supporting bodies, the subject was made a mass of false or frivolous assertions, which the utmost ingenuity could not reconcile with facts, and could still less deduce from the asserted doctrines any new practical truths.

In the case of Statics and Hydrostatics, the most important condition of their advance was undoubtedly the distinct apprehension of these two *appropriate Ideas*—*Statical Pressure*, and *Hydrostatical Pressure* as included in the idea of Fluidity. For the Ideas being once clearly possessed, the experimental laws which they served to express (that the whole pressure of a body downwards was always the same; and that water, and the like, were fluids according to the above idea of fluidity), were so obvious, that there was no doubt nor difficulty about them. These two ideas lie at the root of all mechanical science; and the firm possession of them is, to this day, the first requisite for a student of the subject. After being clearly awakened in the mind of Archimedes, these ideas slept for many centuries, till they were again called up in Galileo, and more remarkably in Stevinus. This time, they were not destined again to slumber; and the results of their activity have been the formation of two Sciences, which are as certain and severe in their demonstrations as geometry itself, and as copious and interesting in their conclusions; but which, besides this recommendation, possess one of a different order,—that they exhibit the exact impress of the laws of the physical world, and unfold a portion of the rules according to which the phenomena of nature take place, and must take place, till nature herself shall alter.

CHAPTER II.

EARLIEST STAGES OF OPTICS.

THE progress made by the ancients in Optics was nearly proportional to that which they made in Statics. As they discovered the true grounds of the doctrine of Equilibrium, without obtaining any sound principles concerning Motion, so they discovered the law of the Reflection of light, but had none but the most indistinct notions concerning Refraction.

The extent of the principles which they really possessed is easily stated. They knew that vision is performed by *rays* which proceed in straight lines, and that these rays are *reflected* by certain surfaces (mirrors) in such manner that the angles which they make with the surface on each side are equal. They drew various conclusions from these premises by the aid of geometry; as, for instance, the convergence of rays which fall on a concave speculum.

It may be observed that the *Idea* which is here introduced, is that of visual *rays*, or lines along which vision is produced and light carried. This idea once clearly apprehended, it was not difficult to show that these lines are straight lines, both in the case of light and of sight. In the beginning of Euclid's "Treatise on Optics," some of the arguments are mentioned by which this was established. We are told in the Proem, "In explaining what concerns the sight, he adduced certain arguments from which he inferred that all light is carried in straight lines. The greatest proof of this is shadows, and the bright spots which are produced by light coming through windows and cracks, and which could not be, except the rays of the sun were carried in straight lines. So in fires, the shadows are greater than the bodies if the fire be small, but less than the bodies if the fire be greater." A clear comprehension of the principle would lead to the perception of innumerable proofs of its truth on every side.

The Law of Equality of Angles of Incidence and Reflection was not quite so easy to verify; but the exact resemblance of the object and its image in a plane mirror (as the surface of still water, for instance), which is a consequence of this law, would afford convincing evidence of its truth in that case, and would be confirmed by the examination of other cases.

With these true principles was mixed much error and indistinctness, even in the best writers. Euclid, and the Platonists, maintained that vision is exercised by rays proceeding *from* the eye, not *to* it; so that when we see objects, we learn their form as a blind man would do, by feeling it out with his staff. This mistake, however, though Montucla speaks severely of it, was neither very discreditable nor very injurious; for the mathematical conclusions on each supposition are necessarily the same. Another curious and false assumption is, that these visual rays are not close together, but separated by intervals, like the fingers when the hand is spread. The motive for this invention was the wish to account for the fact, that in looking for a small object, as a needle, we often cannot see it when it is under our nose; which it was conceived would be impossible if the visual rays reached to all points of the surface before us.

These errors would not have prevented the progress of the science. But the Aristotelian physics, as usual, contained speculations more essentially faulty. Aristotle's views led him to try to describe the kind of causation by which vision is produced, instead of the laws by which it is exercised; and the attempt consisted, as in other subjects, of indistinct principles, and ill-combined facts. According to him, vision must be produced by a Medium,—by something *between* the object and the eye,—for if we press the object on the eye, we do not see it; this Medium is Light, or "the transparent in action;" darkness occurs when the transparency is potential, not actual; color is not the "absolute visible," but something which is *on* the absolute visible; color has the power of setting the transparent in action; it is not, however, all colors that are seen by means of light, but only the proper color of each object; for some things, as the heads, and scales, and eyes of fish, are seen in the dark; but then they are not seen with their proper color."[1]

In all this there is no steady adherence either to one notion, or to one class of facts. The distinction of Power and Act is introduced to modify the Idea of Transparency, according to the formula of the school; then Color is made to be something unknown in addition to Visibility; and the distinction of "proper" and "improper" colors is assumed, as sufficient to account for a phenomenon. Such classifications have in them nothing of which the mind can take steady hold; nor is it difficult to see that they do not come under those

[1] De Anim. ii. 6.

conditions of successful physical speculation, which we have laid down.

It is proper to notice more distinctly the nature of the Geometrical Propositions contained in Euclid's work. The *Optica* contains Propositions concerning Vision and Shadows, derived from the principle that the rays of light are rectilinear: for instance, the Proposition that the shadow is greater than the object, if the illuminating body be less, and *vice versa.* The *Catoptrica* contains Propositions concerning the effects of Reflection, derived from the principle that the Angles of Incidence and Reflection are equal: as, that in a convex mirror the object appears convex, and smaller than the object. We see here an example of the promptitude of the Greeks in deduction. When they had once obtained a knowledge of a principle, they followed it to its mathematical consequences with great acuteness. The subject of concave mirrors is pursued further in Ptolemy's *Optics.*

The Greek writers also cultivated the subject of *Perspective* speculatively, in mathematical treatises, as well as practically, in pictures. The whole of this theory is a consequence of the principle that vision takes place in straight lines drawn from the object to the eye.

"The ancients were in some measure acquainted with the Refraction as well as the Reflection of Light," as I have shown in Book IX. Chap. 2 [2d Ed.] of the *Philosophy.* The current knowledge on this subject must have been very slight and confused; for it does not appear to have enabled them to account for one of the simplest results of Refraction, the magnifying effect of convex transparent bodies. I have noticed in the passage just referred to, Seneca's crude notions on this subject; and in like manner Ptolemy in his *Optics* asserts that an object placed in water must always appear larger then when taken out. Aristotle uses the term ἀνακλάσις (*Meteorol.* iii. 2), but apparently in a very vague manner. It is not evident that he distinguished Refraction from Reflection. His Commentators however do distinguish these as διακλάσις and ἀνακλάσις. See Olympiodorusin Schneider's *Eclogæ Physicæ*, vol. i. p. 397. And Refraction had been the subject of special attention among the Greek Mathematicians. Archimedes had noticed (as we learn from the same writer) that in certain cases, a ring which cannot be seen over the edge of the empty vessel in which it is placed, becomes visible when the vessel is filled with water. The same fact is stated in the *Optics* of Euclid. We do not find this fact explained in that work as we now have it; but in Ptolemy's *Optics* the fact is explained by a flexure of the visual ray: it is noticed that this flexure is different at different angles from the perpendicular, and there is an elaborate collection of measures of the flexure at different angles, made by means of an instrument devised for the purpose. There is also a collection of similar measures of the refraction when the ray passes from air to glass, and when it passes from glass to water. This part of Ptolemy's work is, I think, the oldest extant example of a collection of experimental measures in any other subject than astronomy; and in astronomy our measures are the result of *observation*, rather than of *experiment*. As Delambre says (*Astron. Anc.* vol. ii. p. 427), "On y voit des expériences de physique bien faites, ce qui est sans exemple chez les anciens."

Ptolemy's Optical work was known only by Roger Bacon's references to it (*Opus Majus*, p. 286, &c.) till 1816; but copies of Latin translations of it were known to exist in the Royal Library at Paris, and in the Bodleian at Oxford. Delambre has given an account of the contents of the Paris copy in his *Astron. Anc.* ii. 414, and in the *Connoissance des Temps* for 1816; and Prof. Rigaud's account of the Oxford copy is given in the article *Optics*, in the *Encyclopædia Britannica.* Ptolemy shows great sagacity in applying the notion of Refraction to the explanation of the displacement of astronomical objects which is produced by the atmosphere,—*Astronomical Refraction*, as it is commonly called. He represents the visual ray as refracted in passing from the *ether*, which is above the air, into the air; the air being bounded by a spherical surface which has for its centre "the centre of all the elements, the centre of the earth;" and the refraction being a flexure towards the line drawn perpendicular to this surface. He thus constructs, says Delambre, the same figure on which Cassini afterwards founded the whole of his theory; and gives a theory more complete than that of any astronomer previous to him. Tycho, for instance, believed that astronomical refraction was caused only by the *vapors* of the atmosphere, and did not exist above the altitude of 45°.

Cleomedes, about the time of Augustus, had guessed at Refraction, as an explanation of an eclipse in which the sun and moon are both seen at the same time. "Is it not possible," he says, "that the ray which proceeds from the eye and traverses moist and cloudy air may bend downwards to the sun, even when he is below the horizon?" And Sextus Empiricus, a century later, says, "The air being dense, by the refraction of the visual ray, a constellation may be seen above the horizon when it is yet below the horizon." But from what follows, it appears doubtful whether he clearly distinguished Refraction and Reflection.

In order that we may not attach too much value to the vague expressions of Cleomedes and Sextus Empiricus, we may remark that Cleomedes conceives such an eclipse as he describes not to be possible, though he offers an explanation of it if it be: (the fact must really occur whenever the moon is seen in the horizon in the middle of an eclipse:) and that Sextus Empiricus gives his suggestion of the effect of refraction as an argument why the Chaldean astrology cannot be true, since the constellation which appears to be rising at the moment of a birth is not the one which is truly rising. The Chaldeans might have answered, says Delambre, that the star begins to shed its influence, not when it is really in the horizon, but when its light is seen. (*Ast. Anc.* vol. i. p. 231, and vol. ii. p. 548.)

It has been said that Vitellio, or Vitello, whom we shall hereafter have to speak of in the history of Optics, took his Tables of Refractions from Ptolemy. This is contrary to what Delambre states. He says that Vitello may be accused of plagiarism from Alhazen, and that Alhazen did not borrow his Tables from Ptolemy. Roger Bacon had said (*Opus Majus*, p. 288), "Ptolemæus in libro de Opticis, id est, de Aspectibus, seu in Perspectivâ suâ, qui prius quam Alhazen dedit hanc sententiam, quam a Ptolemæo acceptam Alhazen exposuit." This refers only to the opinion that visual rays proceed from the eye. But this also is erroneous; for Alhazen maintains the contrary: "Visio fit radiis a visibili extrinsecus ad visum manantibus." (*Opt.* Lib. i. cap. 5.) Vitello says of his Table of Refractions, "Acceptis instrumentaliter, prout potuimus propinquius, angulis omnium refractionum . . . invenimus quod semper iidem sunt anguli refractionum: . . . secundum hoc fecimus has tabulas." "Having measured, by means of instruments, as exactly as we could, the whole range of the angles of refraction, we found that the refraction is always the same for the same angle; and hence we have constructed these Tables."

CHAPTER III.

Earliest Stages of Harmonics.

AMONG the ancients, the science of Music was an application of Arithmetic, as Optics and Mechanics were of Geometry. The story which is told concerning the origin of their arithmetical music, is the following, as it stands in the Arithmetical Treatise of Nicomachus.

Pythagoras, walking one day, meditating on the means of measuring musical notes, happened to pass near a blacksmith's shop, and had his attention arrested by hearing the hammers, as they struck the anvil, produce the sounds which had a musical relation to each other. On listening further, he found that the intervals were a Fourth, a Fifth, and an Octave; and on weighing the hammers, it appeared that the one which gave the Octave was *one-half* the heaviest, the one which gave the Fifth was *two-thirds*, and the one which gave the Fourth was *three-quarters*. He returned home, reflected upon this phenomenon, made trials, and finally discovered, that if he stretched musical strings of equal lengths, by weights which have the proportion of one-half, two-thirds, and three-fourths, they produced intervals which were an Octave, a Fifth, and a Fourth. This observation gave an arithmetical measure of the principal Musical Intervals, and made Music an arithmetical subject of speculation.

This story, if not entirely a philosophical fable, is undoubtedly inaccurate; for the musical intervals thus spoken of would not be produced by striking with hammers of the weights there stated. But it is true that the notes of strings have a definite relation to the forces which stretch them; and this truth is still the groundwork of the theory of musical concords and discords.

Nicomachus says that Pythagoras found the weights to be, as I have mentioned, in the proportion of 12, 6, 8, 9; and the intervals, an Octave, corresponding to the proportion 12 to 6, or 2 to 1; a Fifth, corresponding to the proportion 12 to 8, or 3 to 2; and a Fourth, corresponding to the proportion 12 to 9, or 4 to 3. There is no doubt that this statement of the ancient writer is inexact as to the physical fact, for the rate of vibration of a string, on which its note depends, is, other things being equal, not as the weight, but as the square root of the weight. But he is right as to the essential point, that those ratios of 2 to 1, 3 to 2, and 4 to 3, are the characteristic ratios of the Octave, Fifth, and Fourth. In order to produce these intervals, the appended weights must be, not as 12, 9, 8, and 6, but as 12, $6\frac{3}{4}$, $5\frac{1}{3}$, and 3.

The numerical relations of the other intervals of the musical scale, as well as of the Octave, Fifth, and Fourth, were discovered by the Greeks. Thus they found that the proportion in a Major Third was 5 to 4; in a Minor Third, 6 to 5; in a Major Tone, 9 to 8; in a Semitone or *Diesis*, 16 to 15. They even went so far as to determine the *Comma*, in which the interval of two notes is so small that they are in the proportion of 81 to 80. This is the interval between two notes, each of which may be called the Seventeenth above the key-note;—the one note being obtained by ascending a Fifth four times over; the other being obtained by ascending through two Octaves and a Major Third. The want of exact coincidence between these two notes is an inherent arithmetical imperfection in the musical scale, of which the consequences are very extensive.

The numerical properties of the musical scale were worked out to a very great extent by the Greeks, and many of their Treatises on this subject remain to us. The principal ones are the seven authors published by Meibomius.[1] These arithmetical elements of Music are to the present day important and fundamental portions of the Science of Harmonics.

It may at first appear that the truth, or even the possibility of this history, by referring the discovery to accident, disproves our doctrine, that this, like all other fundamental discoveries, required a distinct and well-pondered Idea as its condition. In this, however, as in all cases of supposed accidental discoveries in science, it will be found, that it was exactly the possession of such an Idea which made the accident possible.

Pythagoras, assuming the truth of the tradition, must have had an exact and ready apprehension of those relations of musical sounds, which are called respectively an Octave, a Fifth, and a Fourth. If he had not been able to conceive distinctly this relation, and to apprehend it when heard, the sounds of the anvil would have struck his ears to no more purpose than they did those of the smiths themselves. He

[1] *Antiquæ Musicæ Scriptores septem*, 1652.

must have had, too, a ready familiarity with numerical ratios; and, moreover (that in which, probably, his superiority most consisted), a disposition to connect one notion with the other—the musical relation with the arithmetical, if it were found possible. When the connection was once suggested, it was easy to devise experiments by which it might be confirmed.

"The philosophers of the Pythagorean School,[2] and in particular, Lasus of Hermione, and Hippasus of Metapontum, made many such experiments upon strings; varying both their lengths and the weights which stretched them; and also upon vessels filled with water, in a greater or less degree." And thus was established that connection of the Idea with the Fact, which this Science, like all others, requires.

I shall quit the Physical Sciences of Ancient Greece, with the above brief statement of the discovery of the fundamental principles which they involved; not only because such initial steps must always be the most important in the progress of science, but because, in reality, the Greeks made no advances beyond these. There took place among them no additional inductive processes, by which new facts were brought under the dominion of principles, or by which principles were presented in a more comprehensive shape than before. Their advance terminated in a single stride. Archimedes had stirred the intellectual world, but had not put it in progressive motion: the science of Mechanics stopped where he left it. And though, in some subjects, as in Harmonics, much was written, the works thus produced consisted of deductions from the fundamental principles, by means of arithmetical calculations; occasionally modified, indeed, by reference to the pleasures which music, as an art, affords, but not enriched by any new scientific truths.

[3d Ed.] We should, however, quit the philosophy of the ancient Greeks without a due sense of the obligations which Physical Science in all succeeding ages owes to the acute and penetrating spirit in which their inquiries in that region of human knowledge were conducted, and to the large and lofty aspirations which were displayed, even in their failure, if we did not bear in mind both the multifarious and comprehensive character of their attempts, and some of the causes which limited their progress in positive science. They speculated and

---

[2] Montucla, iii. 10.

theorized under a lively persuasion that a Science of every part of nature was possible, and was a fit object for the exercise of man's best faculties; and they were speedily led to the conviction that such a science must clothe its conclusions in the language of mathematics. This conviction is eminently conspicuous in the writings of Plato. In the *Republic*, in the *Epinomis*, and above all in the *Timæus*, this conviction makes him return, again and again, to a discussion of the laws which had been established or conjectured in his time, respecting Harmonics and Optics, such as we have seen, and still more, respecting Astronomy, such as we shall see in the next Book. Probably no succeeding step in the discovery of the Laws of Nature was of so much importance as the full adoption of this pervading conviction, that there must be Mathematical Laws of Nature, and that it is the business of Philosophy to discover these Laws. This conviction continues, through all the succeeding ages of the history of science, to be the animating and supporting principle of scientific investigation and discovery. And, especially in Astronomy, many of the erroneous guesses which the Greeks made, contain, if not the germ, at least the vivifying life-blood, of great truths, reserved for future ages.

Moreover, the Greeks not only sought such theories of special parts of nature, but a general Theory of the Universe. An essay at such a theory is the *Timæus* of Plato; too wide and too ambitious an attempt to succeed at that time; or, indeed, on the scale on which he unfolds it, even in our time; but a vigorous and instructive example of the claim which man's Intellect feels that it may make to understand the universal frame of things, and to render a reason for all that is presented to it by the outward senses.

Further; we see in Plato, that one of the grounds of the failure in this attempt, was the assumption that the *reason why* every thing is what it is and as it is, must be that so it is *best*, according to some view of better or worse attainable by man. Socrates, in his dying conversation, as given in the *Phædo*, declares this to have been what he sought in the philosophy of his time; and tells his friends that he turned away from the speculations of Anaxagoras because they did not give him such reasons for the constitution of the world; and Plato's *Timæus* is, in reality, an attempt to supply this deficiency, and to present a Theory of the Universe, in which every thing is accounted for by such reasons. Though this is a failure, it is a noble as well as an instructive failure.

BOOK III.

HISTORY

OF

GREEK ASTRONOMY.

Τόδε δὲ μηδέις ποτε φοβηθῇ τῶν Ἑλλήνων, ὡς οὐ χρη περὶ τὰ θεῖα ποτὲ πραγματεύεσθαι θνητοὺς ὄντας· πᾶν δε τούτου διανοηθῆναι τοὐναντίον, ὡς οὔτε ἄφρον ἔστι ποτὲ τὸ θεῖον, οὔτε ἀγνοεῖ που τὴν ἀνθρωπίνην φυσιν· ἀλλ᾽ οἶδεν ὅτι, διδάσκοντος αὐτοῦ, ξυνακολουθήσει καὶ μαθήσεται τὰ διδάσκομενα.—PLATO, *Epinomis*, p. 988.

Nor should any Greek have any misgiving of this kind; that it is not fitting for us to inquire narrowly into the operations of Superior Powers, such as those by which the motions of the heavenly bodies are produced: but, on the contrary, men should consider that the Divine Powers never act without purpose, and that they know the nature of man: they know that by their guidance and aid, man may follow and comprehend the lessons which are vouchsafed him on such subjects.

INTRODUCTION.

THE earliest and fundamental conceptions of men respecting the objects with which Astronomy is concerned, are formed by familiar processes of thought, without appearing to have in them any thing technical or scientific. Days, Years, Months, the Sky, the Constellations, are notions which the most uncultured and incurious minds possess. Yet these are elements of the Science of Astronomy. The reasons why, in this case alone, of all the provinces of human knowledge, men were able, at an early and unenlightened period, to construct a science out of the obvious facts of observation, with the help of the common furniture of their minds, will be more apparent in the course of the philosophy of science: but I may here barely mention two of these reasons. They are, first, that the familiar act of thought, exercised for the common purposes of life, by which we give to an assemblage of our impressions such a unity as is implied in the above notions and terms, a Month, a Year, the Sky, and the like, is, in reality, an *inductive act*, and shares the nature of the processes by which all sciences are formed; and, in the next place, that the ideas appropriate to the induction in this case, are those which, even in the least cultivated minds, are very clear and definite; namely, the ideas of Space and Figure, Time and Number, Motion and Recurrence. Hence, from their first origin, the modifications of those ideas assume a scientific form.

We must now trace in detail the peculiar course which, in consequence of these causes, the knowledge of man respecting the heavenly bodies took, from the earliest period of his history.

CHAPTER I.

Earliest Stages of Astronomy.

Sect. 1.—*Formation of the Notion of a Year.*

THE notion of a *Day* is early and obviously impressed upon man in almost any condition in which we can imagine him. The recurrence of light and darkness, of comparative warmth and cold, of noise and silence, of the activity and repose of animals;—the rising, mounting, descending, and setting of the sun;—the varying colors of the clouds, generally, notwithstanding their variety, marked by a daily progression of appearances;—the calls of the desire of food and of sleep in man himself, either exactly adjusted to the period of this change, or at least readily capable of being accommodated to it;—the recurrence of these circumstances at intervals, equal, so far as our obvious judgment of the passage of time can decide; and these intervals so short that the repetition is noticed with no effort of attention or memory;—this assemblage of suggestions makes the notion of a Day necessarily occur to man, if we suppose him to have the conception of Time, and of Recurrence. He naturally marks by a term such a portion of time, and such a cycle of recurrence; he calls each portion of time, in which this series of appearances and occurrences come round, *a Day;* and such a group of particulars are considered as appearing or happening *in* the same day.

A Year is a notion formed in the same manner; implying in the same way the notion of recurring facts; and also the faculty of arranging facts in time, and of appreciating their recurrence. But the notion of a Year, though undoubtedly very obvious, is, on many accounts, less so than that of a Day. The repetition of similar circumstances, at equal intervals, is less manifest in this case, and the intervals being much longer, some exertion of memory becomes requisite in order that the recurrence may be perceived. A child might easily be persuaded that successive years were of unequal length; or, if the summer were cold, and the spring and autumn warm, might be made to believe, if all who spoke in its hearing agreed to support the delusion, that one year was two. It would be impossible to practise such a deception with regard to the day, without the use of some artifice beyond mere words.

Still, the recurrence of the appearances which suggest the notion of a Year is so obvious, that we can hardly conceive man without it. But though, in all climes and times, there would be a recurrence, and at the same interval in all, the recurring appearances would be extremely different in different countries; and the contrasts and resemblances of the seasons would be widely varied. In some places the winter utterly alters the face of the country, converting grassy hills, deep leafy woods of various hues of green, and running waters, into snowy and icy wastes, and bare snow-laden branches; while in others, the field retains its herbage, and the tree its leaves, all the year; and the rains and the sunshine alone, or various agricultural employments quite different from ours, mark the passing seasons. Yet in all parts of the world the yearly cycle of changes has been singled out from all others, and designated by a peculiar name. The inhabitant of the equatorial regions has the sun vertically over him at the end of every period of six months, and similar trains of celestial phenomena fill up each of these intervals, yet we do not find years of six months among such nations. The Arabs alone,[1] who practise neither agriculture nor navigation, have a year depending upon the moon only; and borrow the word from other languages, when they speak of the solar year.

In general, nations have marked this portion of time by some word which has a reference to the returning circle of seasons and employments. Thus the Latin *annus* signified a ring, as we see in the derivative *annulus:* the Greek term ἐνιαυτὸς implies something which *returns into itself:* and the word as it exists in Teutonic languages, of which our word *year* is an example, is said to have its origin in the word *yra*, which means a ring in Swedish, and is perhaps connected with the Latin *gyrus.*

Sect. 2.—*Fixation of the Civil Year.*

The year, considered as a recurring cycle of seasons and of general appearances, must attract the notice of man as soon as his attention and memory suffice to bind together the parts of a succession of the length of several years. But to make the same term imply a certain fixed number of days, we must know how many days the cycle of the seasons occupies; a knowledge which requires faculties and artifices beyond what we have already mentioned. For instance, men cannot reckon as far as any number at all approaching the number of days in the year, without possessing a system of numeral terms, and methods

[1] Ideler, *Berl. Trans.* 1813, p. 51.

of practical numeration on which such a system of terms is always founded.[2] The South American Indians, the Koussa Caffres and Hottentots, and the natives of New Holland, all of whom are said to be unable to reckon further than the fingers of their hands and feet,[3] cannot, as we do, include in their notion of a year the fact of its consisting of 365 days. This fact is not likely to be known to any nation except those which have advanced far beyond that which may be considered as the earliest scientific process which we can trace in the history of the human race, the formation of a method of designating the successive numbers to an indefinite extent, by means of names, framed according to the decimal, quinary, or vigenary scale.

But even if we suppose men to have the habit of recording the passage of each day, and of counting the score thus recorded, it would be by no means easy for them to determine the exact number of days in which the cycle of the seasons recurs; for the indefiniteness of the appearances which mark the same season of the year, and the changes to which they are subject as the seasons are early or late, would leave much uncertainty respecting the duration of the year. They would not obtain any accuracy on this head, till they had attended for a considerable time to the motions and places of the sun; circumstances which require more precision of notice than the general facts of the degrees of heat and light. The motions of the sun, the succession of the places of his rising and setting at different times of the year, the greatest heights which he reaches, the proportion of the length of day and night, would all exhibit several cycles. The turning back of the sun, when he had reached the greatest distance to the south or to the north, as shown either by his rising or by his height at noon, would perhaps be the most observable of such circumstances. Accordingly the *τροπαὶ ἠελίοιο*, the turnings of the sun, are used repeatedly by Hesiod as a mark from which he reckons the seasons of various employments. "Fifty days," he says, "after the turning of the sun, is a seasonable time for beginning a voyage."[4]

The phenomena would be different in different climates, but the recurrence would be common to all. Any one of these kinds of phenomena, noted with moderate care for a year, would show what was the number of days of which a year consisted; and if several years

[2] *Arithmetic* in *Encyc. Metrop.* (by Dr. Peacock), Art. 8. [3] Ibid. Art. 32.

[4] Ἤματα πεντήκοντα μετὰ τροπὰς ἠελίοιο
Ες τέλος ἐλθόντος θέρεος.—*Op. et Dies*, 661.

were included in the interval through which the scrutiny extended, the knowledge of the length of the year so acquired would be proportionally more exact.

Besides those notices of the sun which offered exact indications of the seasons, other more indefinite natural occurrences were used; as the arrival of the swallow (*χελιδών*) and the kite (*ἰκτίν*). The birds, in Aristophanes' play of that name, mention it as one of their offices to mark the seasons; Hesiod similarly notices the cry of the crane as an indication of the departure of winter.[5]

Among the Greeks the seasons were at first only summer and winter (*θέρος* and *χειμών*), the latter including all the rainy and cold portion of the year. The winter was then subdivided into the *χειμών* and *ἔαρ* (winter proper and spring), and the summer, less definitely, into *θέρος* and *ὀπώρα* (summer and autumn). Tacitus says that the Germans knew neither the blessings nor the name of autumn, "Autumni perinde nomen ac bona ignorantur." Yet *harvest*, *herbst*, is certainly an old German word.[6]

In the same period in which the sun goes through his cycle of positions, the stars also go through a cycle of appearances belonging to them; and these appearances were perhaps employed at as early a period as those of the sun, in determining the exact length of the year. Many of the groups of fixed stars are readily recognized, as exhibiting always the same configuration; and particular bright stars are singled out as objects of attention. These are observed, at particular seasons, to appear in the west after sunset; but it is noted that when they do this, they are found nearer and nearer to the sun every successive evening, and at last disappear in his light. It is observed also, that at a certain interval after this, they rise visibly before the dawn of day renders the stars invisible; and after they are seen to do this, they rise every day at a longer interval before the sun. The risings and settings of the stars under these circumstances, or under others which are easily recognized, were, in countries where the sky is usually clear, employed at an early period to mark the seasons of the year. Eschylus[7] makes Prometheus mention this among the benefits of which

[5] Ideler, i. 240. [6] Ib. i. 243.

[7] Οὐκ ἦν γαρ αυτοῖς οὔτε χείματος τέκμαρ,
Οὔτ' ἀνθεμώδους ἦρος, οὔδε καρπίμου
Θέρους βέβαιον· ἀλλ' ἄτερ γνώμης τὸ πᾶν
Επρασσον, ἔστε δή σφιν ἀνατολὰς ἐγὼ
Αστρων ἔδειξα, τάς τε δυσκρίτους δύσεις.—*Prom. V.* 454.

he, the teacher of arts to the earliest race of men, was the communicator.

Thus, for instance, the rising[8] of the Pleiades in the evening was a mark of the approach of winter. The rising of the waters of the Nile in Egypt coincided with the heliacal rising of Sirius, which star the Egyptians called Sothis. Even without any artificial measure of time or position, it was not difficult to carry observations of this kind to such a degree of accuracy as to learn from them the number of days which compose the year; and to fix the precise season from the appearance of the stars.

A knowledge concerning the stars appears to have been first cultivated with the last-mentioned view, and makes its first appearance in literature with this for its object. Thus Hesiod directs the husbandman when to reap by the rising, and when to plough by the setting of the Pleiades.[9] In like manner Sirius,[10] Arcturus,[11] the Hyades and Orion,[12] are noticed.

---

[8] Ideler (Chronol. i. 242) says that *this* rising of the Pleiades took place at a time of the year which corresponds to our 11th May, and the setting to the 20th October; but this does not agree with the forty days of their being "concealed," which, from the context, must mean, I conceive, the interval between their setting and rising. Pliny, however, says, "Vergiliarum exortu æstas incipit, occasu hiems; *semestri* spatio intra se messes vindemiasque et omnium maturitatem complexæ." (H. N. xviii. 69.)

The autumn of the Greeks, ὀπώρα, was earlier than our autumn, for Homer calls Sirius ἀστὴρ ὀπωρινός, which rose at the end of July.

[9] Πληίαδων 'Ατλαγενέων ἐπιτελλομενάων.
Ἄρχεσθ' ἀμητοῦ· ἀρότοιο δὲ, δυσομενάων.
Αἳ δή τοι νύκτας τε καὶ ἤματα τεσσεράκοντα
Κεκρύφαται, αὖτις δὲ περιπλομένου ἐνιαυτοῦ
Φάινονται. *Op. et Dies*, l. 381.

[10] Ib. l. 413.

[11] Εὖτ' ἂν δ' ἑξήκοντα μετὰ τροπὰς ἠελίοιο
Χειμέρι', ἐκτελέσῃ Ζεὺς ἤματα, δή ῥα τότ' ἀστὴρ
'Αρκτοῦρος, προλιπὼν ἱερον ῥόον 'Ωκεανοῖο
Πρῶτον παμφαίνων ἐπιτέλλεται ἀκροκνέφαιος.
Op. et Dies, l. 562.

'Εὖτ' ἂν δ' 'Ωρίων καὶ Σείριος ἐς μέσον ἔλθῃ
Οὐρανὸν, Αρκτοῦρον δ' ἐσίδῃ ῥοδοδάκτυλος ἠώς.
Ib. 607.

[12] . . . . αὐτὰρ ἐπὴν δὴ
Πληϊάδες 'Υάδες τε τὸ τε σθένος 'Ωρίωνος
Δύνωσιν. Ib. 612.

These methods were employed to a late period, because the Greek months, being lunar, did not correspond to the seasons. Tables of such motions were called παραπήγματά.—Ideler, *Hist. Untersuchungen*, p. 209.

By such means it was determined that the year consisted, at least, nearly, of 365 days. The Egyptians, as we learn from Herodotus,[13] claimed the honor of this discovery. The priests informed him, he says, "that the Egyptians were the first men who discovered the year, dividing it into twelve equal parts; and this they asserted that they discovered from the stars." Each of these parts or months consisted of 30 days, and they added 5 days more at the end of the year, "and thus the circle of the seasons come round." It seems, also, that the Jews, at an early period, had a similar reckoning of time, for the Deluge which continued 150 days (Gen. vii. 24), is stated to have lasted from the 17th day of the second month (Gen. vii. 11) to the 17th day of the seventh month (Gen. viii. 4), that is, 5 months of 30 days.

A year thus settled as a period of a certain number of days is called a *Civil Year*. It is one of the earliest discoverable institutions of States possessing any germ of civilization; and one of the earliest portions of human systematic knowledge is the discovery of the length of the civil year, so that it should agree with the natural year, or year of the seasons.

Sect. 3.—*Correction of the Civil Year.* (*Julian Calendar.*)

In reality, by such a mode of reckoning as we have described, the circle of the seasons would not come round exactly. The real length of the year is very nearly 365 days and a quarter. If a year of 365 days were used, in four years the year would begin a day too soon, when considered with reference to the sun and stars; and in 60 years it would begin 15 days too soon: a quantity perceptible to the loosest degree of attention. The civil year would be found not to coincide with the year of the seasons; the beginning of the former would take place at different periods of the latter; it would *wander* into various seasons, instead of remaining fixed to the same season; the term *year*, and any number of years, would become ambiguous: some correction, at least some comparison, would be requisite.

We do not know by whom the insufficiency of the year of 365 days was first discovered;[14] we find this knowledge diffused among all civilized nations, and various artifices used in making the correction. The method which we employ, and which consists in reckoning an addi-

[13] Ib. ii. 4.

[14] Syncellus (*Chronographia*, p. 123) says that according to the legend, it was King Aseth who first added the 5 additional days to 360, for the year, in the eighteenth century, B. C.

tional day at the end of February every fourth or *leap* year, is an example of the principle of *intercalation*, by which the correction was most commonly made. Methods of intercalation for the same purpose were found to exist in the new world. The Mexicans added 13 days at the end of every 52 years. The method of the Greeks was more complex (by means of the *octaëteris* or cycle of 8 years); but it had the additional object of accommodating itself to the motions of the moon, and therefore must be treated of hereafter. The Egyptians, on the other hand, knowingly permitted their civil year to *wander*, at least so far as their religious observances were concerned. "They do not wish," says Geminus,[15] "the same sacrifices of the gods to be made perpetually at the same time of the year, but that they should go through all the seasons, so that the same feast may happen in summer and winter, in spring and autumn." The period in which any festival would thus pass through all the seasons of the year is 1461 years; for 1460 years of 365¼ days are equal to 1461 years of 365 days. This period of 1461 years is called the *Sothic* Period, from Sothis, the name of the Dog-star, by which their *fixed* year was determined; and for the same reason it is called the *Canicular* Period.[16]

Other nations did not regulate their civil year by intercalation at short intervals, but rectified it by a *reform* when this became necessary. The Persians are said to have added a month of 30 days every 120 years. The Roman calendar, at first very rude in its structure, was reformed by Numa, and was directed to be kept in order by the perpetual interposition of the augurs. This, however, was, from various causes, not properly done; and the consequence was, that the reckoning fell into utter disorder, in which state it was found by Julius Cæsar, when he became dictator. By the advice of Sosigenes, he adopted the mode of intercalation of one day in 4 years, which we still retain; and in order to correct the derangement which had already been produced, he added 90 days to a year of the usual length, which thus became what was called *the year of confusion*. The *Julian Calendar*, thus reformed, came into use, January 1, B. C. 45.

Sect. 4.—Attempts at the Fixation of the Month.

THE circle of changes through which the moon passes in about thirty days, is marked, in the earliest stages of language, by a word which implies the space of time which one such circle occupies; just

[15] *Uranol.* p. 33.

[16] Censorinus *de Die Natali*, c. 18.

as the circle of changes of the seasons is designated by the word *year*. The lunar changes are, indeed, more obvious to the sense, and strike a more careless person, than the annual; the moon, when the sun is absent, is almost the sole natural object which attracts our notice; and we look at her with a far more tranquil and agreeable attention than we bestow on any other celestial object. Her changes of form and place are definite and striking to all eyes; they are uninterrupted, and the duration of their cycle is so short as to require no effort of memory to embrace it. Hence it appears to be more easy, and in earlier stages of civilization more common, to count time by *moons* than by years.

The words by which this period of time is designated in various languages, seem to refer us to the early history of language. Our word *month* is connected with the word *moon*, and a similar connection is noticeable in the other branches of the Teutonic. The Greek word μὴν in like manner is related to μήνη, which though not the common word for the moon, is found in Homer with that signification. The Latin word *mensis* is probably connected with the same group.[17]

The month is not any exact number of days, being more than 29, and less than 30. The latter number was first tried, for men more readily select numbers possessing some distinction of regularity. It existed for a long period in many countries. A very few months of 30 days, however, would suffice to derange the agreement between the days of the months and the moon's appearance. A little further trial would show that months of 29 and 30 days alternately, would preserve, for a considerable period, this agreement.

The Greeks adopted this calendar, and, in consequence, considered the days of their month as representing the changes of the moon: the last day of the month was called ἕνη καὶ νέα, "the old and new," as belonging to both the waning and the reappearing moon:[18] and their

---

[17] Cicero derives this word from the verb *to measure:* "quia *mensa* spatia conficiunt, *menses* nominantur;" and other etymologists, with similar views, connect the above-mentioned words with the Hebrew *manah*, to measure (with which the Arabic word *almanach* is connected). Such a derivation would have some analogy with that of *annus*, &c., noticed above: but if we are to attempt to ascend to the earliest condition of language, we must conceive it probable that men would have a name for a most conspicuous visible object, *the moon*, before they would have a verb denoting the very abstract and general notion, *to measure.*

[18] Aratus says of the moon, in a passage quoted by Geminus, p. 83:

> Ἄιει δ' ἄλλοθεν ἄλλα παρακλίνουσα μετωπὰ
> Ἔιρῃ, ὁπποστάιη μήνος περιτέλλεται ἠῶς.
> As still her shifting visage changing turns,
> By her we count the monthly round of morns.

festivals and sacrifices, as determined by the calendar, were conceived to be necessarily connected with the same periods of the cycles of the sun and moon. "The laws and the oracles," says Geminus, "which directed that they should in sacrifices observe three things, months, days, years, were so understood." With this persuasion, a correct system of intercalation became a religious duty.

The above rule of alternate months of 29 and 30 days, supposes the length of the months 29 days and a half, which is not exactly the length of a lunar month. Accordingly the Months and the Moon were soon at variance. Aristophanes, in "The Clouds," makes the Moon complain of the disorder when the calendar was deranged.

Οὐκ ἄγειν τὰς ἡμέρας
Οὐδὲν ὀρθῶς, ἀλλ' ἄνω τε καὶ κάτω κυδοιδοπᾶν
Ὥστ' ἀπειλεῖν φησὶν αὐτῇ τοὺς θεοὺς ἑκάστοτε
Ἡνίκ' ἂν ψευσθῶσι δείπνου κἀπίωσιν οἴκαδε
Τῆς ἑορτῆς μὴ τυχόντες κατὰ λόγον τῶν ἡμερῶν.

Nubes, 615–19.

CHORUS OF CLOUDS.

The Moon by us to you her greeting sends,
But bids us say that she's an ill-used moon,
And takes it much amiss that you should still
Shuffle her days, and turn them topsy-turvy:
And that the gods (who know their feast-days well)
By your false count are sent home supperless,
And scold and storm at her for your neglect.[19]

The correction of this inaccuracy, however, was not pursued separately, but was combined with another object, the securing a correspondence between the lunar and solar years, the main purpose of all early cycles.

Sect. 5.—Invention of Lunisolar Years.

THERE are 12 complete lunations in a year; which according to the above rule (of $29\frac{1}{2}$ days to a lunation) would make 354 days, leaving $12\frac{1}{4}$ days of difference between such a lunar year and a solar year. It is said that, at an early period, this was attempted to be corrected by interpolating a month of 30 days every alternate year; and Herodotus[20] relates a conversation of Solon, implying a still ruder mode of

---

[19] This passage is supposed by the commentators to be intended as a satire upon those who had introduced the cycle of Meton (spoken of in Sect. 5), which had been done at Athens a few years before "The Clouds" was acted.

[20] B. i. c. 15.

intercalation. This can hardly be considered as an improvement in the Greek calendar already described.

The first cycle which produced any near correspondence of the reckoning of the moon and the sun, was the *Octaëteris*, or period of 8 years: 8 years of 354 days, together with 3 months of 30 days each, making up (in 99 lunations) 2922 days; which is exactly the amount of 8 years of 365¼ days each. Hence this period would answer its purpose, so far as the above lengths of the lunar and solar cycles are exact; and it might assume various forms, according to the manner in which the three intercalary months were distributed. The customary method was to add a thirteenth month at the end of the third, fifth, and eighth year of the cycle. This period is ascribed to various persons and times; probably different persons proposed different forms of it. Dodwell places its introduction in the 59th Olympiad, or in the 6th century, B. C.: but Ideler thinks the astronomical knowledge of the Greeks of that age was too limited to allow of such a discovery.

This cycle, however, was imperfect. The duration of 99 lunations is something more than 2922 days; it is more nearly 2923½; hence in 16 years there was a deficiency of 3 days, with regard to the motions of the moon. This cycle of 16 years (*Heccædecaëteris*), with 3 interpolated days at the end, was used, it is said, to bring the calculation right with regard to the moon; but in this way the origin of the year was displaced with regard to the sun. After 10 revolutions of this cycle, or 160 years, the interpolated days would amount to 30, and hence the end of the lunar year would be a month in advance of the end of the solar. By terminating the lunar year at the end of the preceding month, the two years would again be brought into agreement: and we have thus a cycle of 160 years.[21]

This cycle of 160 years, however, was calculated from the cycle of 16 years; and it was probably never used in civil reckoning; which the others, or at least that of 8 years, appear to have been.

The cycles of 16 and 160 years were corrections of the cycle of 8 years; and were readily suggested, when the length of the solar and lunar periods became known with accuracy. But a much more exact cycle, independent of these, was discovered and introduced by Meton,[22] 432 years B. C. This cycle consisted of 19 years, and is so correct and convenient, that it is in use among ourselves to this day. The time occupied by 19 years, and by 235 lunations, is very nearly the same;

---

[21] Geminus. Ideler.

[22] Ideler, *Hist. Unters.* p. 208.

(the former time is less than 6940 days by 9½ hours, the latter, by 7½ hours). Hence, if the 19 years be divided into 235 months, so as to agree with the changes of the moon, at the end of that period the same succession may begin again with great exactness.

In order that 235 months, of 30 and 29 days, may make up 6940 days, we must have 125 of the former, which were called *full* months, and 110 of the latter, which were termed *hollow*. An artifice was used in order to distribute 110 hollow months among 6940 days. It will be found that there is a hollow month for each 63 days nearly. Hence if we reckon 30 days to every month, but at every 63d day leap over a day in the reckoning, we shall, in the 19 years, omit 110 days; and this accordingly was done. Thus the 3d day of the 3d month, the 6th day of the 5th month, the 9th day of the 7th, must be omitted, so as to make these months "hollow." Of the 19 years, seven must consist of 13 months; and it does not appear to be known according to what order these seven years were selected. Some say they were the 3d, 6th, 8th, 11th, 14th, 17th, and 19th; others, the 3d, 5th, 8th, 11th, 13th, 16th, and 19th.

The near coincidence of the solar and lunar periods in this cycle of 19 years, was undoubtedly a considerable discovery at the time when it was first accomplished. It is not easy to trace the way in which such a discovery was made at that time; for we do not even know the manner in which men then recorded the agreement or difference between the calendar day and the celestial phenomenon which ought to correspond to it. It is most probable that the length of the month was obtained with some exactness by the observation of eclipses, at considerable intervals of time from each other; for eclipses are very noticeable phenomena, and must have been very soon observed to occur only at new and full moon.[23]

The exact length of a certain number of months being thus known, the discovery of a cycle which should regulate the calendar with sufficient accuracy would be a business of arithmetical skill, and would depend, in part, on the existing knowledge of arithmetical methods; but in making the discovery, a natural arithmetical sagacity was probably more efficacious than method. It is very possible that the *Cycle of Meton* is correct more nearly than its author was aware, and more

[23] Thucyd. vii. 50. 'Η *σελήνη ἐκλείπει· ἐτύγχανε γὰρ πανσέληνος οὖσα.* iv. 52. Τοῦ *ἡλίου ἐκλιπές τι ἐγένετο περὶ νουμηνίαν.* ii. 28. Νουμηνίᾳ *κατὰ σελήνην (ὥσπερ καὶ μόνον δοκεῖ εἶναι γίγνεσθαι δυνατὸν) ὁ ἥλιος ἐξέλιπε μετὰ μεσημβρίαν καὶ πάλιν ἀν ἐπληρώθη, γενόμενος μηνοειδὴς καὶ ἀστέρων τινῶν ἐκφανέντων.*

nearly than he could ascertain from any evidence and calculation known to him. It is so exact that it is still used in calculating the new moon for the time of Easter; and the *Golden Number*, which is spoken of in stating such rules, is the number of this Cycle corresponding to the current year.[24]

Meton's Cycle was corrected a hundred years later (330 B. C.), by Calippus, who discovered the error of it by observing an eclipse of the moon six years before the death of Alexander.[25] In this corrected period, four cycles of 19 years were taken, and a day left out at the end of the 76 years, in order to make allowance for the hours by which, as already observed, 6940 days are greater than 19 years, and than 235 lunations: and this *Calippic period* is used in Ptolemy's Almagest, in stating observations of eclipses.

The Metonic and Calippic periods undoubtedly imply a very considerable degree of accuracy in the knowledge which the astronomers, to whom they are due, had of the length of the month; and the first is a very happy invention for bringing the solar and lunar calendars into agreement.

The Roman Calendar, from which our own is derived, appears to have been a much less skilful contrivance than the Greek; though scholars are not agreed on the subject of its construction, we can hardly doubt that months, in this as in other cases, were intended originally to have a reference to the moon. In whatever manner the solar and lunar motions were intended to be reconciled, the attempt seems altogether to have failed, and to have been soon abandoned. The Roman months, both before and after the Julian correction, were portions of the year, having no reference to full and new moons; and we, having adopted this division of the year, have thus, in our common calendar, the traces of one of the early attempts of mankind to seize the law of the succession of celestial phenomena, in a case where the attempt was a complete failure.

Considered as a part of the progress of our astronomical knowledge, improvements in the calendar do not offer many points to our observation, but they exhibit a few very important steps. Calendars which, belonging apparently to unscientific ages and nations, possess a great degree of accordance with the true motions of the sun and moon (like

---

[24] The same cycle of 19 years has been used by the Chinese for a very great length of time; their civil year consisting, like that of the Greeks, of months of 29 and 30 days. The Siamese also have this period. (*Astron.* Lib. U. K.)

[25] Delamb. *A. A.* p. 17.

the solar calendar of the Mexicans, and the lunar calendar of the Greeks), contain the only record now extant of discoveries which must have required a great deal of observation, of thought, and probably of time. The later improvements in calendars, which take place when astronomical observation has been attentively pursued, are of little consequence to the history of science; for they are generally founded on astronomical determinations, and are posterior in time, and inferior in accuracy, to the knowledge on which they depend. But cycles of correction, which are both short and close to exactness, like that of Meton, may perhaps be the original form of the knowledge which they imply; and certainly require both accurate facts and sagacious arithmetical reasonings. The discovery of such a cycle must always have the appearance of a happy guess, like other discoveries of laws of nature. Beyond this point, the interest of the study of calendars, as bearing on our subject, ceases: they may be considered as belonging rather to Art than to Science; rather as an application of a part of our knowledge to the uses of life, than a means or an evidence of its extension.

Sect. 6.—The Constellations.

Some tendency to consider the stars as formed into groups, is inevitable when men begin to attend to them; but how men were led to the fanciful system of names of Stars and of Constellations, which we find to have prevailed in early times, it is very difficult to determine. Single stars, and very close groups, as the Pleiades, were named in the time of Homer and Hesiod, and at a still earlier period, as we find in the book of Job.[26]

Two remarkable circumstances with respect to the Constellations are, first, that they appear in most cases to be arbitrary combinations; the artificial figures which are made to include the stars, not having any resemblance to their obvious configurations; and second, that these figures, in different countries, are so far similar, as to imply some communication. The arbitrary nature of these figures shows that they

---

[26] Job xxxviii. 31. "Canst thou bind the sweet influences of Chima (the Pleiades), or loose the bands of Kesil (Orion)? Canst thou bring forth Mazzaroth (Sirius) in his season? or canst thou guide Ash (or Aisch) (Arcturus) with his sons?"

And ix. 9. "Which maketh Arcturus, Orion, and Pleiades, and the chambers of the south."

Dupuis, vi. 545, thinks that Aisch was αἴξ, the goat and kids. See Hyde, *Ulugh-beigh.*

were rather the work of the imaginative and mythological tendencies of man, than of mere convenience and love of arrangement. "The constellations," says an astronomer of our own time,[27] "seem to have been almost purposely named and delineated to cause as much confusion and inconvenience as possible. Innumerable snakes twine through long and contorted areas of the heavens, where no memory can follow them: bears, lions, and fishes, large and small, northern and southern, confuse all nomenclature. A better system of constellations might have been a material help as an artificial memory." When men indicate the stars by figures, borrowed from obvious resemblances, they are led to combinations quite different from the received constellations. Thus the common people in our own country find a wain or wagon, or a plough, in a portion of the great bear.[28]

The similarity of the constellations recognized in different countries is very remarkable. The Chaldean, the Egyptian, and the Grecian skies have a resemblance which cannot be overlooked. Some have conceived that this resemblance may be traced also in the Indian and Arabic constellations, at least in those of the zodiac.[29] But while the figures are the same, the names and traditions connected with them are different, according to the histories and localities of each country;[30] the river among the stars which the Greeks called the Eridanus, the Egyptians asserted to be the Nile. Some conceive that the Signs of the *Zodiac*, or path along which the sun and moon pass, had its divisions marked by signs which had a reference to the course of the seasons, to the motion of the sun, or the employments of the husbandman. If we take the position of the heavens, which, from the knowledge we now possess, we are sure they must have had 15,000 years ago, the significance of the signs of the zodiac, in which the sun was, as referred to the Egyptian year, becomes very marked,[31] and has led some to suppose that the zodiac was invented at such a period. Others have rejected this as an improbably great antiquity, and have thought it more likely that the constellation assigned to each season was that which, at that season, rose at the beginning of the night:

---

[27] Sir J. Herschel.

[28] So also the Greeks, Homer, *Il.* XVIII. 487.

> *Ἄρκτον ἣν καὶ ἄμαξαν ἐπίκλησιν καλέουσιν.*
> The Northern Bear which oft the Wain they call.

Ἄρκτος was the traditional name; *ἄμαξα*, that suggested by the form.

[29] Dupuis, vi. 548. The Indian zodiac contains, in the place of our Capricorn, a ram *and* a fish, which proves the resemblance without chance of mistake. Bailly, i. p. 157.

[30] Dupuis, vi. 549.

[31] Laplace, *Hist. Astron.* p. 8.

thus the balance (which is conceived to designate the equality of days and nights) was placed among the stars which rose in the evening when the spring began: this would fix the origin of these signs 2500 years before our era.

It is clear, as has already been said, that Fancy, and probably Superstition, had a share in forming the collection of constellations. It is certain that, at an early period, superstitious notions were associated with the stars.[32] Astrology is of very high antiquity in the East. The stars were supposed to influence the character and destiny of man, and to be in some way connected with superior natures and powers.

We may, I conceive, look upon the formation of the constellations, and the notions thus connected with them, as a very early attempt to find a meaning in the relations of the stars; and as an utter failure. The first effort to associate the appearances and motions of the skies by conceptions implying unity and connection, was made in a wrong direction, as may very easily be supposed. Instead of considering the appearances only with reference to space, time, number, in a manner purely rational, a number of other elements, imagination, tradition, hope, fear, awe of the supernatural, belief in destiny, were called into action. Man, still young, as a philosopher at least, had yet to learn what notions his successful guesses on these subjects must involve, and what they must exclude. At that period, nothing could be more natural or excusable than this ignorance; but it is curious to see how long and how obstinately the belief lingered (if indeed it be yet extinct) that the motions of the stars, and the dispositions and fortunes of men, may come under some common conceptions and laws, by which a connection between the one and the other may be established.

We cannot, therefore, agree with those who consider Astrology in the early ages as "only a degraded Astronomy, the abuse of a more ancient science."[33] It was the first step to astronomy by leading to habits and means of grouping phenomena; and, after a while, by showing that pictorial and mythological relations among the stars had no very obvious value. From that time, the inductive process went on steadily in the true road, under the guidance of ideas of space, time, and number.

Sect. 7.—The Planets.

WHILE men were becoming familiar with the fixed stars, the planets must have attracted their notice. Venus, from her brightness, and

---

[32] Dupuis, vi. 546. [33] Ib. vi. 546.

from her accompanying the sun at no great distance, and thus appearing as the morning and evening star, was very conspicuous. Pythagoras is said to have maintained that the evening and morning star are the same body, which certainly must have been one of the earliest discoveries on this subject; and indeed we can hardly conceive men noticing the stars for a year or two without coming to this conclusion. Jupiter and Mars, sometimes still brighter than Venus, were also very noticeable. Saturn and Mercury were less so, but in fine climates they and their motion would soon be detected by persons observant of the heavens. To reduce to any rule the movements of these luminaries must have taken time and thought; probably before this was done, certainly very early, these heavenly bodies were brought more peculiarly under those views which we have noticed as leading to astrology.

At a time beyond the reach of certain history, the planets, along with the sun and moon, had been arranged in a certain recognized order by the Egyptians or some other ancient nation. Probably this arrangement had been made according to the slowness of their motions among the stars; for though the motion of each is very variable, the gradation of their velocities is, on the whole, very manifest; and the different rate of travelling of the different planets, and probably other circumstances of difference, led, in the ready fancy of early times, to the attribution of a peculiar character to each luminary. Thus Saturn was held to be of a cold and gelid nature; Jupiter, who, from his more rapid motion, was supposed to be lower in place, was temperate; Mars, fiery, and the like.[34]

It is not necessary to dwell on the details of these speculations, but we may notice a very remarkable evidence of their antiquity and generality in the structure of one of the most familiar of our measures of time, the *Week*. This distribution of time according to periods of seven days, comes down to us, as we learn from the Jewish scriptures, from the beginning of man's existence on the earth. The same usage is found over all the East; it existed among the Arabians, Assyrians,

---

[34] Achilles Tatius (*Uranol.* pp. 135, 136), gives the Grecian and Egyptian names of the planets.

	Egyptian.	Greek.	
Saturn	Νεμεσέως	Κρόνου ἀστὴρ	*φαίνων*
Jupiter	'Οσίριδος	Δῖος	*φαέθων*
Mars	‘Ηρακλεοῦς	'Αρέος	*πυρόεις*
Venus		'Αφροδίτης	*ἑώσφορος*
Mercury	'Απόλλωνος	‘Ερμοῦ	*στίλβων*

Egyptians.[35] The same week is found in India among the Bramins; it has there, also, its days marked by those of the heavenly bodies; and it has been ascertained that the same day has, in that country, the name corresponding with its designation in other nations.

The notion which led to the usual designations of the days of the week is not easily unravelled. The days each correspond to one of the heavenly bodies, which were, in the earliest systems of the world, conceived to be the following, enumerating them in the order of their remoteness from the earth:[36] Saturn, Jupiter, Mars, the Sun, Venus, Mercury, the Moon. At a later period, the received systems placed the seven luminaries in *the seven spheres*. The knowledge which was implied in this view, and the time when it was obtained, we must consider hereafter. The order in which the names are assigned to the days of the week (beginning with Saturday) is, Saturn, the Sun, the Moon, Mars, Mercury, Jupiter, Venus; and various accounts are given of the manner in which one of these orders is obtained from the other; all the methods proceeding upon certain arbitrary arithmetical processes, connected in some way with astrological views. It is perhaps not worth our while here to examine further the steps of this process; it would be difficult to determine with certainty why the former order of the planets was adopted, and how and why the latter was deduced from it. But there is something very remarkable in the universality of the notions, apparently so fantastic, which have produced this result; and we may probably consider the Week, with Laplace,[37] as "the most ancient monument of astronomical knowledge." This period has gone on without interruption or irregularity from the earliest recorded times to our own days, traversing the extent of ages and the revolutions of empires; the names of the ancient deities which were associated with the stars have been replaced by those of the objects of the worship of our Teutonic ancestors, according to their views of the correspondence of the two mythologies; and the Quakers, in rejecting these names of days, have cast aside the most ancient existing relic of astrological as well as idolatrous superstition.

Sec. 8.—The Circles of the Sphere.

THE inventions hitherto noticed, though undoubtedly they were steps in astronomical knowledge, can hardly be considered as purely abstract and scientific speculations; for the exact reckoning of time is one of

[35] Laplace, *Hist. Astron.* p. 16. [36] *Philol. Mus.* No. 1. [37] *Hist. Ast.* p. 17.

the wants, even of the least civilized nations. But the distribution of the places and motions of the heavenly bodies by means of a celestial sphere with imaginary lines drawn upon it, is a step in *speculative* astronomy, and was occasioned and rendered important by the scientific propensities of man.

It is not easy to say with whom this notion originated. Some parts of it are obvious. The appearance of the sky naturally suggests the idea of a concave Sphere, with the stars fixed on its surface. Their motions during any one night, it would be readily seen, might be represented by supposing this Sphere to turn round a Pole or Axis; for there is a conspicuous star in the heavens which apparently stands still (the Pole-star); all the others travel round this in circles, and keep the same positions with respect to each other. This stationary star is every night the same, and in the same place; the other stars also have the same relative position; but their general position at the same time of night varies gradually from night to night, so as to go through its cycle of appearances once a year. All this would obviously agree with the supposition that the sky is a concave sphere or dome, that the stars have fixed places on this sphere, and that it revolves perpetually and uniformly about the Pole or fixed point.

But this supposition does not at all explain the way in which the appearances of different nights succeed each other. This, however, may be explained, it appears, by supposing the *sun* also *to move among the stars* on the surface of the concave sphere. The sun by his brightness makes the stars invisible which are on his side of the heavens: this we can easily believe; for the moon, when bright, also puts out all but the largest stars; and we see the stars appearing in the evening, each in its place, according to their degree of splendor, as fast as the declining light of day allows them to become visible. And as the sun brings day, and his absence night, if he move through the circuit of the stars in a year, we shall have, in the course of that time, every part of the starry sphere in succession presented to us as our nocturnal sky.

This notion, that *the sun moves round among the stars in a year*, is the basis of astronomy, and a considerable part of the science is only the development and particularization of this general conception. It is not easy to ascertain either the exact method by which the path of the sun among the stars was determined, or the author and date of the discovery. That there is some difficulty in tracing the course of the sun among the stars will be clearly seen, when it is considered that no star can ever be seen at the same time with the sun. If the whole circuit of the sky be divided into twelve parts or *signs*, it is estimated by Autolycus, the oldest writer on these subjects whose works remain to us,[38] that the stars which occupy one of these parts are absorbed by the solar rays, so that they cannot be seen. Hence the stars which are seen nearest to the place of the setting and the rising sun in the evening and in the morning, are distant from him by the half of a sign: the evening stars being to the west, and the morning stars to the east of him. If the observer had previously obtained a knowledge of the places of all the principal stars, he might in this way determine the position of the sun each night, and thus trace his path in a year.

In this, or some such way, the sun's path was determined by the early astronomers of Egypt. Thales, who is mentioned as the father of Greek astronomy, probably learnt among the Egyptians the results of such speculations, and introduced them into his own country. His knowledge, indeed, must have been a great deal more advanced than that which we are now describing, if it be true, as is asserted, that he predicted an eclipse. But his having done so is not very consistent with what we are told of the steps which his successors had still to make.

The Circle of the Signs, in which the sun moves among the stars, is obliquely situated with regard to the circles in which the stars move about the poles. Pliny[39] states that Anaximander,[40] a scholar of Thales, was the first person who pointed out this obliquity, and thus, as he says, "opened the gate of nature." Certainly, the person who first had a clear view of the nature of the sun's path in the celestial sphere, made that step which led to all the rest; but it is difficult to conceive that the Egyptians and Chaldeans had not already advanced so far.

The diurnal motion of the celestial sphere, and the motion of the moon in the circle of the signs, gave rise to a mathematical science, *the Doctrine of the Sphere*, which was one of the earliest branches of applied mathematics. A number of technical conceptions and terms were soon introduced. The *Sphere* of the heavens was conceived to be complete, though we see but a part of it; it was supposed to turn about the visible *pole* and another pole opposite to this, and these poles were connected by an imaginary *Axis*. The circle which divided the sphere exactly midway between these poles was called the *Equator* (ἰσημέρινος).

---

[38] Delamb. *A. A.* p. xiii. [39] Lib. ii. c. (viii.)

[40] Plutarch, *De Plac. Phil.* lib. ii. cap. xii. says Pythagoras was the author of this discovery.

The two circles parallel to this which bounded the sun's path among the stars were called *Tropics* (τροπικαί), because the sun *turns* back again towards the equator when he reaches them. The stars which never set are bounded by a circle called the *Arctic Circle* (ἄρκτικος, from ἄρκτος, the Bear, the constellation to which some of the principal stars within that circle belong.) A circle about the opposite pole is called *Antarctic*, and the stars which are within it can never rise to us,[41] The sun's path or circle of the signs is called the *Zodiac*, or circle of animals; the points where this circle meets the equator are the *Equinoctial Points*, the days and nights being equal when the sun is in them; the *Solstitial Points* are those where the sun's path touches the tropics; his motion to the south or to the north ceases when he is there, and he appears in that respect to stand still. The *Colures* (κόλουροι, *mutilated*) are circles which pass through the poles and through the equinoctial and solstitial points; they have their name because they are only visible in part, a portion of them being below the horizon.

The *Horizon* (ὁρίζων) is commonly understood as the boundary of the visible earth and heaven. In the doctrine of the sphere, this boundary is *a great circle*, that is, a circle of which the plane passes through the centre of the sphere; and, therefore, an entire hemisphere is always above the horizon. The term occurs for the first time in the work of Euclid, called *Phænomena* (Φαινόμενα). We possess two treatises written by Autolycus[42] (who lived about 300 B. C.) which trace *deductively* the results of the doctrine of the sphere. Supposing its diurnal motion to be uniform, in a work entitled Περὶ Κινουμένης Σφαίρας, "On the Moving Sphere," he demonstrates various properties of the diurnal risings, settings, and motions of the stars. In another work, Περὶ 'Επιτολῶν καὶ Δύσεων, "On Risings and Settings,"[43] *tacitly* assuming the sun's motion in his circle to be uniform, he proves certain propositions, with regard to those risings and settings of the stars, which take place at the same time when the sun rises and sets,[44] or *vice versâ*;[45] and also their *apparent* risings and settings when they cease to be visible after sunset, or begin to be visible after sunrise.[46]

---

41 The Arctic and Antarctic Circles of modern astronomers are different from these.

42 Delambre, *Astron. Ancienne*, p. 19.

43 Delambre, *Astron. Anc.* p. 25.

44 *Cosmical* rising and setting.

45 *Acronycal* rising and setting; (ἀκρονύκιος, happening at the extremity of the night.)

46 *Heliacal* rising and setting.

Several of the propositions contained in the former of these treatises are still necessary to be understood, as fundamental parts of astronomy.

The work of Euclid, just mentioned, is of the same kind. Delambre[47] finds in it evidence that Euclid was merely a book-astronomer, who had never observed the heavens.

We may here remark the first instance of that which we shall find abundantly illustrated in every part of the history of science; that man is *prone* to become a deductive reasoner;—that as soon as he obtains principles which can be traced to details by logical consequence, he sets about forming a body of science, by making a system of such reasonings. Geometry has always been a favorite mode of exercising this propensity: and that science, along with Trigonometry, Plane and Spherical, to which the early problems of astronomy gave rise, have, up to the present day, been a constant field for the exercise of mathematical ingenuity; a few simple astronomical truths being assumed as the basis of the reasoning.

Sect. 9.—The Globular Form of the Earth.

The establishment of the globular form of the earth is an important step in astronomy, for it is the first of those convictions, directly opposed to the apparent evidence of the senses, which astronomy irresistibly proves. To make men believe that *up* and *down* are different directions in different places; that the sea, which seems so level, is, in fact, convex; that the earth, which appears to rest on a solid foundation, is, in fact, not supported at all; are great triumphs both of the power of discovering and the power of convincing. We may readily allow this, when we recollect how recently the doctrine of the *antipodes*, or the existence of inhabitants of the earth, who stand on the opposite side of it, with their feet turned towards ours, was considered both monstrous and heretical.

Yet the different positions of the horizon at different places, necessarily led the student of spherical astronomy towards this notion of the earth as a round body. Anaximander[48] is said by some to have held the earth to be globular, and to be detached or suspended; he is also stated to have constructed a sphere, on which were shown the extent of land and water. As, however, we do not know the arguments upon which he maintained the earth's globular form, we cannot judge of the

---

[47] *Ast. Anc.* p. 53.

[48] See Brucker, *Hist. Phil.* vol. i. p. 486.

value of his opinion; it may have been no better founded than a different opinion ascribed to him by Laertius, that the earth had the shape of a pillar. Probably, the authors of the doctrine of the globular form of the earth were led to it, as we have said, by observing the different height of the pole at different places. They would find that the space which they passed over from north to south on the earth, was proportional to the change of place of the horizon in the celestial sphere; and as the horizon is, at every place, in the direction of the earth's apparently level surface, this observation would naturally suggest to them the opinion that the earth is placed within the celestial sphere, as a small globe in the middle of a much larger one.

We find this doctrine so distinctly insisted on by Aristotle, that we may almost look on him as the establisher of it.[49] "As to the figure of the earth, it must necessarily be spherical." This he proves, first by the tendency of things, in all places, downwards. He then adds,[50] "And, moreover, from the phenomena according to the sense: for if it were not so, the eclipses of the moon would not have such sections as they have. For in the configurations in the course of a month, the deficient part takes all different shapes; it is straight, and concave, and convex; but in eclipses it always has the line of division convex; wherefore, since the moon is eclipsed in consequence of the interposition of the earth, the periphery of the earth must be the cause of this by having a spherical form. And again, from the appearances of the stars, it is clear, not only that the earth is round, but that its size is not very large: for when we make a small removal to the south or the north, the circle of the horizon becomes palpably different, so that the stars overhead undergo a great change, and are not the same to those that travel to the north and to the south. For some stars are seen in Egypt or at Cyprus, but are not seen in the countries to the north of these; and the stars that in the north are visible while they make a complete circuit, there undergo a setting. So that from this it is manifest, not only that the form of the earth is round, but also that it is a part of not a very large sphere: for otherwise the difference would not be so obvious to persons making so small a change of place. Wherefore we may judge that those persons *who connect the region in the neighborhood of the pillars of Hercules with that towards India, and who assert that in this way the sea is* ONE, do not assert things very improbable. They confirm this conjecture moreover by the

---

49 Arist. *de Cœlo*, lib. ii. cap. xiv. ed. Casaub. p. 290.

50 p. 291 C.

elephants, which are said to be of the same species (γένος) towards each extreme; as if this circumstance was a consequence of the conjunction of the extremes. The mathematicians, who try to calculate the measure of the circumference, make it amount to 400,000 stadia; whence we collect that the earth is not only spherical, but is not large compared with the magnitude of the other stars."

When this notion was once suggested, it was defended and confirmed by such arguments as we find in later writers: for instance,[51] that the tendency of all things was to fall to the place of heavy bodies, and that this place being the centre of the earth, the whole earth had no such tendency; that the inequalities on the surface were so small as not materially to affect the shape of so vast a mass; that drops of water naturally form themselves into figures with a convex surface; that the end of the ocean would fall if it were not rounded off; that we see ships, when they go out to sea, disappearing downwards, which shows the surface to be convex. These are the arguments still employed in impressing the doctrines of astronomy upon the student of our own days; and thus we find that, even at the early period of which we are now speaking, truths had begun to accumulate which form a part of our present treasures.

Sect. 10.—*The Phases of the Moon.*

When men had formed a steady notion of the Moon as a solid body, revolving about the earth, they had only further to conceive it spherical, and to suppose the sun to be beyond the region of the moon, and they would find that they had obtained an explanation of the varying forms which the bright part of the moon assumes in the course of a month. For the convex side of the crescent-moon, and her full edge when she is gibbous, are always turned towards the sun. And this explanation, once suggested, would be confirmed, the more it was examined. For instance, if there be near us a spherical stone, on which the sun is shining, and if we place ourselves so that this stone and the moon are seen in the same direction (the moon appearing just over the top of the stone), we shall find that the visible part of the stone, which is then illuminated by the sun, is exactly similar in form to the moon, at whatever period of her changes she may be. The stone and the moon being in the same position with respect to us, and both being enlightened by the sun, the bright parts are the same in figure;

---

[51] Pliny, *Nat. Hist.* ii. LXV.

the only difference is, that the dark part of the moon is usually not visible at all.

This doctrine is ascribed to Anaximander. Aristotle was fully aware of it.[52] It could not well escape the Chaldeans and Egyptians, if they speculated at all about the causes of the appearances in the heavens.

Sect. 11.—*Eclipses.*

Eclipses of the sun and moon were from the earliest times regarded with a peculiar interest. The notions of superhuman influences and relations, which, as we have seen, were associated with the luminaries of the sky, made men look with alarm at any sudden and striking change in those objects; and as the constant and steady course of the celestial revolutions was contemplated with a feeling of admiration and awe, any marked interruption and deviation in this course, was regarded with surprise and terror. This appears to be the case with all nations at an early stage of their civilization.

This impression would cause Eclipses to be noted and remembered; and accordingly we find that the records of Eclipses are the earliest astronomical information which we possess. When men had discovered some of the laws of succession of other astronomical phenomena, for instance, of the usual appearances of the moon and sun, it might then occur to them that these unusual appearances also might probably be governed by some rule.

The search after this rule was successful at an early period. The Chaldeans were able to predict Eclipses of the Moon. This they did, probably, by means of their Cycle of 223 months, or about 18 years; for at the end of this time, the eclipses of the moon begin to return, at the same intervals and in the same order as at the beginning.[53] Probably this was the first instance of the prediction of peculiar astronomical phenomena. The Chinese have, indeed, a legend, in which it is related that a solar eclipse happened in the reign of Tchongkang, above 2000 years before Christ, and that the emperor was so much irritated against two great officers of state, who had neglected to predict this eclipse, that he put them to death. But this cannot be accepted as a real event: for, during the next ten centuries, we find no single observation or fact connected with astronomy in the Chinese

---

[52] Probl. Cap. xv. Art. 7.

[53] The eclipses of the sun are more difficult to calculate; since they depend upon the place of the spectator on the earth.

histories; and their astronomy has never advanced beyond a very rude and imperfect condition.

We can only conjecture the mode in which the Chaldeans discovered their Period of 18 years; and we may make very different suppositions with regard to the degree of science by which they were led to it. We may suppose, with Delambre,[54] that they carefully recorded the eclipses which happened, and then, by the inspection of their registers, discovered that those of the moon recurred after a certain period. Or we may suppose, with other authors, that they sedulously determined the motions of the moon, and having obtained these with considerable accuracy, sought and found a period which should include cycles of these motions. This latter mode of proceeding would imply a considerable degree of knowledge.

It appears probable rather that such a period was discovered by noticing the *recurrence* of eclipses, than by studying the moon's *motions*. After $6585\frac{1}{3}$ days, or 223 lunations, the same eclipses nearly will recur. It is not contested that the Chaldeans were acquainted with this period, which they called *Saros;* or that they calculated eclipses by means of it.

Sect. 12.—*Sequel to the Early Stages of Astronomy.*

Every stage of science has its train of practical applications and systematic inferences, arising both from the demands of convenience and curiosity, and from the pleasure which, as we have already said, ingenuous and active-minded men feel in exercising the process of deduction. The earliest condition of astronomy, in which it can be looked upon as a science, exhibits several examples of such applications and inferences, of which we may mention a few.

Prediction of Eclipses.—The Cycles which served to keep in order the Calendar of the early nations of antiquity, in some instances enabled them also, as has just been stated, to predict Eclipses; and this application of knowledge necessarily excited great notice. Cleomedes, in the time of Augustus, says, "We never see an eclipse happen which has not been predicted by those who made use of the Tables." (*ὑπὸ τῶν κανονικῶν.*)

Terrestrial Zones.—The globular form of the earth being assented to, the doctrine of the sphere was applied to the earth as well as the heavens; and the earth's surface was divided by various imaginary

[54] *A. A.* p. 212.

circles; among the rest, the equator, the tropics, and circles, at the same distance from the poles as the tropics are from the equator. One of the curious consequences of this division was the *assumption* that there must be some marked difference in the stripes or *zones* into which the earth's surface was thus divided. In going to the south, Europeans found countries hotter and hotter, in going to the north, colder and colder; and it was supposed that the space between the tropical circles must be uninhabitable from heat, and that within the polar circles, again, uninhabitable from cold. This fancy was, as we now know, entirely unfounded. But the principle of the globular form of the earth, when dealt with by means of spherical geometry, led to many true and important propositions concerning the lengths of days and nights at different places. These propositions still form a part of our Elementary Astronomy.

Gnomonic.—Another important result of the doctrine of the sphere was *Gnomonic* or *Dialling.* Anaximenes is said by Pliny to have first taught this art in Greece; and both he and Anaximander are reported to have erected the first dial at Lacedemon. Many of the ancient dials remain to us; some of these are of complex forms, and must have required great ingenuity and considerable geometrical knowledge in their construction.

Measure of the Sun's Distance.—The explanation of the phases of the moon led to no result so remarkable as the attempt of Aristarchus of Samos to obtain from this doctrine a measure of the Distance of the Sun as compared with that of the Moon. If the moon was a perfectly smooth sphere, when she was exactly midway between the new and full in position (that is, a quadrant from the sun), she would be somewhat more than a half moon; and the place when she was *dichotomized*, that is, was an exact semicircle, the bright part being bounded by a straight line, would depend upon the sun's distance from the earth. Aristarchus endeavored to fix the exact place of this Dichotomy; but the irregularity of the edge which bounds the bright part of the moon, and the difficulty of measuring with accuracy, by means then in use, either the precise time when the boundary was most nearly a straight line, or the exact distance of the moon from the sun at that time, rendered his conclusion false and valueless. He collected that the sun is at 18 times the distance of the moon from us; we now know that he is at 400 times the moon's distance.

It would be easy to dwell longer on subjects of this kind; but we have already perhaps entered too much in detail. We have been tempted to do this by the interest which the mathematical spirit of the Greeks gave to the earliest astronomical discoveries, when these were the subjects of their reasonings; but we must now proceed to contemplate them engaged in a worthier employment, namely, in adding to these discoveries.

CHAPTER II.

Prelude to the Inductive Epoch of Hipparchus.

WITHOUT pretending that we have exhausted the consequences of the elementary discoveries which we have enumerated, we now proceed to consider the nature and circumstances of the next great discovery which makes an Epoch in the history of Astronomy; and this we shall find to be the Theory of Epicycles and Eccentrics. Before, however, we relate the establishment of this theory, we must, according to the general plan we have marked out, notice some of the conjectures and attempts by which it was preceded, and the growing acquaintance with facts, which made the want of such an explanation felt.

In the steps previously made in astronomical knowledge, no ingenuity had been required to devise the view which was adopted. The motions of the stars and sun were most naturally and almost irresistibly conceived as the results of motion in a revolving sphere; the indications of position which we obtain from different places on the earth's surface, when clearly combined, obviously imply a globular shape. In these cases, the first conjectures, the supposition of the simplest form, of the most uniform motion, required no after-correction. But this manifest simplicity, this easy and obvious explanation, did not apply to the movement of all the heavenly bodies. The Planets, the "wandering stars," could not be so easily understood; the motion of each, as Cicero says, "undergoing very remarkable changes in its course, going before and behind, quicker and slower, appearing in the evening, but gradually lost there, and emerging again in the morning."[1] A continued attention to these stars would, however,

---

[1] Cic. *de Nat. D.* lib. ii. p. 450. "Ea quæ Saturni stella dicitur, φαίνωνque a Græcis nominatur, quæ a terrâ abest plurimum, xxx fere annis cursum suum con-

detect a kind of intricate regularity in their motions, which might naturally be described as "a dance." The Chaldeans are stated by Diodorus[2] to have observed assiduously the risings and settings of the planets, from the top of the temple of Belus. By doing this, they would find the times in which the forward and backward movements of Saturn, Jupiter, and Mars recur; and also the time in which they come round to the same part of the heavens.[3] Venus and Mercury never recede far from the sun, and the intervals which elapse while either of them leaves its greatest distance from the sun and returns again to the greatest distance on the same side, would easily be observed.

Probably the manner in which the motions of the planets were originally reduced to rule was something like the following :—In about 30 of our years, Saturn goes 29 times through his *Anomaly*, that is, the succession of varied motions by which he sometimes goes forwards and sometimes backwards among the stars. During this time, he goes once round the heavens, and returns nearly to the same place. This is the cycle of his apparent motions.

Perhaps the eastern nations contented themselves with thus referring these motions to cycles of time, so as to determine their recurrence. Something of this kind was done at an early period, as we have seen.

But the Greeks soon attempted to frame to themselves a sensible image of the mechanism by which these complex motions were produced; nor did they find this difficult. Venus, for instance, who, upon the whole, moves from west to east among the stars, is seen, at certain intervals, to return or move *retrograde* a short way back from east to west, then to become for a short time *stationary*, then to turn again and resume her *direct* motion westward, and so on. Now this can be explained by supposing that she is placed in the rim of a wheel, which is turned edgeways to us, and of which the centre turns round in the heavens from west to east, while the wheel, carrying the planet in its motion, moves round its own centre. In this way the motion of the wheel about its centre, would, in some situations, counterbalance the general motion of the centre, and make the planet retrograde, while, on the whole, the westerly motion would prevail. Just as if we suppose that a person, holding a lamp in his hand in the dark, and at a

ficit; in quo cursu multa mirabiliter efficiens, tum antecedendo, tum retardando, tum vespertinis temporibus delitescendo, tum matutinis se rursum aperiendo, nihil immutat sempiternis sæculorum ætatibus, quin eadem iisdem temporibus efficiat." And so of the other planets.

[2] Del. *A. A.* i. p. 4.

[3] Plin. *H. N.* ii. p. 204.

distance, so that the lamp alone is visible, should run on turning himself round; we should see the light sometimes stationary, sometimes retrograde, but on the whole progressive.

A mechanism of this kind was imagined for each of the planets, and the wheels of which we have spoken were in the end called *Epicycles.*

The application of such mechanism to the planets appears to have arisen in Greece about the time of Aristotle. In the works of Plato we find a strong taste for this kind of mechanical speculation. In the tenth book of the "Polity," we have the apologue of Alcinus the Pamphylian, who, being supposed to be killed in battle, revived when he was placed on the funeral pyre, and related what he had seen during his trance. Among other revelations, he beheld the machinery by which all the celestial bodies revolve. The axis of these revolutions is the adamantine distaff which Destiny holds between her knees; on this are fixed, by means of different sockets, flat rings, by which the planets are carried. The order and magnitude of these spindles are minutely detailed. Also, in the "Epilogue to the Laws" (*Epinomis*), he again describes the various movements of the sky, so as to show a distinct acquaintance with the general character of the planetary motions; and, after speaking of the Egyptians and Syrians as the original cultivators of such knowledge, he adds some very remarkable exhortations to his countrymen to prosecute the subject. "Whatever we Greeks," he says, "receive from the barbarians, we improve and perfect; there is good hope and promise, therefore, that Greeks will carry this knowledge far beyond that which was introduced from abroad." To this task, however, he looks with a due appreciation of the qualities and preparation which it requires. "An astronomer must be," he says, "the wisest of men; his mind must be duly disciplined in youth; especially is mathematical study necessary; both an acquaintance with the doctrine of number, and also with that other branch of mathematics, which, closely connected as it is with the science of the *heavens*, we very absurdly call *geometry*, the measurement of the *earth*."[4]

These anticipations were very remarkably verified in the subsequent career of the Greek Astronomy.

The theory, once suggested, probably made rapid progress. Simplicius[5] relates, that Eudoxus of Cnidus introduced the hypothesis of revolving circles or spheres. Calippus of Cyzicus, having visited Pole-

[4] *Epinomis*, pp. 988, 990.

[5] Lib. ii. *de Cœlo.* Bullialdus, p. 18.

marchus, an intimate friend of Eudoxus, they went together to Athens, and communicated to Aristotle the invention of Eudoxus, and with his help improved and corrected it.

Probably at first this hypothesis was applied only to account for the general phenomena of the progressions, retrogradations, and stations of the planet; but it was soon found that the motions of the sun and moon, and the circular motions of the planets, which the hypothesis supposed, had other *anomalies* or irregularities, which made a further extension of the hypothesis necessary.

The defect of uniformity in these motions of the sun and moon, though less apparent than in the planets, is easily detected, as soon as men endeavor to obtain any accuracy in their observations. We have already stated (Chap. I.) that the Chaldeans were in possession of a period of about eighteen years, which they used in the calculation of eclipses, and which might have been discovered by close observation of the moon's motions; although it was probably rather hit upon by noting the recurrence of eclipses. The moon moves in a manner which is not reducible to regularity without considerable care and time. If we trace her path among the stars, we find that, like the path of the sun, it is oblique to the equator, but it does not, like that of the sun, pass over the same stars in successive revolutions. Thus its *latitude*, or distance from the equator, has a cycle different from its revolution among the stars; and its *Nodes*, or the points where it cuts the equator, are perpetually changing their position. In addition to this, the moon's motion in her own path is not uniform; in the course of each lunation, she moves alternately slower and quicker, passing gradually through the intermediate degrees of velocity; and goes through the cycle of these changes in something less than a month; this is called a revolution of *Anomaly*. When the moon has gone through a complete number of revolutions of Anomaly, and has, in the same time, returned to the same position with regard to the sun, and also with regard to her Nodes, her motions with respect to the sun will thenceforth be the same as at the first, and all the circumstances on which lunar eclipses depend being the same, the eclipses will occur in the same order. In 6585⅓ days there are 239 revolutions of anomaly, 241 revolutions with regard to one of the Nodes, and, as we have said, 223 lunations or revolutions with regard to the sun. Hence this Period will bring about a succession of the same lunar eclipses.

If the Chaldeans observed the moon's motion among the stars with any considerable accuracy, so as to detect this period by that means, they could hardly avoid discovering the anomaly or unequal motion of the moon; for in every revolution, her daily progression in the heavens varies from about twenty-two to twenty-six times her own diameter. But there is not, in their knowledge of this Period, any evidence that they had measured the amount of this variation; and Delambre[6] is probably right in attributing all such observations to the Greeks.

The sun's motion would also be seen to be irregular as soon as men had any exact mode of determining the lengths of the four seasons, by means of the passage of the sun through the equinoctial and solstitial points. For spring, summer, autumn, and winter, which would each consist of an equal number of days if the motions were uniform, are, in fact, found to be unequal in length.

It was not very difficult to see that the mechanism of epicycles might be applied so as to explain irregularities of this kind. A wheel travelling round the earth, while it revolved upon its centre, might produce the effect of making the sun or moon fixed in its rim go sometimes faster and sometimes slower in appearance, just in the same way as the same suppositions would account for a planet going sometimes forwards and sometimes backwards: the epicycles of the sun and moon would, for this purpose, be less than those of the planets. Accordingly, it is probable that, at the time of Plato and Aristotle, philosophers were already endeavoring to apply the hypothesis to these cases, though it does not appear that any one fully succeeded before Hipparchus.

The problem which was thus present to the minds of astronomers, and which Plato is said to have proposed to them in a distinct form, was, "To reconcile the celestial phenomena by the combination of equable circular motions." That the circular motions should be equable as well as circular, was a condition, which, if it had been merely tried at first, as the most simple and definite conjecture, would have deserved praise. But this condition, which is, in reality, inconsistent with nature, was, in the sequel, adhered to with a pertinacity which introduced endless complexity into the system. The history of this assumption is one of the most marked instances of that love of simplicity and symmetry which is the source of all general truths, though it so often produces and perpetuates error. At present we can easily see how fancifully the notion of simplicity and perfection was interpreted, in the arguments by which the opinion was defended, that the

---

[6] *Astronomie Ancienne*, i. 212.

real motions of the heavenly bodies must be circular and uniform. The Pythagoreans, as well as the Platonists, maintained this dogma. According to Geminus, "They supposed the motions of the sun, and the moon, and the five planets, to be circular and equable: for they would not allow of such disorder among divine and eternal things, as that they should sometimes move quicker, and sometimes slower, and sometimes stand still; for no one would tolerate such anomaly in the movements, even of a man, who was decent and orderly. The occasions of life, however, are often reasons for men going quicker or slower, but in the incorruptible nature of the stars, it is not possible that any cause can be alleged of quickness and slowness. Whereupon they propounded this question, how the phenomena might be represented by equable and circular motions."

These conjectures and assumptions led naturally to the establishment of the various parts of the Theory of Epicycles. It is probable that this theory was adopted with respect to the Planets at or before the time of Plato. And Aristotle gives us an account of the system thus devised.[7] "Eudoxus," he says, "attributed four spheres to each Planet: the first revolved with the fixed stars (and this produced the diurnal motion); the second gave the planet a motion along the ecliptic (the mean motion in longitude); the third had its axis perpendicular[8] to the ecliptic (and this gave the inequality of each planetary motion, really arising from its special motion about the sun); the fourth produced the oblique motion transverse to this (the motion in latitude)." He is also said to have attributed a motion in latitude and a corresponding sphere to the Sun as well as to the Moon, of which it is difficult to understand the meaning, if Aristotle has reported rightly of the theory; for it would be absurd to ascribe to Eudoxus a knowledge of the motions by which the sun deviates from the ecliptic. Calippus conceived that two additional spheres must be given to the sun and to the moon, in order to explain the phenomena: probably he was aware of the inequalities of the motions of these luminaries. He also proposed an additional sphere for each planet, to account, we may suppose, for the results of the eccentricity of the orbits.

The hypothesis, in this form, does not appear to have been reduced to measure, and was, moreover, unnecessarily complex. The resolution

---

[7] Metaph. xi. 8.

[8] Aristotle says "has its poles in the ecliptic," but this must be a mistake of his. He professes merely to receive these opinions from the mathematical astronomers, ‘ἐκ τῆς οἰκειοτάτης φιλοσοφίας των μαθηματικῶν.

of the oblique motion of the moon into two separate motions, by Eudoxus, was not the simplest way of conceiving it; and Calippus imagined the connection of these spheres in some way which made it necessary nearly to double their number; in this manner his system had no less than 55 spheres.

Such was the progress which the *Idea* of the hypothesis of epicycles had made in men's minds, previously to the establishment of the theory by Hipparchus. There had also been a preparation for this step, on the other side, by the collection of *Facts*. We know that observations of the Eclipses of the Moon were made by the Chaldeans 367 B. C. at Babylon, and were known to the Greeks; for Hipparchus and Ptolemy founded their Theory of the Moon on these observations. Perhaps we cannot consider, as equally certain, the story that, at the time of Alexander's conquest, the Chaldeans possessed a series of observations, which went back 1903 years, and which Aristotle caused Callisthenes to bring to him in Greece. All the Greek observations which are of any value, begin with the school of Alexandria. Aristyllus and Timocharis appear, by the citations of Hipparchus, to have observed the Places of Stars and Planets, and the Times of the Solstices, at various periods from B. C. 295 to B. C. 269. Without their observations, indeed, it would not have been easy for Hipparchus to establish either the Theory of the Sun or the Precession of the Equinoxes.

In order that observations at distant intervals may be compared with each other, they must be referred to some common era. The Chaldeans dated by the era of Nabonassar, which commenced 749 B. C. The Greek observations were referred to the Calippic periods of 76 years, of which the first began 331 B. C. These are the dates used by Hipparchus and Ptolemy.

CHAPTER III.

INDUCTIVE EPOCH OF HIPPARCHUS.

Sect. 1.—Establishment of the Theory of Epicycles and Eccentrics.

ALTHOUGH, as we have already seen, at the time of Plato, the Idea of Epicycles had been suggested, and the problem of its general application proposed, and solutions of this problem offered by his followers; we still consider Hipparchus as the real discoverer and founder of that theory; inasmuch as he not only guessed that it *might*, but showed that it *must*, account for the phenomena, both as to their nature and as to their quantity. The assertion that "he only discovers who proves," is just; not only because, until a theory is proved to be the true one, it has no pre-eminence over the numerous other guesses among which it circulates, and above which the proof alone elevates it; but also because he who takes hold of the theory so as to apply calculation to it, possesses it with a distinctness of conception which makes it peculiarly his.

In order to establish the Theory of Epicycles, it was necessary to assign the magnitudes, distances, and positions of the circles or spheres in which the heavenly bodies were moved, in such a manner as to account for their apparently irregular motions. We may best understand what was the problem to be solved, by calling to mind what we now know to be the real motions of the heavens. The true motion of the earth round the sun, and therefore the apparent annual motion of the sun, is performed, not in a circle of which the earth is the centre, but in an ellipse or oval, the earth being nearer to one end than to the other; and the motion is most rapid when the sun is at the nearer end of this oval. But instead of an oval, we may suppose the sun to move uniformly in a circle, the earth being now, not in the centre, but nearer to one side; for on this supposition, the sun will appear to move most quickly when he is nearest to the earth, or in his *Perigee,* as that point is called. Such an orbit is called an *Eccentric,* and the distance of the earth from the centre of the circle is called the *Eccentricity.* It may easily be shown by geometrical reasoning, that the inequality of apparent motion so produced, is exactly the same in detail, as the inequality which follows from the hypothesis of a small *Epicycle*, turning uniformly on its axis, and carrying the sun in its circumference, while the centre of this epicycle moves uniformly in a circle of which the earth is the centre. This identity of the results of the hypothesis of the Eccentric and the Epicycle is proved by Ptolemy in the third book of the "Almagest."

The Sun's Eccentric.—When Hipparchus had clearly conceived these hypotheses, as *possible* ways of accounting for the sun's motion, the task which he had to perform, in order to show that they deserved to be adopted, was to assign a place to the *Perigee,* a magnitude to the *Eccentricity*, and an *Epoch* at which the sun was at the perigee; and to show that, in this way, he had produced a true representation of the motions of the sun. This, accordingly, he did; and having thus determined, with considerable exactness, both the law of the solar irregularities, and the numbers on which their amount depends, he was able to assign the motions and places of the sun for any moment of future time with corresponding exactness; he was able, in short, to construct *Solar Tables*, by means of which the sun's place with respect to the stars could be correctly found at any time. These tables (as they are given by Ptolemy)[1] give the *Anomaly*, or inequality of the sun's motion; and this they exhibit by means of the *Prosthapheresis*, the quantity of which, at any distance of the sun from the *Apogee*, it is requisite to add to or subtract from the arc, which he would have described if his motion had been equable.

The reader might perhaps expect that the calculations which thus exhibited the motions of the sun for an indefinite future period must depend upon a considerable number of observations made at all seasons of the year. That, however, was not the case; and the genius of the discoverer appeared, as such genius usually does appear, in his perceiving how small a number of facts, rightly considered, were sufficient to form a foundation for the theory. The number of days contained in two seasons of the year sufficed for this purpose to Hipparchus. "Having ascertained," says Ptolemy, "that the time from the vernal equinox to the summer tropic is 94½ days, and the time from the summer tropic to the autumnal equinox 92½ days, from these phenomena alone he demonstrates that the straight line joining the centre of the sun's eccentric path with the centre of the zodiac (the spectator's eye) is nearly the 24th part of the radius of the eccentric path; and that

[1] Syntax. l. iii.

its *apogee* precedes the summer solstice by 24½ degrees nearly, the zodiac containing 360."

The exactness of the Solar Tables, or *Canon*, which was founded on these data, was manifested, not only by the coincidence of the sun's calculated place with such observations as the Greek astronomers of this period were able to make (which were indeed very rude), but by its enabling them to calculate solar and lunar eclipses; phenomena which are a very precise and severe trial of the accuracy of such tables, inasmuch as a very minute change in the apparent place of the sun or moon would completely alter the obvious features of the eclipse. Though the tables of this period were by no means perfect, they bore with tolerable credit this trying and perpetually recurring test; and thus proved the soundness of the theory on which the tables were calculated.

The Moon's Eccentric.—The moon's motions have many irregularities; but when the hypothesis of an Eccentric or an Epicycle had sufficed in the case of the sun, it was natural to try to explain, in the same way, the motions of the moon; and it was shown by Hipparchus that such hypotheses would account for the more obvious anomalies. It is not very easy to describe the several ways in which these hypotheses were applied, for it is, in truth, very difficult to explain in words even the mere facts of the moon's motion. If she were to leave a visible bright line behind her in the heavens wherever she moved, the path thus exhibited would be of an extremely complex nature; the circle of each revolution slipping away from the preceding, and the traces of successive revolutions forming a sort of band of net-work running round the middle of the sky.[2] In each revolution, the motion in longitude is affected by an anomaly of the same nature as the sun's anomaly already spoken of; but besides this, the path of the moon deviates from the ecliptic to the north and to the south of the ecliptic, and thus she has a motion in latitude. This motion in latitude would be sufficiently known if we knew the period of its *restoration*, that is, the time which the moon occupies in moving from any latitude till she is restored to the same latitude; as, for instance, from the ecliptic on one side of the heavens to the ecliptic on the same side of the heavens again. But it is found that the period of the restoration of the latitude is not the same as the period of the restoration of the longitude, that is, as the period of the moon's revolution among the

[2] The reader will find an attempt to make the nature of this path generally intelligible in the *Companion to the British Almanac* for 1814.

stars; and thus the moon describes a different path among the stars in every successive revolution, and her path, as well as her velocity, is constantly variable.

Hipparchus, however, reduced the motions of the moon to rule and to Tables, as he did those of the sun, and in the same manner. He determined, with much greater accuracy than any preceding astronomer, the *mean* or average equable motions of the moon in longitude and in latitude; and he then represented the anomaly of the motion in longitude by means of an eccentric, in the same manner as he had done for the sun.

But here there occurred still an additional change, besides those of which we have spoken. The Apogee of the Sun was always in the same place in the heavens; or at least so nearly so, that Ptolemy could detect no error in the place assigned to it by Hipparchus 250 years before. But the Apogee of the Moon was found to have a motion among the stars. It had been observed before the time of Hipparchus, that in 6585⅓ days, there are 241 revolutions of the moon with regard to the stars, but only 239 revolutions with regard to the anomaly. This difference could be suitably represented by supposing the eccentric, in which the moon moves, to have itself an angular motion, perpetually carrying its apogee in the same direction in which the moon travels; but this supposition being made, it was necessary to determine, not only the eccentricity of the orbit, and place of the apogee at a certain time, but also the rate of motion of the apogee itself, in order to form tables of the moon.

This task, as we have said, Hipparchus executed; and in this instance, as in the problem of the reduction of the sun's motion to tables, the data which he found it necessary to employ were very few. He deduced all his conclusions from six eclipses of the moon.[3] Three of these, the records of which were brought from Babylon, where a register of such occurrences was kept, happened in the 366th and 367th years from the era of Nabonassar, and enabled Hipparchus to determine the eccentricity and apogee of the moon's orbit at that time. The three others were observed at Alexandria, in the 547th year of Nabonassar, which gave him another position of the orbit at an interval of 180 years; and he thus became acquainted with the motion of the orbit itself, as well as its form.[4]

---

[3] Ptol. *Syn*. iv. 10.

[4] Ptolemy uses the hypothesis of an epycicle for the moon's first inequality; but Hipparchus employs an eccentric.

The moon's motions are really affected by several other inequalities, of very considerable amount, besides those which were thus considered by Hipparchus; but the lunar paths, constructed on the above data, possessed a considerable degree of correctness, and especially when applied, as they were principally, to the calculation of eclipses; for the greatest of the additional irregularities which we have mentioned disappear at new and full moon, which are the only times when eclipses take place.

The numerical explanation of the motions of the sun and moon, by means of the Hypothesis of Eccentrics, and the consequent construction of tables, was one of the great achievements of Hipparchus. The general explanation of the motions of the planets, by means of the hypothesis of epicycles, was in circulation previously, as we have seen. But the special motions of the planets, in their epicycles, are, in reality, affected by anomalies of the same kind as those which render it necessary to introduce eccentrics in the cases of the sun and moon.

Hipparchus determined, with great exactness, the *Mean Motions* of the Planets; but he was not able, from want of data, to explain the planetary *Irregularities* by means of Eccentrics. The whole mass of good observations of the planets which he received from preceding ages, did not contain so many, says Ptolemy, as those which he has transmitted to us of his own. "Hence[5] it was," he adds, "that while he labored, in the most assiduous manner to represent the motions of the sun and moon by means of equable circular motions; with respect to the planets, so far as his works show, he did not even make the attempt, but merely put the extant observations in order, added to them himself more than the whole of what he received from preceding ages, and showed the insufficiency of the hypothesis current among astronomers to explain the phenomena." It appears that preceding mathematicians had already pretended to construct "a Perpetual Canon," that is, Tables which should give the places of the planets at any future time; but these being constructed without regard to the eccentricity of the orbits, must have been very erroneous.

Ptolemy declares, with great reason, that Hipparchus showed his usual love of truth, and his right sense of the responsibility of his task, in leaving this part of it to future ages. The Theories of the Sun and Moon, which we have already described, constitute him a great astronomical discoverer, and justify the reputation he has always

---

[5] *Synt.* ix. 2.

possessed. There is, indeed, no philosopher who is so uniformly spoken of in terms of admiration. Ptolemy, to whom we owe our principal knowledge of him, perpetually couples with his name epithets of praise: he is not only an excellent and careful observer, but "a[6] most truth-loving and labor-loving person," one who had shown extraordinary sagacity and remarkable desire of truth in every part of science. Pliny, after mentioning him and Thales, breaks out into one of his passages of declamatory vehemence: "Great men! elevated above the common standard of human nature, by discovering the laws which celestial occurrences obey, and by freeing the wretched mind of man from the fears which eclipses inspired—Hail to you and to your genius, interpreters of heaven, worthy recipients of the laws of the universe, authors of principles which connect gods and men!" Modern writers have spoken of Hipparchus with the same admiration; and even the exact but severe historian of astronomy, Delambre, who bestows his praise so sparingly, and his sarcasm so generally;—who says[7] that it is unfortunate for the memory of Aristarchus that his work has come to us entire, and who cannot refer[8] to the statement of an eclipse rightly predicted by Halicon of Cyzicus without adding, that if the story be true, Halicon was more lucky than prudent;—loses all his bitterness when he comes to Hipparchus.[9] "In Hipparchus," says he, "we find one of the most extraordinary men of antiquity; the *very greatest,* in the sciences which require a combination of observation with geometry." Delambre adds, apparently in the wish to reconcile this eulogium with the depreciating manner in which he habitually speaks of all astronomers whose observations are inexact, "a long period and the continued efforts of many industrious men are requisite to produce good instruments, but energy and assiduity depend on the man himself."

Hipparchus was the author of other great discoveries and improvements in astronomy, besides the establishment of the Doctrine of Eccentrics and Epicycles; but this, being the greatest advance in the *theory* of the celestial motions which was made by the ancients, must be the leading subject of our attention in the present work; our object being to discover in what the progress of real theoretical knowledge consists, and under what circumstances it has gone on.

---

6 *Syn.* ix. 2.
7 *Astronomie Ancienne*, i. 75.
8 Ib. i. 17.
9 Ib. i. 186.

Sect. 2.—Estimate of the Value of the Theory of Eccentrics and Epicycles.

It may be useful here to explain the value of the theoretical step which Hipparchus thus made; and the more so, as there are, perhaps, opinions in popular circulation, which might lead men to think lightly of the merit of introducing or establishing the Doctrine of Epicycles. For, in the first place, this doctrine is now acknowledged to be false; and some of the greatest men in the more modern history of astronomy owe the brightest part of their fame to their having been instrumental in overturning this hypothesis. And, moreover, in the next place, the theory is not only false, but extremely perplexed and entangled, so that it is usually looked upon as a mass of arbitrary and absurd complication. Most persons are familiar with passages in which it is thus spoken of.[10]

> . . . . . He his fabric of the heavens
> Hath left to their disputes, perhaps to move
> His laughter at their quaint opinions wide;
> Hereafter, when they come to model heaven
> And calculate the stars, how will they wield
> The mighty frame! how build, unbuild, contrive,
> To save appearances! how gird the sphere
> With centric and eccentric scribbled o'er,
> Cycle in epicycle, orb in orb!

And every one will recollect the celebrated saying of Alphonso X., king of Castile,[11] when this complex system was explained to him; that "if God had consulted him at the creation, the universe should have been on a better and simpler plan." In addition to this, the system is represented as involving an extravagant conception of the nature of the orbs which it introduces; that they are crystalline spheres, and that the vast spaces which intervene between the celestial luminaries are a solid mass, formed by the fitting together of many masses perpetually in motion; an imagination which is presumed to be incredible and monstrous.

We must endeavor to correct or remove these prejudices, not only in order that we may do justice to the Hipparchian, or, as it is usually called, Ptolemaic system of astronomy, and to its founder; but for another reason, much more important to the purpose of this work;

---

[10] *Paradise Lost*, viii. [11] A. D. 1252.

namely, that we may see how theories may be highly estimable, though they contain false representations of the real state of things, and may be extremely useful, though they involve unnecessary complexity. In the advance of knowledge, the value of the true part of a theory may much outweigh the accompanying error, and the use of a rule may be little impaired by its want of simplicity. The first steps of our progress do not lose their importance because they are not the last; and the outset of the journey may require no less vigor and activity than its close.

That which is true in the Hipparchian theory, and which no succeeding discoveries have deprived of its value, is the *Resolution* of the apparent motions of the heavenly bodies into an assemblage of circular motions. The test of the truth and reality of this Resolution is, that it leads to the construction of theoretical Tables of the motions of the luminaries, by which their places are given at any time, agreeing nearly with their places as actually observed. The assumption that these circular motions, thus introduced, are all exactly uniform, is the fundamental principle of the whole process. This assumption is, it may be said, false; and we have seen how fantastic some of the arguments were, which were originally urged in its favor. But *some* assumption is necessary, in order that the motions, at different points of a revolution, may be somehow connected, that is, in order that we may have *any* theory of the motions; and no assumption more simple than the one now mentioned can be selected. The merit of the theory is this; —that obtaining the amount of the eccentricity, the place of the apogee, and, it may be, other elements, from a *few* observations, it deduces from these, results agreeing with *all* observations, however numerous and distant. To express an inequality by means of an epicycle, implies, not only that there is an inequality, but further,—that the inequality is at its greatest value at a certain known place,—diminishes in proceeding from that place by a known law,—continues its diminution for a known portion of the revolution of the luminary,—then increases again; and so on: that is, the introduction of the epicycle represents the inequality of motion, as completely as it can be represented with respect to its *quantity*.

We may further illustrate this, by remarking that such a Resolution of the unequal motions of the heavenly bodies into equable circular motions, is, in fact, equivalent to the most recent and improved processes by which modern astronomers deal with such motions. Their universal method is to resolve all unequal motions into a series of *terms*, or expressions of partial motions; and these terms involve *sines* and *cosines*, that is, certain technical modes of measuring circular motion, the circular motion having some constant relation to the time. And thus the problem of the resolution of the celestial motions into equable circular ones, which was propounded above two thousand years ago in the school of Plato, is still the great object of the study of modern astronomers, whether observers or calculators.

That Hipparchus should have succeeded in the first great steps of this resolution for the sun and moon, and should have seen its applicability in other cases, is a circumstance which gives him one of the most distinguished places in the roll of great astronomers. As to the charges or the sneers against the complexity of his system, to which we have referred, it is easy to see that they are of no force. As a system of *calculation*, his is not only good, but, as we have just said, in many cases no better has yet been discovered. If, when the actual motions of the heavens are calculated in the best possible way, the process is complex and difficult, and if we are discontented at this, nature, and not the astronomer, must be the object of our displeasure. This plea of the astronomers must be allowed to be reasonable. "We must not be repelled," says Ptolemy,[12] "by the complexity of the hypotheses, but explain the phenomena as well as we can. If the hypotheses satisfy each apparent inequality separately, the combination of them will represent the truth; and why should it appear wonderful to any that such a complexity should exist in the heavens, when we know nothing of their nature which entitles us to suppose that any inconsistency will result?"

But it may be said, we now know that the motions are more simple than they were thus represented, and that the Theory of Epicycles was false, as a conception of the real construction of the heavens. And to this we may reply, that it does not appear that the best astronomers of antiquity conceived the cycles and epicycles to have a material existence. Though the dogmatic philosophers, as the Aristotelians, appear to have taught that the celestial spheres were real solid bodies, they are spoken of by Ptolemy as imaginary;[13] and it is clear, from his proof of the identity of the results of the hypothesis of an eccentric and an epicycle, that they are intended to pass for no more than geometrical conceptions, in which view they are true representations of the apparent motions.

---

[12] *Synt.* xiii. 2. [13] Ibid. iii. 3.

It is true, that the real motions of the heavenly bodies are simpler than the apparent motions; and that we, who are in the habit of representing to our minds their real arrangement, become impatient of the seeming confusion and disorder of the ancient hypotheses. But this real arrangement never could have been detected by philosophers, if the apparent motions had not been strictly examined and successfully analyzed. How far the connection between the facts and the true theory is from being obvious or easily traced, any one may satisfy himself by endeavoring, from a general conception of the moon's real motions, to discover the rules which regulate the occurrences of eclipses; or even to explain to a learner, of what nature the apparent motions of the moon among the stars will be.

The unquestionable evidence of the merit and value of the Theory of Epicycles is to be found in this circumstance;—that it served to embody all the most exact knowledge then extant, to direct astronomers to the proper methods of making it more exact and complete, to point out new objects of attention and research; and that, after doing this at first, it was also able to take in, and preserve, all the new results of the active and persevering labors of a long series of Greek, Latin, Arabian, and modern European astronomers, till a new theory arose which could discharge this office. It may, perhaps, surprise some readers to be told, that the author of this next *great* step in astronomical theory, Copernicus, adopted the theory of epicycles; that is, he employed that which we have spoken of as its really valuable characteristic. "We[14] must confess," he says, "that the celestial motions are circular, or compounded of several circles, since their inequalities observe a fixed law and recur in value at certain intervals, which could not be, except that they were circular; for a circle alone can make that which has been, recur again."

In this sense, therefore, the Hipparchian theory was a real and indestructible truth, which was not rejected, and replaced by different truths, but was adopted and incorporated into every succeeding astronomical theory; and which can never cease to be one of the most important and fundamental parts of our astronomical knowledge.

A moment's reflection will show that, in the events just spoken of, the introduction and establishment of the Theory of Epicycles, those characteristics were strictly exemplified, which we have asserted to be the conditions of every real advance in progressive science; namely,

---

[14] Copernicus. *De Rev.* l. i. c. 4.

the application of distinct and appropriate Ideas to a real series of Facts. The distinctness of the geometrical conceptions which enabled Hipparchus to assign the Orbits of the Sun and Moon, requires no illustration; and we have just explained how these ideas combined into a connected whole the various motions and places of those luminaries. To make this step in astronomy, required diligence and care, exerted in collecting observations, and mathematical clearness and steadiness of view, exercised in seeing and showing that the theory was a successful analysis of them.

Sect. 3.—*Discovery of the Precession of the Equinoxes.*

The same qualities which we trace in the researches of Hipparchus already examined,—diligence in collecting observations, and clearness of idea in representing them,—appear also in other discoveries of his, which we must not pass unnoticed. The Precession of the Equinoxes, in particular, is one of the most important of these discoveries.

The circumstance here brought into notice was a Change of Longitude of the Fixed Stars. The longitudes of the heavenly bodies, being measured from the point where the sun's annual path cuts the equator, will change if that path changes. Whether this happens, however, is not very easy to decide; for the sun's path among the stars is made out, not by merely looking at the heavens, but by a series of inferences from other observable facts. Hipparchus used for this purpose eclipses of the moon; for these, being exactly opposite to the sun, afford data in marking out his path. By comparing the eclipses of his own time with those observed at an earlier period by Timocharis, he found that the bright star, Spica Virginis, was six degrees behind the equinoctial point in his own time, and had been eight degrees behind the same point at an earlier epoch. The suspicion was thus suggested, that the longitudes of all the stars increase perpetually; but Hipparchus had too truly philosophical a spirit to take this for granted. He examined the places of Regulus, and those of other stars, as he had done those of Spica; and he found, in all these instances, a change of place which could be explained by a certain alteration of position in the circles to which the stars are referred, which alteration is described as the Precession of the Equinoxes.

The distinctness with which Hipparchus conceived this change of relation of the heavens, is manifested by the question which, as we are told by Ptolemy, he examined and decided;—that this motion of the heavens takes place about the poles of the ecliptic, and not about those of the equator. The care with which he collected this motion from the stars themselves, may be judged of from this, that having made his first observations for this purpose on Spica and Regulus, zodiacal stars, his first suspicion was that the stars of the zodiac alone changed their longitude, which suspicion he disproved by the examination of other stars. By his processes, the idea of the nature of the motion, and the evidence of its existence, the two conditions of a discovery, were fully brought into view. The scale of the facts which Hipparchus was thus able to reduce to law, may be in some measure judged of, by recollecting that the precession, from his time to ours, has only carried the stars through one sign of the zodiac; and that, to complete one revolution of the sky by the motion thus discovered, would require a period of 25,000 years. Thus this discovery connected the various aspects of the heavens at the most remote periods of human history; and, accordingly, the novel and ingenious views which Newton published in his chronology, are founded on this single astronomical fact, the Precession of the Equinoxes.

The two discoveries which have been described, the mode of constructing Solar and Lunar Tables, and the Precession, were advances of the greatest importance in astronomy, not only in themselves, but in the new objects and undertakings which they suggested to astronomers. The one discovery detected a constant law and order in the midst of perpetual change and apparent disorder; the other disclosed mutation and movement perpetually operating where every thing had been supposed fixed and stationary. Such discoveries were well adapted to call up many questionings in the minds of speculative men; for, after this, nothing could be supposed constant till it had been ascertained to be so by close examination; and no apparent complexity or confusion could justify the philosopher in turning away in despair from the task of simplification. To answer the inquiries thus suggested, new methods of observing the facts were requisite, more exact and uniform than those hitherto employed. Moreover, the discoveries which were made, and others which could not fail to follow in their train, led to many consequences, required to be reasoned upon, systematized, completed, enlarged. In short, the *Epoch of Induction* led, as we have stated that such epochs must always lead, to a *Period of Development, of Verification, Application, and Extension.*

CHAPTER IV.

Sequel to the Inductive Epoch of Hipparchus.

Sect. 1.—Researches which verified the Theory.

THE discovery of the leading Laws of the Solar and Lunar Motions, and the detection of the Precession, may be considered as the great positive steps in the Hipparchian astronomy;—the parent discoveries, from which many minor improvements proceeded. The task of pursuing the collateral and consequent researches which now offered themselves,—of bringing the other parts of astronomy up to the level of its most improved portions,—was prosecuted by a succession of zealous observers and calculators, first, in the school of Alexandria, and afterwards in other parts of the world. We must notice the various labors of this series of astronomers; but we shall do so very briefly; for the ulterior development of doctrines once established is not so important an object of contemplation for our present purpose, as the first conception and proof of those fundamental truths on which systematic doctrines are founded. Yet Periods of Verification, as well as Epochs of Induction, deserve to be attended to; and they can nowhere be studied with so much advantage as in the history of astronomy.

In truth, however, Hipparchus did not leave to his successors the task of pursuing into detail those views of the heavens to which his discoveries led him. He examined with scrupulous care almost every part of the subject. We must briefly mention some of the principal points which were thus settled by him.

The verification of the laws of the changes which he assigned to the skies, implied that the condition of the heavens was constant, except so far as it was affected by those changes. Thus, the doctrine that the changes of position of the stars were rightly represented by the precession of the equinoxes, supposed that the stars were fixed with regard to each other; and the doctrine that the unequal number of days, in certain subdivisions of months and years, was adequately explained by the theory of epicycles, assumed that years and days were always of constant lengths. But Hipparchus was not content with assuming these bases of his theory, he endeavored to prove them.

1. *Fixity of the Stars.*—The question necessarily arose after the discovery of the precession, even if such a question had never suggested itself before, whether the stars which were called *fixed*, and to which the motions of the other luminaries are referred, do really retain constantly the same relative position. In order to determine this fundamental question, Hipparchus undertook to construct a *Map* of the heavens; for though the result of his survey was expressed in words, we may give this name to his Catalogue of the positions of the most conspicuous stars. These positions are described by means of *alineations;* that is, three or more such stars are selected as can be touched by an apparent straight line drawn in the heavens. Thus Hipparchus observed that the southern claw of Cancer, the bright star in the same constellation which precedes the head of the Hydra, and the bright star Procyon, were nearly in the same line. Ptolemy quotes this and many other of the configurations which Hipparchus had noted, in order to show that the positions of the stars had not changed in the intermediate time; a truth which the catalogue of Hipparchus thus gave astronomers the means of ascertaining. It contained 1080 stars.

The construction of this catalogue of the stars by Hipparchus is an event of great celebrity in the history of astronomy. Pliny,[1] who speaks of it with admiration as a wonderful and superhuman task ("ausus rem etiam Deo improbam, annumerare posteris stellas"), asserts the undertaking to have been suggested by a remarkable astronomical event, the appearance of a new star; "novam stellam et aliam in ævo suo genitam deprehendit; ejusque motu, qua die fulsit, ad dubitationem est adductus anne hoc sæpius fieret, moverenturque et eæ quas putamus affixas." There is nothing inherently improbable in this tradition, but we may observe, with Delambre,[2] that we are not informed whether this new star remained in the sky, or soon disappeared again. Ptolemy makes no mention of the star or the story; and his catalogue contains no *bright* star which is not found in the "Catasterisms" of Eratosthenes. These Catasterisms were an enumeration of 475 of the principal stars, according to the constellations in which they are, and were published about sixty years before Hipparchus.

2. *Constant Length of Years.*—Hipparchus also attempted to ascertain whether successive years are all of the same length; and though, with his scrupulous love of accuracy,[3] he does not appear to have

---

[1] *Nat. Hist.* lib. ii. (xxvi.) [2] *A. A.* i. 290. [3] Ptolem. *Synt.* iii. 2.

thought himself justified in asserting that the years were always exactly equal, he showed, both by observations of the time when the sun passed the equinoxes, and by eclipses, that the difference of successive years, if there were any difference, must be extremely slight. The observations of succeeding astronomers, and especially of Ptolemy, confirmed this opinion, and proved, with certainty, that there is no progressive increase or diminution in the duration of the year.

3. *Constant Length of Days. Equation of Time.*—The equality of days was more difficult to ascertain than that of years; for the year is measured, as on a natural scale, by the number of days which it contains; but the day can be subdivided into hours only by artificial means; and the mechanical skill of the ancients did not enable them to attain any considerable accuracy in the measure of such portions of time; though clepsydras and similar instruments were used by astronomers. The equality of days could only be proved, therefore, by the consequences of such a supposition; and in this manner it appears to have been assumed, as the fact really is, that the apparent revolution of the stars is accurately uniform, never becoming either quicker or slower. It followed, as a consequence of this, that the solar days (or rather the *nycthemers*, compounded of a night and a day) would be unequal, in consequence of the sun's unequal motion, thus giving rise to what we now call the *Equation of Time*,—the interval by which the time, as marked on a dial, is before or after the time, as indicated by the accurate timepieces which modern skill can produce. This inequality was fully taken account of by the ancient astronomers; and they thus in fact assumed the equality of the sidereal days.

Sect. 2.—Researches which did not verify the Theory.

Some of the researches of Hipparchus and his followers fell upon the weak parts of his theory; and if the observations had been sufficiently exact, must have led to its being corrected or rejected.

Among these we may notice the researches which were made concerning the *Parallax* of the heavenly bodies, that is, their apparent displacement by the alteration of position of the observer from one part of the earth's surface to the other. This subject is treated of at length by Ptolemy; and there can be no doubt that it was well examined by Hipparchus, who invented a *parallactic instrument* for that purpose. The idea of parallax, as a geometrical possibility, was indeed too obvious to be overlooked by geometers at any time; and when the doctrine of the sphere was established, it must have appeared strange to the student, that every place on the earth's surface might alike be considered as the centre of the celestial motions. But if this was true with respect to the motions of the fixed stars, was it also true with regard to those of the sun and moon? The displacement of the sun by parallax is so small, that the best observers among the ancients could never be sure of its existence; but with respect to the moon, the case is different. She may be displaced by this cause to the amount of twice her own breadth, a quantity easily noticed by the rudest process of instrumental observation. The law of the displacement thus produced is easily obtained by theory, the globular form of the earth being supposed known; but the amount of the displacement depends upon the distance of the moon from the earth, and requires at least one good observation to determine it. Ptolemy has given a table of the effects of parallax, calculated according to the apparent altitude of the moon, assuming certain supposed distances; these distances, however, do not follow the real law of the moon's distances, in consequence of their being founded upon the Hypothesis of the Eccentric and Epicycle.

In fact this Hypothesis, though a very close representation of the truth, so far as the *positions* of the luminaries are concerned, fails altogether when we apply it to their *distances*. The radius of the epicycle, or the eccentricity of the eccentric, are determined so as to satisfy the observations of the apparent *motions* of the bodies; but, inasmuch as the hypothetical motions are different altogether from the real motions, the Hypothesis does not, at the same time, satisfy the observations of the *distances* of the bodies, if we are able to make any such observations.

Parallax is one method by which the distances of the moon, at different times, may be compared; her Apparent Diameters afford another method. Neither of these modes, however, is easily capable of such accuracy as to overturn at once the Hypothesis of epicycles; and, accordingly, the Hypothesis continued to be entertained in spite of such measures; the measures being, indeed, in some degree falsified in consequence of the reigning opinion. In fact, however, the imperfection of the methods of measuring parallax and magnitude, which were in use at this period, was such, their results could not lead to any degree of conviction deserving to be set in opposition to a theory which was so satisfactory with regard to the more certain observations, namely, those of the motions.

The Eccentricity, or the Radius of the Epicycle, which would satisfy the inequality of the *motions* of the moon, would, in fact, double the inequality of the *distances*. The Eccentricity of the moon's orbit is determined by Ptolemy as $\frac{1}{12}$ of the radius of the orbit; but its real amount is only half as great; this difference is a necessary consequence of the supposition of uniform circular motions, on which the Epicyclic Hypothesis proceeds.

We see, therefore, that this part of the Hipparchian theory carries in itself the germ of its own destruction. As soon as the art of celestial measurement was so far perfected, that astronomers could be sure of the apparent diameter of the moon within $\frac{1}{30}$ or $\frac{1}{40}$ of the whole, the inconsistency of the theory with itself would become manifest. We shall see, hereafter, the way in which this inconsistency operated; in reality a very long period elapsed before the methods of observing were sufficiently good to bring it clearly into view.

Sect. 3.—Methods of Observation of the Greek Astronomers.

We must now say a word concerning the Methods above spoken of. Since one of the most important tasks of verification is to ascertain with accuracy the magnitude of the quantities which enter, as elements, into the theory which occupies men during the period; the improvement of instruments, and the methods of observing and experimenting, are principal features in such periods. We shall, therefore, mention some of the facts which bear upon this point.

The estimation of distances among the stars by the eye, is an extremely inexact process. In some of the ancient observations, however, this appears to have been the method employed; and stars are described as being *a cubit* or *two cubits* from other stars. We may form some notion of the scale of this kind of measurement, from what Cleomedes remarks,[4] that the sun appears to be about a foot broad; an opinion which he confutes at length.

A method of determining the positions of the stars, susceptible of a little more exactness than the former, is the use of *alineations*, already noticed in speaking of Hipparchus's catalogue. Thus, a straight line passing through two stars of the Great Bear passes also through the pole-star; this is, indeed, even now a method usually employed to enable us readily to fix on the pole-star; and the two stars β and α of Ursa Major, are hence often called "the pointers."

---

[4] Del. *A. A.* i. 222.

But nothing like accurate measurements of any portions of the sky were obtained, till astronomers adopted the method of making visual coincidences of the objects with the instruments, either by means of *shadows* or of *sights*.

Probably the oldest and most obvious measurements of the positions of the heavenly bodies were those in which the elevation of the sun was determined by comparing the length of the shadow of an upright staff or *gnomon*, with the length of the staff itself. It appears,[5] from a memoir of Gautil, first printed in the *Connaissance des Temps* for 1809, that, at the lower town of Loyang, now called Hon-anfou, Tchon-kong found the length of the shadow of the gnomon, at the summer solstice, equal to one foot and a half, the gnomon itself being eight feet in length. This was about 1100 B.C. The Greeks, at an early period, used the same method. Strabo says[6] that "Byzantium and Marseilles are on the same parallel of latitude, because the shadows at those places have the same proportion to the gnomon, according to the statement of Hipparchus, who follows Pytheas."

But the relations of position which astronomy considers, are, for the most part, angular distances; and these are most simply expressed by the intercepted portion of a circumference described about the angular point. The use of the gnomon might lead to the determination of the angle by the graphical methods of geometry; but the numerical expression of the circumference required some progress in trigonometry; for instance, a table of the tangents of angles.

Instruments were soon invented for measuring angles, by means of circles, which had a border or *limb*, divided into equal parts. The whole circumference was divided into 360 *degrees:* perhaps because the circles, first so divided, were those which represented the sun's annual path; one such degree would be the sun's daily advance, more nearly than any other convenient aliquot part which could be taken. The position of the sun was determined by means of the shadow of one part of the instrument upon the other. The most ancient instrument of this kind appears to be the *Hemisphere of Berosus*. A hollow hemisphere was placed with its rim horizontal, and a style was erected in such a manner that the extremity of the style was exactly at the centre of the sphere. The shadow of this extremity, on the concave surface, had the same position with regard to the lowest point of the sphere which the sun had with regard to the highest point of the heavens.

---

[5] Lib. U. K. *Hist. Ast.* p. 5.

[6] Del. *A. A.* i. 257.

But this instrument was in fact used rather for dividing the day into portions of time than for determining position.

Eratosthenes[7] observed the amount of the obliquity of the sun's path to the equator: we are not informed what instruments he used for this purpose; but he is said to have obtained, from the munificence of Ptolemy Euergetes, two *Armils*, or instruments composed of circles, which were placed in the portico at Alexandria, and long used for observations. If a circular rim or hoop were placed so as to coincide with the plane of the equator, the inner concave edge would be enlightened by the sun's rays which came under the front edge, when the sun was south of the equator, and by the rays which came over the front edge, when the sun was north of the equator: the moment of the transition would be the time of the equinox. Such an instrument appears to be referred to by Hipparchus, as quoted by Ptolemy.[8] "The circle of copper, which stands at Alexandria in what is called the Square Porch, appears to mark, as the day of the equinox, that on which the concave surface begins to be enlightened from the other side." Such an instrument was called an *equinoctial armil.*

A *solstitial armil* is described by Ptolemy, consisting of two circular rims, one sliding round within the other, and the inner one furnished with two pegs standing out from its surface at right angles, and diametrically opposite to each other. These circles being fixed in the plane of the meridian, and the inner one turned, till, at noon, the shadow of the peg in front falls upon the peg behind, the position of the sun at noon would be determined by the degrees on the outer circle.

In calculation, the degree was conceived to be divided into 60 *minutes*, the minute into 60 *seconds*, and so on. But in practice it was impossible to divide the limb of the instrument into parts so small. The armils of Alexandria were divided into no parts smaller than sixths of degrees, or divisions of 10 minutes.

The angles, observed by means of these divisions, were expressed as a fraction of the circumference. Thus Eratosthenes stated the interval between the tropics to be $\frac{11}{83}$ of the circumference.[9]

It was soon remarked that the whole circumference of the circle

---

[7] Delambre, *A. A.* i. 86. [8] Ptol. *Synt.* iii. 2.

[9] Delambre, *A. A.* i. 87. It is probable that his observation gave him $47^2/_3$ degrees. The fraction $\frac{47^2/_3}{360} = \frac{143}{1080} = \frac{11 \cdot 13}{1080} = \frac{11}{83^1/_{13}}$, which is very nearly $\frac{11}{83}$.

was not wanted for such observations. Ptolemy[10] says that he found it more convenient to observe altitudes by means of a square flat piece of stone or wood, with a *quadrant* of a circle described on one of its flat faces, about a centre near one of the angles. A peg was placed at the centre, and one of the extreme radii of the quadrant being perpendicular to the horizon, the elevation of the sun above the horizon was determined by observing the point of the arc of the quadrant on which the shadow of the peg fell.

As the necessity of accuracy in the observations was more and more felt, various adjustments of such instruments were practised. The instruments were placed in the meridian by means of *a meridian line* drawn by astronomical methods on the floor on which they stood. The plane of the instrument was made vertical by means of a plumb-line: the bounding radius, from which angles were measured, was also adjusted by the *plumb-line*.[11]

In this manner, the places of the sun and of the moon could be observed by means of the shadows which they cast. In order to observe the stars,[12] the observer looked along the face of the circle of the armil, so as to see its two edges apparently brought together, and the star apparently touching them.[13]

It was afterwards found important to ascertain the position of the sun with regard to the ecliptic: and, for this purpose, an instrument, called an *astrolabe*, was invented, of which we have a description in Ptolemy.[14] This also consisted of circular rims, movable within one another, or about poles; and contained circles which were to be brought into the position of the ecliptic, and of a plane passing through the sun and the poles of the ecliptic. The position of the moon with regard to the ecliptic, and its position in longitude with regard to the sun or a star, were thus determined.

The astrolabe continued long in use, but not so long as the quadrant described by Ptolemy; this, in a larger form, is the *mural quadrant*, which has been used up to the most recent times.

It may be considered surprising,[15] that Hipparchus, after having

---

[10] *Synt.* i. 1.

[11] The curvature of the plane of the circle, by warping, was noticed. Ptol. iii. 2. p. 155, observes that his equatorial circle was illuminated on the hollow side twice in the same day. (He did not know that this might arise from refraction.)

[12] Delamb. *A. A.* i. 185.

[13] Ptol. *Synt.* i. 1. *Ὥσπερ κεκολλήμενος ἀμφοτέραις αὐτῶν ταῖς ἐπιφανέιαις ὁ ἀστὴρ ἐν τῷ δι' αὐτῶν ἐπιπέδῳ διοπτεύηται.*

[14] *Synt.* v. 1.

[15] Del. *A. A.* 181.

observed, for some time, right ascensions and declinations, quitted equatorial armils for the astrolabe, which immediately refers the stars to the ecliptic. He probably did this because, after the discovery of precession, he found the latitudes of the stars constant, and wanted to ascertain their motion in longitude.

To the above instruments, may be added the *dioptra*, and the *parallactic instrument* of Hipparchus and Ptolemy. In the latter, the distance of a star from the zenith was observed by looking through two sights fixed in a rule, this being annexed to another rule, which was kept in a vertical position by a plumb-line; and the angle between the two rules was measured.

The following example of an observation, taken from Ptolemy, may serve to show the form in which the results of the instruments, just described, were usually stated.[16]

"In the 2d year of Antoninus, the 9th day of Pharmouthi, the sun being near setting, the last division of Taurus being on the meridian (that is, 5½ equinoctial hours after noon), the moon was in 3 degrees of Pisces, by her distance from the sun (which was 92 degrees, 8 minutes); and half an hour after, the sun being set, and the quarter of Gemini on the meridian, Regulus appeared, by the other circle of the astrolabe, 57½ degrees more forwards than the moon in longitude." From these data the longitude of Regulus is calculated.

From what has been said respecting the observations of the Alexandrian astronomers, it will have been seen that their instrumental observations could not be depended on for any close accuracy. This defect, after the general reception of the Hipparchian theory, operated very unfavorably on the progress of the science. If they could have traced the moon's place distinctly from day to day, they must soon have discovered all the inequalities which were known to Tycho Brahe; and if they could have measured her parallax or her diameter with any considerable accuracy, they must have obtained a confutation of the epicycloidal form of her orbit. By the badness of their observations, and the imperfect agreement of these with calculation, they not only were prevented making such steps, but were led to receive the theory with a servile assent and an indistinct apprehension, instead of that rational conviction and intuitive clearness which would have given a progressive impulse to their knowledge.

---

[16] Del. *A. A.* ii. 248.

Sect. 4.—Period from Hipparchus to Ptolemy.

We have now to speak of the cultivators of astronomy from the time of Hipparchus to that of Ptolemy, the next great name which occurs in the history of this science; though even he holds place only among those who verified, developed, and extended the theory of Hipparchus. The astronomers who lived in the intermediate time, indeed, did little, even in this way; though it might have been supposed that their studies were carried on under considerable advantages, inasmuch as they all enjoyed the liberal patronage of the kings of Egypt.[17] The "divine school of Alexandria," as it is called by Synesius, in the fourth century, appears to have produced few persons capable of carrying forwards, or even of verifying, the labors of its great astronomical teacher. The mathematicians of the school wrote much, and apparently they observed sometimes; but their observations are of little value; and their books are expositions of the theory and its geometrical consequences, without any attempt to compare it with observation. For instance, it does not appear that any one verified the remarkable discovery of the precession, till the time of Ptolemy, 250 years after; nor does the statement of this motion of the heavens appear in the treatises of the intermediate writers; nor does Ptolemy quote a single observation of any person made in this long interval of time; while his references to those of Hipparchus are perpetual; and to those of Aristyllus and Timocharis, and of others, as Conon, who preceded Hipparchus, are not unfrequent.

This Alexandrian period, so inactive and barren in the history of science, was prosperous, civilized, and literary; and many of the works which belong to it are come down to us, though those of Hipparchus are lost. We have the "Uranologion" of Geminus,[18] a systematic treatise on Astronomy, expounding correctly the Hipparchian Theories and their consequences, and containing a good account of the use of the various Cycles, which ended in the adoption of the Calippic Period. We have likewise "The Circular Theory of the Celestial Bodies" of Cleomedes,[19] of which the principal part is a development of the doctrine of the sphere, including the consequences of the globular form of the earth. We have also another work on "Spherics" by Theodosius of Bithynia,[20] which contains some of the most important propositions of the subject, and has been used as a book of in-

[17] Delamb. *A. A.* ii. 240. [18] B. C. 70. [19] B. C. 60. [20] B.C. 50.

struction even in modern times. Another writer on the same subject is Menelaus, who lived somewhat later, and whose Three Books on Spherics still remain.

One of the most important kinds of deduction from a geometrical theory, such as that of the doctrine of the sphere, or that of epicycles, is the calculation of its numerical results in particular cases. With regard to the latter theory, this was done in the construction of Solar and Lunar Tables, as we have already seen; and this process required the formation of a *Trigonometry*, or system of rules for calculating the relations between the sides and angles of triangles. Such a science had been formed by Hipparchus, who appears to be the author of every great step in ancient astronomy.[21] He wrote a work in twelve books, "On the Construction of the Tables of Chords of Arcs;" such a table being the means by which the Greeks solved their triangles. The Doctrine of the Sphere required, in like manner, a *Spherical Trigonometry*, in order to enable mathematicians to calculate its results; and this branch of science also appears to have been formed by Hipparchus,[22] who gives results that imply the possession of such a method. Hypsicles, who was a contemporary of Ptolemy, also made some attempts at the solution of such problems: but it is extraordinary that the writers whom we have mentioned as coming after Hipparchus, namely, Theodosius, Cleomedes, and Menelaus, do not even mention the calculation of triangles,[23] either plain or spherical; though the latter writer[24] is said to have written on "the Table of Chords," a work which is now lost.

We shall see, hereafter, how prevalent a disposition in literary ages is that which induces authors to become commentators. This tendency showed itself at an early period in the school of Alexandria. Aratus,[25] who lived 270 B. C. at the court of Antigonus, king of Macedonia, described the celestial constellations in two poems, entitled "Phænomena," and "Prognostics." These poems were little more than a versification of the treatise of Eudoxus on the acronycal and heliacal risings and settings of the stars. The work was the subject of a comment by Hipparchus, who perhaps found this the easiest way of giving connection and circulation to his knowledge. Three Latin translations of this poem gave the Romans the means of becoming acquainted with it: the first is by Cicero, of which we have numerous fragments ex-

[21] Delamb. *A. A.* ii. 87. [22] *A. A.* i. 117. [23] *A. A.* i. 249.
[24] *A. A.* ii. 87. [25] *A. A.* i. 74.

tant;[26] Germanicus Cæsar, one of the sons-in-law of Augustus, also translated the poem, and this translation remains almost entire. Finally, we have a complete translation by Avienus.[27] The "Astronomica" of Manilius, the "Poeticon Astronomicon" of Hyginus, both belonging to the time of Augustus, are, like the work of Aratus, poems which combine mythological ornament with elementary astronomical exposition; but have no value in the history of science. We may pass nearly the same judgment upon the explanations and declamations of Cicero, Seneca, and Pliny, for they do not apprise us of any additions to astronomical knowledge; and they do not always indicate a very clear apprehension of the doctrines which the writers adopt.

Perhaps the most remarkable feature in the two last-named writers, is the declamatory expression of their admiration for the discoverers of physical knowledge; and in one of them, Seneca, the persuasion of a boundless progress in science to which man was destined. Though this belief was no more than a vague and arbitrary conjecture, it suggested other conjectures in detail, some of which, having been verified, have attracted much notice. For instance, in speaking of comets,[28] Seneca says, "The time will come when those things which are now hidden shall be brought to light by time and persevering diligence. Our posterity will wonder that we should be ignorant of what is so obvious." "The motions of the planets," he adds, "complex and seemingly confused, have been reduced to rule; and some one will come hereafter, who will reveal to us the paths of comets." Such convictions and conjectures are not to be admired for their wisdom; for Seneca was led rather by enthusiasm, than by any solid reasons, to entertain this opinion; nor, again, are they to be considered as merely lucky guesses, implying no merit; they are remarkable as showing how the persuasion of the universality of law, and the belief of the probability of its discovery by man, grow up in men's minds, when speculative knowledge becomes a prominent object of attention.

An important practical application of astronomical knowledge was made by Julius Cæsar, in his correction of the calendar, which we have already noticed; and this was strictly due to the Alexandrian School: Sosigenes, an astronomer belonging to that school, came from Egypt to Rome for the purpose.

---

[26] Two copies of this translation, illustrated by drawings of different ages, one set Roman, and the other Saxon, according to Mr. Ottley, are described in the *Archæologia*, vol. xviii.

[27] Montucla, i. 221.

[28] Seneca, *Qu. N.* vii. 25.

Sect. 5.—Measures of the Earth.

There were, as we have said, few attempts made, at the period of which we are speaking, to improve the accuracy of any of the determinations of the early Alexandrian astronomers. One question naturally excited much attention at all times, the *magnitude* of the earth, its figure being universally acknowledged to be a globe. The Chaldeans, at an earlier period, had asserted that a man, walking without stopping, might go round the circuit of the earth in a year; but this might be a mere fancy, or a mere guess. The attempt of Eratosthenes to decide this question went upon principles entirely correct. Syene was situated on the tropic; for there, on the day of the solstice, at noon, objects cast no shadow; and a well was enlightened to the bottom by the sun's rays. At Alexandria, on the same day, the sun was, at noon, distant from the zenith by a fiftieth part of the circumference. These two cities were north and south from each other: and the distance had been determined, by the royal overseers of the roads, to be 5000 stadia. This gave a circumference of 250,000 stadia to the earth, and a radius of about 40,000. Aristotle[29] says that the mathematicians make the circumference 400,000 stadia. Hipparchus conceived that the measure of Eratosthenes ought to be increased by about one-tenth.[30] Posidonius, the friend of Cicero, made another attempt of the same kind. At Rhodes, the star Canopus but just appeared above the horizon; at Alexandria, the same star rose to an altitude of $\frac{1}{48}$th of the circumference; the direct distance on the meridian was 5000 stadia, which gave 240,000 for the whole circuit. We cannot look upon these measures as very precise; the stadium employed is not certainly known; and no peculiar care appears to have been bestowed on the measure of the direct distance.

When the Arabians, in the ninth century, came to be the principal cultivators of astronomy, they repeated this observation in a manner more suited to its real importance and capacity of exactness. Under the Caliph Almamon,[31] the vast plain of Singiar, in Mesopotamia, was the scene of this undertaking. The Arabian astronomers there divided themselves into two bands, one under the direction of Chalid ben Abdolmalic, and the other having at its head Alis ben Isa. These two parties proceeded, the one north, the other south, determining the distance by the actual application of their measuring-rods to the ground,

[29] *De Cœlo*, ii. ad fin. [30] Plin. ii. (cviii.) [31] Montu. 357.

till each was found, by astronomical observation, to be a degree from the place at which they started. It then appeared that these terrestrial degrees were respectively 56 miles, and 56 miles and two-thirds, the mile being 4000 cubits. In order to remove all doubt concerning the scale of this measure, we are informed that the cubit is that called the black cubit, which consists of 27 inches, each inch being the thickness of six grains of barley.

Sect. 6.—*Ptolemy's Discovery of Evection.*

By referring, in this place, to the last-mentioned measure of the earth, we include the labors of the Arabian as well as the Alexandrian astronomers, in the period of mere detail, which forms the sequel to the great astronomical revolution of the Hipparchian epoch. And this period of verification is rightly extended to those later times; not merely because astronomers were then still employed in determining the magnitude of the earth, and the amount of other elements of the theory,—for these are some of their employments to the present day,—but because no great intervening discovery marks a new epoch, and begins a new period;—because no great revolution in the theory added to the objects of investigation, or presented them in a new point of view. This being the case, it will be more instructive for our purpose to consider the general character and broad intellectual features of this period, than to offer a useless catalogue of obscure and worthless writers, and of opinions either borrowed or unsound. But before we do this, there is one writer whom we cannot leave undistinguished in the crowd; since his name is more celebrated even than that of Hipparchus; his works contain ninety-nine hundredths of what we know of the Greek astronomy; and though he was not the author of a new theory, he made some very remarkable steps in the verification, correction, and extension of the theory which he received. I speak of Ptolemy, whose work, "The Mathematical Construction" (of the heavens), contains a complete exposition of the state of astronomy in his time, the reigns of Adrian and Antonine. This book is familiarly known to us by a term which contains the record of our having received our first knowledge of it from the Arabic writers. The "*Megiste* Syntaxis," or Great Construction, gave rise, among them, to the title *Al Magisti*, or *Almagest*, by which the work is commonly described. As a mathematical exposition of the Theory of Epicycles and Eccentrics, of the observations and calculations which were employed in order to apply this theory to the sun, moon, and planets, and of the other calculations which are requisite, in order to deduce the consequences of this theory, the work is a splendid and lasting monument of diligence, skill, and judgment. Indeed, all the other astronomical works of the ancients hardly add any thing whatever to the information we obtain from the Almagest; and the knowledge which the student possesses of the ancient astronomy must depend mainly upon his acquaintance with Ptolemy. Among other merits, Ptolemy has that of giving us a very copious account of the manner in which Hipparchus established the main points of his theories; an account the more agreeable, in consequence of the admiration and enthusiasm with which this author everywhere speaks of the great master of the astronomical school.

In our present survey of the writings of Ptolemy, we are concerned less with his exposition of what had been done before him, than with his own original labors. In most of the branches of the subject, he gave additional exactness to what Hipparchus had done; but our main business, at present, is with those parts of the Almagest which contain new steps in the application of the Hipparchian hypothesis. There are two such cases, both very remarkable,—that of the moon's *Evection*, and that of the *Planetary Motions.*

The law of the moon's anomaly, that is, of the leading and obvious inequality of her motion, could be represented, as we have seen, either by an eccentric or an epicycle; and the amount of this inequality had been collected by observations of eclipses. But though the hypothesis of an epicycle, for instance, would bring the moon to her proper place, so far as eclipses could show it, that is, at new and full moon, this hypothesis did not rightly represent her motions at other points of her course. This appeared, when Ptolemy set about measuring her distances from the sun at different times. "These," he[32] says, "sometimes agreed, and sometimes disagreed." But by further attention to the facts, a rule was detected in these differences. "As my knowledge became more complete and more connected, so as to show the order of this new inequality, I perceived that this difference was small, or nothing, at new and full moon; and that at both the *dichotomies* (when the moon is half illuminated) it was small, or nothing, if the moon was at the apogee or perigee of the epicycle, and was greatest when she was in the middle of the interval, and therefore when the first inequal-

---

[32] *Synth.* v. 2.

ity was greatest also." He then adds some further remarks on the circumstances according to which the moon's place, as affected by this new inequality, is before or behind the place, as given by the epicyclical hypothesis.

Such is the announcement of the celebrated discovery of the moon's second inequality, afterwards called (by Bullialdus) the *Evection*. Ptolemy soon proceeded to represent this inequality by a combination of circular motions, uniting, for this purpose, the hypothesis of an epicycle, already employed to explain the first inequality, with the hypothesis of an eccentric, in the circumference of which the centre of the epicycle was supposed to move. The mode of combining these was somewhat complex; more complex we may, perhaps, say, than was absolutely requisite;[33] the apogee of the eccentric moved backwards, or contrary to the order of the signs, and the centre of the epicycle moved forwards nearly twice as fast upon the circumference of the eccentric, so as to reach a place nearly, but not exactly, the same, as if it had moved in a concentric instead of an eccentric path. Thus the centre of the epicycle went twice round the eccentric in the course of one month: and in this manner it satisfied the condition that it should vanish at new and full moon, and be greatest when the moon was in the quarters of her monthly course.[34]

The discovery of the Evection, and the reduction of it to the epi-

---

[33] If Ptolemy had used the hypothesis of an eccentric instead of an epicycle for the first inequality of the moon, an epicycle would have represented the second inequality more simply than his method did.

[34] I will insert here the explanation which my German translator, the late distinguished astronomer Littrow, has given of this point. The Rule of this Inequality, the Evection, may be most simply expressed thus. If a denote the excess of the Moon's Longitude over the Sun's, and b the Anomaly of the Moon reckoned from her Perigee, the Evection is equal to $1^\circ.\ 3 \,.\, \sin\,(2a-b)$. At New and Full Moon, a is 0 or 180°, and thus the Evection is $-1^\circ.\ 3 \,.\, \sin\, b$. At both quarters, or dichotomies, a is 90° or 270°, and consequently the Evection is $+\ 1^\circ.\ 3 \,.\, \sin\, b$. The Moon's Elliptical Equation of the centre is at all points of her orbit equal to $6^\circ.\ 3 \,.\, \sin\, b$. The Greek Astronomers before Ptolemy observed the moon only at the time of eclipses; and hence they necessarily found for the sum of these two greatest inequalities of the moon's motion the quantity $6^\circ.\ 3 \,.\, \sin\, b - 1^\circ.\ 3 \,.\, \sin\, b$, or $5^\circ.\ \sin\, b$: and as they took this for the moon's equation of the centre, which depends upon the eccentricity of the moon's orbit, we obtain from this too small equation of the centre, an eccentricity also smaller than the truth. Ptolemy, who first observed the moon in her quarters, found for the sum of those Inequalities at those points the quantity $6^\circ.\ 3 \,.\, \sin\, b + 1^\circ.\ 3 \,.\, \sin\, b$, or $7^\circ.\ 6 \,.\, \sin\, b$; and thus made the eccentricity of the moon as much too great at the quarters as the observers of eclipses had made it too small. He hence concluded that the eccentricity of the Moon's orbit is variable, which is not the case.

cyclical theory, was, for several reasons, an important step in astronomy; some of these reasons may be stated.

1. It obviously suggested, or confirmed, the suspicion that the motions of the heavenly bodies might be subject to *many* inequalities:—that when one set of anomalies had been discovered and reduced to rule, another set might come into view;—that the discovery of a rule was a step to the discovery of deviations from the rule, which would require to be expressed in other rules;—that in the application of theory to observation, we find, not only the *stated phenomena*, for which the theory does account, but also *residual phenomena*, which remain unaccounted for, and stand out beyond the calculation;—that thus nature is not simple and regular, by conforming to the simplicity and regularity of our hypotheses, but leads us forwards to apparent complexity, and to an accumulation of rules and relations. A fact like the Evection, explained by an Hypothesis like Ptolemy's, tended altogether to discourage any disposition to guess at the laws of nature from mere ideal views, or from a few phenomena.

2. The discovery of Evection had an importance which did not come into view till long afterwards, in being the first of a numerous series of inequalities of the moon, which results from the *Disturbing Force* of the sun. These inequalities were successfully discovered; and led finally to the establishment of the law of universal gravitation. The moon's first inequality arises from a different cause;—from the same cause as the inequality of the sun's motion;—from the motion in an ellipse, so far as the central attraction is undisturbed by any other. This first inequality is called the Elliptic Inequality, or, more usually, the *Equation of the Centre.*[35] All the planets have such inequalities, but the Evection is peculiar to the moon. The discovery of other inequalities of the moon's motion, the Variation and Annual Equation, made an immediate sequel in the order of the subject to

[35] The Equation of the Centre is the difference between the place of the Planet in its elliptical orbit, and that place which a Planet would have, which revolved uniformly round the Sun as a centre in a circular orbit in the same time. An imaginary Planet moving in the manner last described, is called the *mean* Planet, while the actual Planet which moves in the ellipse is called the *true* Planet. The Longitude of the mean Planet at a given time is easily found, because its motion is uniform. By adding to it the Equation of the Centre, we find the Longitude of the true Planet, and thus, its place in its orbit.—*Littrow's Note.*

I may add that the word *Equation*, used in such cases, denotes in general a quantity which must be added to or subtracted from a mean quantity, to make it *equal* to the true quantity; or rather, a quantity which must be added to or subtracted from a variably increasing quantity to make it increase *equably*.

the discoveries of Ptolemy, although separated by a long interval of time; for these discoveries were only made by Tycho Brahe in the sixteenth century. The imperfection of astronomical instruments was the great cause of this long delay.

3. The Epicyclical Hypothesis was found capable of accommodating itself to such new discoveries. These new inequalities could be represented by new combinations of eccentrics and epicycles: all the real and imaginary discoveries by astronomers, up to Copernicus, were actually embodied in these hypotheses; Copernicus, as we have said, did not reject such hypotheses; the lunar inequalities which Tycho detected might have boen similarly exhibited; and even Newton[36] represents the motion of the moon's apogee by means of an epicycle. As a mode of expressing the law of the irregularity, and of calculating its results in particular cases, the epicyclical theory was capable of continuing to render great service to astronomy, however extensive the progress of the science might be. It was, in fact, as we have already said, the modern process of representing the motion by means of a series of circular functions.

4. But though the doctrine of eccentrics and epicycles was thus admissible as an Hypothesis, and convenient as a means of expressing the laws of the heavenly motions, the successive occasions on which it was called into use, gave no countenance to it as a Theory; that is, as a true view of the nature of these motions, and their causes. By the steps of the progress of this Hypothesis, it became more and more complex, instead of becoming more simple, which, as we shall see, was the course of the true Theory. The notions concerning the position and connection of the heavenly bodies, which were suggested by one set of phenomena, were not confirmed by the indications of another set of phenomena; for instance, those relations of the epicycles which were adopted to account for the Motions of the heavenly bodies, were not found to fall in with the consequences of their apparent Diameters and Parallaxes. In reality, as we have said, if the relative distances of the sun and moon at different times could have been accurately determined, the Theory of Epicycles must have been forthwith overturned. The insecurity of such measurements alone maintained the theory to later times.[37]

---

[36] *Principia*, lib. iii. prop. xxxv.

[37] The alteration of the apparent diameter of the moon is so great that it cannot escape us, even with very moderate instruments. This apparent diameter contains, when the moon is nearest the earth, 2010 seconds; when she is farthest off

Sect. 7.—Conclusion of the History of Greek Astronomy.

I MIGHT now proceed to give an account of Ptolemy's other great step, the determination of the Planetary Orbits; but as this, though in itself very curious, would not illustrate any point beyond those already noticed, I shall refer to it very briefly. The planets all move in ellipses about the sun, as the moon moves about the earth; and as the sun apparently moves about the earth. They will therefore each have an Elliptic Inequality or Equation of the centre, for the same reason that the sun and moon have such inequalities. And this inequality may be represented, in the cases of the planets, just as in the other two, by means of an eccentric; the epicycle, it will be recollected, had already been used in order to represent the more obvious changes of the planetary motions. To determine the amount of the Eccentricities and the places of the Apogees of the planetary orbits, was the task which Ptolemy undertook; Hipparchus, as we have seen, having been destitute of the observations which such a process required. The determination of the Eccentricities in these cases involved some peculiarities which might not at first sight occur to the reader. The ecliptical motion of the planets takes place about the sun; but Ptolemy considered their movements as altogether independent of the sun, and referred them to the earth alone; and thus the apparent eccentricities which he had to account for, were the compound result of the Eccentricity of the earth's orbit, and of the proper eccentricity of the orbit of the Planet. He explained this result by the received mechanism of an eccentric *Deferent*, carrying an Epicycle; but the motion in the Deferent is uniform, not about the centre of the circle, but about another point, the *Equant*. Without going further into detail, it may be sufficient to state that, by a combination of Eccentrics and Epicycles, he did account for the leading features of these motions; and by using his own observations, compared with more ancient ones (for instance, those of Timocharis for Venus), he was able to determine the Dimensions and Positions of the orbits.[38]

1762 seconds; that is, 248 seconds, or 4 minutes 8 seconds, less than in the former case. [The two qantities are in the proportion of 8 to 7, nearly.]—*Littrow's Note.*

38 Ptolemy determined the Radius and the Periodic Time of his two circles for each Planet in the following manner: For the *inferior* Planets, that is, Mercury and Venus, he took the Radius of the Deferent equal to the Radius of the Earth's orbit, and the Radius of the Epicycle equal to that of the Planet's orbit. For these Planets, according to his assumption, the Periodic Time of the Planet in its Epi-

I shall here close my account of the astronomical progress of the Greek School. My purpose is only to illustrate the principles on which the progress of science depends, and therefore I have not at all pretended to touch upon every part of the subject. Some portion of the ancient theories, as, for instance, the mode of accounting for the motions of the moon and planets in latitude, are sufficiently analogous to what has been explained, not to require any more especial notice. Other parts of Greek astronomical knowledge, as, for instance, their acquaintance with refraction, did not assume any clear or definite form, and can only be considered as the prelude to modern discoveries on the same subject. And before we can with propriety pass on to these, there is a long and remarkable, though unproductive interval, of which some account must be given.

Sect. 8.—*Arabian Astronomy.*

The interval to which I have just alluded may be considered as extending from Ptolemy to Copernicus; we have no advance in Greek astronomy after the former; no signs of a revival of the power of discovery till the latter. During this interval of 1350 years,[39] the principal cultivators of astronomy were the Arabians, who adopted this science from the Greeks whom they conquered, and from whom the conquerors of western Europe again received back their treasure, when the love of science and the capacity for it had been awakened in their minds. In the intervening time, the precious deposit had undergone little change. The Arab astronomer had been the scrupulous but unprofitable servant, who kept his talent without apparent danger of loss, but also without prospect of increase. There is little in Ara-

---

cycle was to the Periodic Time of the Epicyclical Centre on the Deferent, as the *synodical* Revolution of the Planet to the *tropical* Revolution of the Earth above the Sun. For the three *superior* Planets, Mars, Jupiter, and Saturn, the Radius of the Deferent was equal to the Radius of the Planet's orbit, and the Radius of the Epicycle was equal to the Radius of the Earth's orbit; the Periodic Time on the Planet in its Epicycle was to the Periodic Time of the Epicyclical Centre on the Deferent, as the *synodical* Revolution of the Planet to the *tropical* Revolution of the same Planet.

Ptolemy might obviously have made the geometrical motions of *all* the Planets correspond with the observations by one of these two modes of construction; but he appears to have adopted this double form of the theory, in order that in the inferior, as well as in the superior Planets, he might give the smaller of the two Radii to the Epicycle: that is, in order that he might make the smaller circle move round the larger, not *vice versâ*.—*Littrow's Notes.*

[39] Ptolemy died about A. D. 150. Copernicus was living A. D. 1500.

bic literature which bears upon the *progress* of astronomy; but as the little that there is must be considered as a sequel to the Greek science, I shall notice one or two points before I treat of the stationary period in general.

When the sceptre of western Asia had passed into the hands of the Abasside caliphs,[40] Bagdad, "the city of peace," rose to splendor and refinement, and became the metropolis of science under the successors of Almansor the Victorious, as Alexandria had been under the successors of Alexander the Great. Astronomy attracted peculiarly the favor of the powerful as well as the learned; and almost all the culture which was bestowed upon the science, appears to have had its source in the patronage, often also in the personal studies, of Saracen princes. Under such encouragement, much was done, in those scientific labors which money and rank can command. Translations of Greek works were made, large instruments were erected, observers were maintained; and accordingly as observation showed the defects and imperfection of the extant tables of the celestial motions, new ones were constructed. Thus under Almansor, the Grecian works of science were collected from all quarters, and many of them translated into Arabic.[41] The translation of the "Megiste Syntaxis" of Ptolemy, which thus became the Almagest, is ascribed to Isaac ben Homain in this reign.

The greatest of the Arabian Astronomers comes half a century later. This is Albategnius, as he is commonly called; or more exactly, Muhammed ben Geber Albatani, the last appellation indicating that he was born at Batan, a city of Mesopotamia.[42] He was a Syrian prince, whose residence was at Aracte or Racha in Mesopotamia: a part of his observations were made at Antioch. His work still remains to us in Latin. "After having read," he says, "the Syntaxis of Ptolemy, and learnt the methods of calculation employed by the Greeks, his observations led him to conceive that some improvements might be made in their results. He found it necessary to add to Ptolemy's observations as Ptolemy had added to those of Abrachis" (Hipparchus). He then published Tables of the motions of the sun, moon, and planets, which long maintained a high reputation.

These, however, did not prevent the publication of others. Under the Caliph Hakem (about A. D. 1000), Ebon Iounis published Tables of the Sun, Moon, and Planets, which were hence called the *Hakemite* Tables. Not long after, Arzachel of Toledo published the *Toletan* Ta-

[40] Gibbon, x. 31. [41] Id. x. 36. [42] Del. *Astronomie du Moyen Age*, 4.

bles. In the 13th century, Nasir Eddin published Tables of the Stars, dedicated to Ilchan, a Tartar prince, and hence termed the *Ilchanic* Tables. Two centuries later, Ulugh Beigh, the grandson of Tamerlane, and prince of the countries beyond the Oxus, was a zealous practical astronomer; and his Tables, which were published in Europe by Hyde in 1665, are referred to as important authority by modern astronomers. The series of Astronomical Tables which we have thus noticed, in which, however, many are omitted, leads us to the *Alphonsine* Tables, which were put forth in 1488, and in succeeding years, under the auspices of Alphonso, king of Castile; and thus brings us to the verge of modern astronomy.

For all these Tables, the Ptolemaic hypotheses were employed; and, for the most part, without alteration. The Arabs sometimes felt the extreme complexity and difficulty of the doctrine which they studied; but their minds did not possess that kind of invention and energy by which the philosophers of Europe, at a later period, won their way into a simpler and better system.

Thus Alpetragius states, in the outset of his "Planetarum Theorica," that he was at first astonished and stupefied with this complexity, but that afterwards "God was pleased to open to him the occult secret in the theory of his orbs, and to make known to him the truth of their essence, and the rectitude of the quality of their motion." His system consists, according to Delambre,[43] in attributing to the planets a spiral motion from east to west, an idea already refuted by Ptolemy. Geber of Seville criticises Ptolemy very severely,[44] but without introducing any essential alteration into his system. The Arabian observations are in many cases valuable; both because they were made with more skill and with better instruments than those of the Greeks; and also because they illustrate the permanence or variability of important elements, such as the obliquity of the ecliptic and the inclination of the moon's orbit.

We must, however, notice one or two peculiar Arabian doctrines. The most important of these is the discovery of the Motion of the Sun's Apogee by Albategnius. He found the Apogee to be in longitude 82 degrees; Ptolemy had placed it in longitude 65 degrees. The difference of 17 degrees was beyond all limit of probable error of calculation, though the process is not capable of great precision; and the inference of the Motion of the Apogee was so obvious, that we cannot

---

[43] Delambre, *M. A.* p. 7.

[44] *M. A.* p. 180, &c.

agree with Delambre, in doubting or extenuating the claim of Albategnius to this discovery, on the ground of his not having expressly stated it.

In detecting this motion, the Arabian astronomers reasoned rightly from facts well observed: they were not always so fortunate. Arzachel, in the 11th century, found the apogee of the sun to be less advanced than Albategnius had found it, by some degrees; he inferred that it had receded in the intermediate time; but we now know, from an acquaintance with its real rate of moving, that the true inference would have been, that Albategnius, whose method was less trustworthy than that of Arzachel, had made an error to the amount of the difference thus arising. A curious, but utterly false hypothesis was founded on observations thus erroneously appreciated; namely, the *Trepidation of the fixed stars.* Arzachel conceived that a uniform Precession of the equinoctial points would not account for the apparent changes of position of the stars, and that for this purpose, it was necessary to conceive two circles of about eight degrees radius described round the equinoctial points of the immovable sphere, and to suppose the first points of Aries and Libra to describe the circumference of these circles in about 800 years. This would produce, at one time a progression, and at another a regression, of the apparent equinoxes, and would moreover change the latitude of the stars. Such a motion is entirely visionary; but the doctrine made a sect among astronomers, and was adopted in the first edition of the Alphonsine Tables, though afterwards rejected.

An important exception to the general unprogressive character of Arabian science has been pointed out recently by M. Sedillot.[45] It appears that Mohammed-Aboul Wefa-al-Bouzdjani, an Arabian astronomer of the tenth century, who resided at Cairo, and observed at Bagdad in 975, discovered a third inequality of the moon, in addition to the two expounded by Ptolemy, the Equation of the Centre, and the Evection. This third inequality, the *Variation*, is usually supposed to have been discovered by Tycho Brahe, six centuries later. It is an inequality of the moon's motion, in virtue of which she moves quickest when she is at new or full, and slowest at the first and third quarter; in consequence of this, from the first quarter to the full, she is behind her *mean* place; at the full, she does not differ from her mean place; from the full to the third quarter, she is before her true

---

[45] Sedillot, Nouvelles Rech. sur l'Hist. de l'Astron. chez les Arabes. *Nouveau Journal Asiatique.* 1836.

place; and so on; and the greatest effect of the inequality is in the *octants*, or points half-way between the four quarters. In an Almagest of Aboul Wefa, a part of which exists in the Royal Library at Paris, after describing the two inequalities of the moon, he has a Section ix., "Of the Third Anomaly of the moon called *Muhazal* or *Prosneusis*." He there says, that taking cases when the moon was in apogee or perigee, and when, consequently, the effect of the two first inequalities vanishes, he found, *by observation of the moon*, when she was nearly *in trine* and *in sextile* with the sun, that she was a degree and a quarter from her calculated place. "And hence," he adds, "I perceived that this anomaly exists independently of the two first: and this can only take place by a declination of the diameter of the epicycle with respect to the centre of the zodiac."

We may remark that we have here this inequality of the moon made out in a really philosophical manner; a residual quantity in the moon's longitude being detected by observation, and the cases in which it occurs selected and grouped by an inductive effort of the mind. The advance is not great; for Aboul Wefa appears only to have detected the existence, and not to have fixed the law or the exact quantity of the inequality; but still it places the scientific capacity of the Arabs in a more favorable point of view than any circumstance with which we were previously acquainted.

But this discovery of Aboul Wefa appears to have excited no notice among his contemporaries and followers: at least it had been long quite forgotten when Tycho Brahe rediscovered the same lunar inequality. We can hardly help looking upon this circumstance as an evidence of a servility of intellect belonging to the Arabian period. The learned Arabians were so little in the habit of considering science as progressive, and looking with pride and confidence at examples of its progress, that they had not the courage to believe in a discovery which they themselves had made, and were dragged back by the chain of authority, even when they had advanced beyond their Greek masters.

As the Arabians took the whole of their theory (with such slight exceptions as we have been noticing) from the Greeks, they took from them also the mathematical processes by which the consequences of the theory were obtained. Arithmetic and Trigonometry, two main branches of these processes, received considerable improvements at their hands. In the former, especially, they rendered a service to the world which it is difficult to estimate too highly, in abolishing the cumbrous Sexagesimal Arithmetic of the Greeks, and introducing the notation by means of the digits 1, 2, 3, 4, 5, 6, 7, 8, 9, 0, which we now employ.[46] These numerals appear to be of Indian origin, as is acknowledged by the Arabs themselves; and thus form no exception to the sterility of the Arabian genius as to great scientific inventions. Another improvement, of a subordinate kind, but of great utility, was Arabian, being made by Albategnius. He introduced into calculation the *sine*, or half-chord of the double arc, instead of the chord of the arc itself, which had been employed by the Greek astronomers. There have been various conjectures concerning the origin of the word *sine;* the most probable appears to be that *sinus* is the Latin translation of the Arabic word *gib*, which signifies a fold, the two halves of the chord being conceived to be folded together.

The great obligation which Science owes to the Arabians, is to have preserved it during a period of darkness and desolation, so that Europe might receive it back again when the evil days were past. We shall see hereafter how differently the European intellect dealt with this hereditary treasure when once recovered.

Before quitting the subject, we may observe that Astronomy brought back, from her sojourn among the Arabs, a few terms which may still be perceived in her phraseology. Such are the *zenith*, and the opposite imaginary point, the *nadir;*—the circles of the sphere termed *almacantars* and *azimuth* circles. The *alidad* of an instrument is its index, which possesses an angular motion. Some of the stars still retain their Arabic names; *Aldebaran*, *Rigel*, *Fomalhaut;* many others were known by such appellations a little while ago. Perhaps the word *almanac* is the most familiar vestige of the Arabian period of astronomy.

It is foreign to my purpose to note any efforts of the intellectual faculties among other nations, which may have taken place independently of the great system of progressive European culture, from which all our existing science is derived. Otherwise I might speak of the astronomy of some of the Orientals, for example, the Chinese, who are said, by Montucla (i. 465), to have discovered the first equation of the moon, and the proper motion of the fixed stars (the Precession), in the third century of our era. The Greeks had made these discoveries 500 years earlier.

---

[46] Mont. i. 376.

BOOK IV.

HISTORY

OF

PHYSICAL SCIENCE IN THE MIDDLE AGES;

OR,

VIEW OF THE STATIONARY PERIOD

OF

INDUCTIVE SCIENCE.

In vain, in vain! the all-composing hour
Resistless falls . . . .

. . . .

As one by one, at dread Medea's strain,
The sickening stars fade off th' ethereal plain;
As Argus' eyes, by Hermes' wand opprest,
Closed one by one to everlasting rest;
Thus at her felt approach and secret might,
Art after art goes out, and all is night.
See skulking Truth to her old cavern fled,
Mountains of casuistry heaped on her head;
Philosophy, that reached the heavens before,
Shrinks to her hidden cause, and is no more.
Physic of Metaphysic begs defence,
And Metaphysic calls for aid to Sense:
See Mystery to Mathematics fly!
In vain! they gaze, turn giddy, rave, and die.

Dunciad, B. iv.

INTRODUCTION.

WE have now to consider more especially a long and barren period, which intervened between the scientific activity of ancient Greece and that of modern Europe; and which we may, therefore, call the Stationary Period of Science. It would be to no purpose to enumerate the various forms in which, during these times, men reproduced the discoveries of the inventive ages; or to trace in them the small successes of Art, void of any principle of genuine Philosophy. Our object requires rather that we should point out the general and distinguishing features of the intellect and habits of those times. We must endeavor to delineate the character of the Stationary Period, and, as far as possible, to analyze its defects and errors; and thus obtain some knowledge of the causes of its barrenness and darkness.

We have already stated, that real scientific progress requires distinct general Ideas, applied to many special and certain Facts. In the period of which we now have to speak, men's Ideas were obscured; their disposition to bring their general views into accordance with Facts was enfeebled. They were thus led to employ themselves unprofitably, among indistinct and unreal notions. And the evil of these tendencies was further inflamed by moral peculiarities in the character of those times;—by an abjectness of thought on the one hand, which could not help looking towards some intellectual superior, and by an impatience of dissent on the other. To this must be added an enthusiastic temper, which, when introduced into speculation, tends to subject the mind's operations to ideas altogether distorted and delusive.

These characteristics of the stationary period, its obscurity of thought, its servility, its intolerant disposition, and its enthusiastic temper, will be treated of in the four following chapters, on the Indistinctness of Ideas, the Commentatorial Spirit, the Dogmatism, and the Mysticism of the Middle Ages.

CHAPTER I.

On the Indistinctness of Ideas of the Middle Ages.

THAT firm and entire possession of certain clear and distinct general ideas which is necessary to sound science, was the character of the minds of those among the ancients who created the several sciences which arose among them. It was indispensable that such inventors should have a luminous and steadfast apprehension of certain general relations, such as those of space and number, order and cause; and should be able to apply these notions with perfect readiness and precision to special facts and cases. It is necessary that such scientific notions should be more definite and precise than those which common language conveys; and in this state of unusual clearness, they must be so familiar to the philosopher, that they are the language in which he thinks. The discoverer is thus led to doctrines which other men adopt and follow out, in proportion as they seize the fundamental ideas, and become acquainted with the leading facts. Thus Hipparchus, conceiving clearly the motions and combinations of motion which enter into his theory, saw that the relative lengths of the seasons were sufficient data for determining the form of the sun's orbit; thus Archimedes, possessing a steady notion of mechanical pressure, was able, not only to deduce the properties of the lever and of the centre of gravity, but also to see the truth of those principles respecting the distribution of pressure in fluids, on which the science of hydrostatics depends.

With the progress of such distinct ideas, the inductive sciences rise and flourish; with the decay and loss of such distinct ideas, these sciences become stationary, languid, and retrograde. When men merely repeat the terms of science, without attaching to them any clear conceptions;—when their apprehensions become vague and dim;—when they assent to scientific doctrines as a matter of tradition, rather than of conviction, on trust rather than on sight;—when science is considered as a collection of opinions, rather than a record of laws by which the universe is really governed;—it must inevitably happen, that men will lose their hold on the knowledge which the great discoverers who preceded them have brought to light. They are not able to push forwards the truths on which they lay so feeble and irresolute a hand; probably they cannot even prevent their sliding back towards the obscurity from which they had been drawn, or from being lost altogether. Such indistinctness and vacillation of thought appear to have prevailed in the stationary period, and to be, in fact, intimately connected with its stationary character. I shall point out some indications of the intellectual peculiarity of which I speak.

1. *Collections of Opinions.*—The fact, that mere Collections of the opinions of physical philosophers came to hold a prominent place in literature, already indicated a tendency to an indistinct and wandering apprehension of such opinions. I speak of such works as Plutarch's five Books "on the Opinions of Philosophers," or the physical opinions which Diogenes Laërtius gives in his "Lives of the Philosophers." At an earlier period still, books of this kind appear; as for instance, a large portion of Pliny's Natural History, a work which has very appropriately been called the Encyclopædia of Antiquity; even Aristotle himself is much in the habit of enumerating the opinions of those who had preceded him. To present such statements as an important part of physical philosophy, shows an erroneous and loose apprehension of its nature. For the only proof of which its doctrines admit, is the possibility of applying the general theory to each particular case; the authority of great men, which in moral and practical matters may or must have its weight, is here of no force; and the technical precision of ideas which the terms of a sound physical theory usually demand, renders a mere statement of the doctrines very imperfectly intelligible to readers familiar with common notions only. To dwell upon such collections of opinions, therefore, both implies, and produces, in writers and readers, an obscure and inadequate apprehension of the full meaning of the doctrines thus collected; supposing there be among them any which really possess such a clearness, solidity, and reality, as to make them important in the history of science. Such diversities of opinion convey no truth; such a multiplicity of statements of what has been *said*, in no degree teaches us what *is*; such accumulations of indistinct notions, however vast and varied, do not make up one distinct idea. On the contrary, the habit of dwelling upon the verbal expressions of the views of other persons, and of being content with such an apprehension of doctrines as a transient notice can give us, is fatal to firm and clear thought: it indicates wavering and feeble conceptions, which are inconsistent with sound physical speculation.

We may, therefore, consider the prevalence of Collections of the kind just referred to, as indicating a deficiency of philosophical talent in the ages now under review. As evidence of the same character, we may add the long train of publishers of Abstracts, Epitomes, Bibliographical Notices, and similar writers. All such writers are worthless for all purposes of *science*, and their labors may be considered as dead works; they have in them no principle of philosophical vitality; they draw their origin and nutriment from the death of true physical knowledge; and resemble the swarms of insects that are born from the perishing carcass of some noble animal.

2. *Indistinctness of Ideas in Mechanics.*—But the indistinctness of thought which is so fatal a feature in the intellect of the stationary period, may be traced more directly in the works, even of the best authors, of those times. We find that they did not retain steadily the ideas on which the scientific success of the previous period had depended. For instance, it is a remarkable circumstance in the history of the science of Mechanics, that it did not make any advance from the time of Archimedes to that of Stevinus and Galileo. Archimedes had established the doctrine of the lever; several persons tried, in the intermediate time, to prove the property of the inclined plane, and none of them succeeded. But let us look to the attempts; for example, that of Pappus, in the eighth Book of his Mathematical Collections, and we may see the reason of the failure. His Problem shows, in the very terms in which it is propounded, the want of a clear apprehension of the subject. "Having given the power which will draw a given weight along the horizontal plane, to find the additional power which will draw the same weight along a given inclined plane." This is proposed without previously defining how Powers, producing such effects, are to be measured; and as if the speed with which the body were drawn, and the nature of the surface of the plane, were of no consequence. The proper elementary Problem is, To find the force which will *support* a body on a smooth inclined plane; and no doubt the solution of Pappus has more reference to this problem than to his own. His reasoning is, however, totally at variance with mechanical ideas on any view of the problem. He supposes the weight to be formed into a sphere; and this sphere being placed in contact with the inclined plane, he assumes that the effect will be the same as if the weight were supported on a horizontal lever, the fulcrum being the point of contact of the sphere with the plane, and the power acting at the circumference of the sphere. Such an assumption implies an entire

absence of those distinct ideas of force and mechanical pressure, on which our perception of the identity or difference of different modes of action must depend;—of those ideas by the help of which Archimedes had been able to demonstrate the properties of the lever, and Stevinus afterwards discovered the true solution of the problem of the inclined plane. The motive to Pappus's assumption was probably no more than this;—he perceived that the additional power, which he thus obtained, vanished when the plane became horizontal, and increased as the inclination became greater. Thus his views were vague; he had no clear conception of mechanical action, and he tried a geometrical conjecture. This is not the way to real knowledge.

Pappus (who lived about A. D. 400) was one of the best mathematicians of the Alexandrian school; and, on subjects where his ideas were so indistinct, it is not likely that any much clearer were to be found in the minds of his contemporaries. Accordingly, on all subjects of speculative mechanics, there appears to have been an entire confusion and obscurity of thought till modern times. Men's minds were busy in endeavoring to systematize the distinctions and subtleties of the Aristotelian school, concerning Motion and Power; and, being thus employed among doctrines in which there was involved no definite meaning capable of real exemplification, they, of course, could not acquire sound physical knowledge. We have already seen that the physical opinions of Aristotle, even as they came from him, had no proper scientific precision. His followers, in their endeavors to perfect and develop his statements, never attempted to introduce clearer ideas than those of their master; and as they never referred, in any steady manner, to facts, the vagueness of their notions was not corrected by any collision with observation. The physical doctrines which they extracted from Aristotle were, in the course of time, built up into a regular system; and though these doctrines could not be followed into a practical application without introducing distinctions and changes, such as deprived the terms of all steady signification, the dogmas continued to be repeated, till the world was persuaded that they were self-evident; and when, at a later period, experimental philosophers, such as Galileo and Boyle, ventured to contradict these current maxims, their new principles sounded in men's ears as strange as they now sound familiar. Thus Boyle promulgated his opinions on the mechanics of fluids, as "Hydrostatical *Paradoxes*, proved and illustrated by experiments." And the opinions which he there opposes, are those which the Aristotelian philosophers habitually propounded as certain and indisputable; such, for instance, as that "in fluids the upper parts do not gravitate on the lower;" that "a lighter fluid will not gravitate on a heavier;" that "levity is a positive quality of bodies as well as gravity." So long as these assertions were left uncontested and untried, men heard and repeated them, without perceiving the incongruities which they involved: and thus they long evaded refutation, amid the vague notions and undoubting habits of the stationary period. But when the controversies of Galileo's time had made men think with more acuteness and steadiness, it was discovered that many of these doctrines were inconsistent with themselves, as well as with experiment. We have an example of the confusion of thought to which the Aristotelians were liable, in their doctrine concerning falling bodies. "Heavy bodies," said they, "must fall quicker than light ones; for weight is the cause of their fall, and the weight of the greater bodies is greater." They did not perceive that, if they considered the weight of the body as a power acting to produce motion, they must consider the body itself as offering a resistance to motion; and that the effect must depend on the proportion of the power to the resistance; in short, they had no clear idea of *accelerating force*. This defect runs through all their mechanical speculations, and renders them entirely valueless.

We may exemplify the same confusion of thought on mechanical subjects in writers of a less technical character. Thus, if men had any distinct idea of mechanical action, they could not have accepted for a moment the fable of the Echineis or Remora, a little fish which was said to be able to stop a large ship merely by sticking to it.[1] Lucan refers to this legend in a poetical manner, and notices this creature only in bringing together a collection of monstrosities; but Pliny relates the tale gravely, and moralizes upon it after his manner. "What," he cries,[2] "is more violent than the sea and the winds? what a greater work of art than a ship? Yet one little fish (the Echineis) can hold back all these when they all strain the same way. The winds may

[1] Lucan is describing one of the poetical compounds produced in incantations.

> Huc quicquid fœtu genuit Natura sinistro
> Miscetur: non spuma canum quibus unda timori est,
> Viscera non lyncis, non duræ nodus hyænæ
> Defuit, et cervi pasti serpente medullæ;
> Non puppes retinens, Euro tendente rudentes
> In mediis *Echineis* aquis, oculique draconum.
> Etc. *Pharsalia*, iv. 670.

[2] Plin. *Hist. N.* xxxii. 5.

blow, the waves may rage; but this small creature controls their fury, and stops a vessel, when chains and anchors would not hold it: and this it does, not by hard labor, but merely by adhering to it. Alas, for human vanity! when the turreted ships which man has built, that he may fight from castle-walls, at sea as well as at land, are held captive and motionless by a fish a foot and a half long! Such a fish is said to have stopped the admiral's ship at the battle of Actium, and compelled Antony to go into another. And in our own memory, one of these animals held fast the ship of Caius, the emperor, when he was sailing from Astura to Antium. The stopping of this ship, when all the rest of the fleet went on, caused surprise; but this did not last long, for some of the men jumped into the water to look for the fish, and found it sticking to the rudder; they showed it to Caius, who was indignant that this animal should interpose its prohibition to his progress, when impelled by four hundred rowers. It was like a slug; and had no power, after it was taken into the ship."

A very little advance in the power of thinking clearly on the force which it exerted in pulling, would have enabled the Romans to see that the ship and its rowers must pull the adhering fish by the hold the oars had upon the water; and that, except the fish had a hold equally strong on some external body, it could not resist this force.

3. *Indistinctness of Ideas shown in Architecture.*—Perhaps it may serve to illustrate still further the extent to which, under the Roman empire, men's notions of mechanical relations became faint, wavered, and disappeared, if we observe the change which took place in architecture. All architecture, to possess genuine beauty, must be mechanically consistent. The decorative members must represent a structure which has in it a principle of support and stability. Thus the Grecian colonnade was a straight horizontal beam, resting on vertical props; and the pediment imitated a frame like a roof, where oppositely inclined beams support each other. These forms of building were, therefore, proper models of art, because they implied supporting forces. But to be content with colonnades and pediments, which, though they imitated the forms of the Grecian ones, were destitute of their mechanical truth, belonged to the decline of art; and showed that men had lost the idea of force, and retained only that of shape. Yet this was what the architects of the Roman empire did. Under their hands, the pediment was severed at its vertex, and divided into separate halves, so that it was no longer a mechanical possibility. The entablature no longer lay straight from pillar to pillar, but, projecting over each

column, turned back to the wall, and adhered to it in the intervening space. The splendid remains of Palmyra, Balbec, Petra, exhibit endless examples of this kind of perverse inventiveness; and show us, very instructively, how the decay of art and of science alike accompany this indistinctness of ideas which we are now endeavoring to illustrate.

4. *Indistinctness of Ideas in Astronomy.*—Returning to the sciences, it may be supposed, at first sight, that, with regard to astronomy, we have not the same ground for charging the stationary period with indistinctness of ideas on that subject, since they were able to acquire and verify, and, in some measure, to apply, the doctrines previously established. And, undoubtedly, it must be confessed that men's notions of the relations of space and number are never very indistinct. It appears to be impossible for these chains of elementary perception ever to be much entangled. The later Greeks, the Arabians, and the earliest modern astronomers, must have conceived the hypotheses of the Ptolemaic system with tolerable completeness. And yet, we may assert, that during the stationary period, men did not possess the notions, even of space and number, in that vivid and vigorous manner which enables them to discover new truths. If they had perceived distinctly that the astronomical theorist had merely to do with *relative* motions, they must have been led to see the possibility, at least, of the Copernican system; as the Greeks, at an earlier period, had already perceived it. We find no trace of this. Indeed, the mode in which the Arabian mathematicians present the solutions of their problems, does not indicate that clear apprehension of the relations of space, and that delight in the contemplation of them, which the Greek geometrical speculations imply. The Arabs are in the habit of giving conclusions without demonstrations, precepts without the investigations by which they are obtained; as if their main object were practical rather than speculative,—the calculation of results rather than the exposition of theory. Delambre[3] has been obliged to exercise great ingenuity, in order to discover the method by which Ibn Iounis proved his solution of certain difficult problems.

5. *Indistinctness of Ideas shown by Skeptics.*—The same unsteadiness of ideas which prevents men from obtaining clear views, and steady and just convictions, on special subjects, may lead them to despair of or deny the possibility of acquiring certainty at all, and may thus make them skeptics with regard to all knowledge. Such skeptics

---

[3] Delamb. *M. A.* p. 125–8.

are themselves men of indistinct views, for they could not otherwise avoid assenting to the demonstrated truths of science; and, so far as they may be taken as specimens of their contemporaries, they prove that indistinct ideas prevail in the age in which they appear. In the stationary period, moreover, the indefinite speculations and unprofitable subtleties of the schools might further impel a man of bold and acute mind to this universal skepticism, because they offered nothing which could fix or satisfy him. And thus the skeptical spirit may deserve our notice as indicative of the defects of a system of doctrine too feeble in demonstration to control such resistance.

The most remarkable of these philosophical skeptics is Sextus Empiricus; so called, from his belonging to that medical sect which was termed the *empirical*, in contradistinction to the *rational* and *methodical* sects. His works contain a series of treatises, directed against all the divisions of the science of his time. He has chapters against the Geometers, against the Arithmeticians, against the Astrologers, against the Musicians, as well as against Grammarians, Rhetoricians, and Logicians; and, in short, as a modern writer has said, his skepticism is employed as a sort of frame-work which embraces an encyclopedical view of human knowledge. It must be stated, however, that his objections are rather to the metaphysical grounds, than to the details of the sciences; he rather denies the possibility of speculative truth in general, than the experimental truths which had been then obtained. Thus his objections to geometry and arithmetic are founded on abstract cavils concerning the nature of points, letters, unities, &c. And when he comes to speak against astrology, he says, "I am not going to consider that perfect science which rests upon geometry and arithmetic; for I have already shown the weakness of those sciences: nor that faculty of prediction (of the motions of the heavens) which belongs to the pupils of Eudoxus, and Hipparchus, and the rest, which some call Astronomy; for that is an observation of phenomena, like agriculture or navigation: but against the Art of Prediction from the time of birth, which the Chaldeans exercise." Sextus, therefore, though a skeptic by profession, was not insensible to the difference between experimental knowledge and mystical dogmas, though even the former had nothing which excited his admiration.

The skepticism which denies the evidence of the truths of which the best established physical sciences consist, must necessarily involve a very indistinct apprehension of those truths; for such truths, properly exhibited, contain their own evidence, and are the best antidote to this skepticism. But an incredulity or contempt towards the asserted truths of physical science may arise also from the attention being mainly directed to the certainty and importance of religious truths. A veneration for revealed religion may thus assume the aspect of a skepticism with regard to natural knowledge. Such appears to be the case with Algazel or Algezeli, who is adduced by Degerando[4] as an example of an Arabian skeptic. He was a celebrated teacher at Bagdad in the eleventh century, and he declared himself the enemy, not only of the mixed Peripatetic and Platonic philosophy of the time, but of Aristotle himself. His work entitled *The Destructions of the Philosophers*, is known to us by the refutation of it which Averrhoes published, under the title of *Destruction of Algazel's Destructions of the Philosophers.* It appears that he contested the fundamental principles both of the Platonic and of the Aristotelian schools, and denied the possibility of a known connection between cause and effect; thus making a prelude, says Degerando, to the celebrated argumentation of Hume.

[2d Ed.] Since the publication of my first edition, an account of Algazel or Algazzali and his works has been published under the title of *Essai sur les Ecoles Philosophiques chez les Arabes, et notamment sur la Doctrine d'Algazzali*, par August Schmölders. Paris. 1842. From this book it appears that Degerando's account of Algazzali is correct, when he says[5] that "his skepticism seems to have essentially for its object to destroy all systems of merely rational theology, in order to open an indefinite career, not only to faith guided by revelation, but also to the free exaltation of a mystical enthusiasm." It is remarked by Dr. Schmölders, following M. de Hammer-Purgstall, that the title of the work referred to in the text ought rather to be *Mutual Refutation of the Philosophers:* and that its object is to show that Philosophy consists of a mass of systems, each of which overturns the others. The work of Algazzali which Dr. Schmölders has published, *On the Errors of Sects, &c.*, contains a kind of autobiographical account of the way in which the author was led to his views. He does not reject the truths of science, but he condemns the mental habits which are caused by laying too much stress upon science. Religious men, he says, are, by such a course, led to reject all science, even what relates to eclipses of the moon and sun; and men of science are led to hate religion.[6]

4 Degerando, *Hist. Comp. de Systèmes*, iv. 224.

5 *Hist. Comp.* iv. p. 227.

6 *Essai*, p. 33.

6. *Neglect of Physical Reasoning in Christendom.*—If the Arabians, who, during the ages of which we are speaking, were the most eminent cultivators of science, entertained only such comparatively feeble and servile notions of its doctrines, it will easily be supposed, that in the Christendom of that period, where physical knowledge was comparatively neglected, there was still less distinctness and vividness in the prevalent ideas on such subjects. Indeed, during a considerable period of the history of the Christian Church, and by many of its principal authorities, the study of natural philosophy was not only disregarded but discommended. The great practical doctrines which were presented to men's minds, and the serious tasks, of the regulation of the will and affections, which religion impressed upon them, made inquiries of mere curiosity seem to be a reprehensible misapplication of human powers; and many of the fathers of the Church revived, in a still more peremptory form, the opinion of Socrates, that the only valuable philosophy is that which teaches us our moral duties and religious hopes.[7] Thus Eusebius says,[8] "It is not through ignorance of the things admired by them, but through contempt of their useless labor, that we think little of these matters, turning our souls to the exercise of better things." When the thoughts were thus intentionally averted from those ideas which natural philosophy involves, the ideas inevitably became very indistinct in their minds; and they could not conceive that any other persons could find, on such subjects, grounds of clear conviction and certainty. They held the whole of their philosophy to be, as Lactantius[9] asserts it to be, "empty and false." "To search," says he, "for the causes of natural things; to inquire whether the sun be as large as he seems, whether the moon is convex or concave, whether the stars are fixed in the sky or float freely in the air; of what size and of what material are the heavens; whether they be at rest or in motion; what is the magnitude of the earth; on what foundations it is suspended and balanced;—to dispute and conjecture on such matters, is just as if we chose to discuss what we think of a city in a remote country, of which we never heard but the name." It is impossible to express more forcibly that absence of any definite notions on physical subjects which led to this tone of thought.

7. *Question of Antipodes.*—With such habits of thought, we are not to be surprised if the relations resulting from the best established theories were apprehended in an imperfect and incongruous manner.

---

[7] Brucker, iii. 317. [8] *Præp. Ev.* xv. 61. [9] *Inst.* l. iii. init.

We have some remarkable examples of this; and a very notable one is the celebrated question of the existence of *Antipodes*, or persons inhabiting the opposite side of the globe of the earth, and consequently having the soles of their feet directly opposed to ours. The doctrine of the globular form of the earth results, as we have seen, by a geometrical necessity, from a clear conception of the various points of knowledge which we obtain, bearing upon that subject. This doctrine was held distinctly by the Greeks; it was adopted by all astronomers, Arabian and European, who followed them; and was, in fact, an inevitable part of every system of astronomy which gave a consistent and intelligible representation of phenomena. But those who did not call before their minds any distinct representation at all, and who referred the whole question to other relations than those of space, might still deny this doctrine; and they did so. The existence of inhabitants on the opposite side of the terraqueous globe, was a fact of which experience alone could teach the truth or falsehood; but the religious relations, which extend alike to all mankind, were supposed to give the Christian philosopher grounds for deciding against the possibility of such a race of men. Lactantius,[10] in the fourth century, argues this matter in a way very illustrative of that impatience of such speculations, and consequent confusion of thought, which we have mentioned. "Is it possible," he says, "that men can be so absurd as to believe that the crops and trees on the other side of the earth hang downwards, and that men there have their feet higher than their heads? If you ask of them how they defend these monstrosities—how things do not fall away from the earth on that side—they reply, that the nature of things is such that heavy bodies tend towards the centre, like the spokes of a wheel, while light bodies, as clouds, smoke, fire, tend from the centre towards the heavens on all sides. Now I am really at a loss what to say of those who, when they have once gone wrong, steadily persevere in their folly, and defend one absurd opinion by another." It is obvious that so long as the writer refused to admit into his thoughts the fundamental conception of their theory, he must needs be at a loss what to say to their arguments without being on that account in any degree convinced of their doctrines.

In the sixth century, indeed, in the reign of Justinian, we find a writer (Cosmas Indicopleustes[11]) who does not rest in this obscurity of

---

[10] *Inst.* l. iii. 23.

[11] Montfaucon, *Collectio Nova Patrum*, t. ii. p. 113. Cosmas Indicopleustes. Christianorum Opiniones de Mundo, sive Topographia Christiana.

representation; but in this case, the distinctness of his pictures only serves to show his want of any clear conception as to what suppositions would explain the phenomena. He describes the earth as an oblong floor, surrounded by upright walls, and covered by a vault, below which the heavenly bodies perform their revolutions, going round a certain high mountain, which occupies the northern parts of the earth, and makes night by intercepting the light of the sun. In Augustin[12] (who flourished A. D. 400) the opinion is treated on other grounds; and without denying the globular form of the earth, it is asserted that there are no inhabitants on the opposite side, because no such race is recorded by Scripture among the descendants of Adam.[13] Considerations of the same kind operated in the well-known instance of Virgil, Bishop of Salzburg, in the eighth century. When he was reported to Boniface, Archbishop of Mentz, as holding the existence of Antipodes, the prelate was shocked at the assumption, as it seemed to him, of a world of human beings, out of the reach of the conditions of salvation; and application was made to Pope Zachary for a censure of the holder of this dangerous doctrine. It does not, however, appear that this led to any severity; and the story of the deposition of Virgil from his bishopric, which is circulated by Kepler and by more modern writers, is undoubtedly altogether false. The same scruples continued to prevail among Christian writers to a later period; and Tostatus[14] notes the opinion of the rotundity of the earth as an "unsafe" doctrine, only a few years before Columbus visited the other hemisphere.

8. *Intellectual Condition of the Religious Orders.*—It must be recollected, however, that though these were the views and tenets of many religious writers, and though they may be taken as indications of the prevalent and characteristic temper of the times of which we speak, they never were universal. Such a confusion of thought affects the minds of many persons, even in the most enlightened times; and in what we call the Dark Ages, though clear views on such subjects might be more rare, those who gave their minds to science, entertained the true opinion of the figure of the earth. Thus Boëthius[15] (in the sixth century) urges the smallness of the globe of the earth, com-

---

[12] *Civ. D.* xvi. 9.

[13] It appears, however, that scriptural arguments were found on the other side. St. Jerome says (*Comm. in Ezech.* i. 6), speaking of the two cherubims with four faces, seen by the prophet, and the interpretation of the vision: "Alii vero qui philosophorum stultam sequuntur sapientiam, duo hemispheria in duobus templi cherubim, nos et antipodes, quasi supinos et cadentes homines suspicantur."

[14] Montfauc. *Patr.* t. ii.

[15] Boëthius, *Cons.* ii. pr. 7.

pared with the heavens, as a reason to repress our love of glory. This work, it will be recollected, was translated into the Anglo-Saxon by our own Alfred. It was also commented on by Bede, who, in what he says on this passage, assents to the doctrine, and shows an acquaintance with Ptolemy and his commentators, both Arabian and Greek. Gerbert, in the tenth century, went from France to Spain to study astronomy with the Arabians, and soon surpassed his masters. He is reported to have fabricated clocks, and an astrolabe of peculiar construction. Gerbert afterwards (in the last year of the first thousand from the birth of Christ) became pope, by the name of Sylvester II. Among other cultivators of the sciences, some of whom, from their proficiency, must have possessed with considerable clearness and steadiness the elementary ideas on which it depends, we may here mention, after Montucla,[16] Adelbold, whose work On the Sphere was addressed to Pope Sylvester, and whose geometrical reasonings are, according to Montucla,[17] vague and chimerical; Hermann Contractus, a monk of St. Gall, who, in 1050, published astronomical works; William of Hirsaugen, who followed his example in 1080; Robert of Lorraine, who was made Bishop of Hereford by William the Conqueror, in consequence of his astronomical knowledge. In the next century, Adelhard Goth, an Englishman, travelled among the Arabs for purposes of study, as Gerbert had done in the preceding age; and on his return, translated the Elements of Euclid, which he had brought from Spain or Egypt. Robert Grostête, Bishop of Lincoln, was the author of an Epitome on the Sphere; Roger Bacon, in his youth the contemporary of Robert, and of his brother Adam Marsh, praises very highly their knowledge in mathematics.

"And here," says the French historian of mathematics, whom I have followed in the preceding relation, "it is impossible not to reflect that all those men who, if they did not augment the treasure of the sciences, at least served to transmit it, were monks, or had been such originally. Convents were, during these stormy ages, the asylum of sciences and letters. Without these religious men, who, in the silence of their monasteries, occupied themselves in transcribing, in studying, and in imitating the works of the ancients, well or ill, those works would have perished; perhaps not one of them would have come down to us. The thread which connects us with the Greeks and Romans would have been snapt asunder; the precious productions of

---

16 Mont. i. 502. 17 Ib. i. 503.

ancient literature would no more exist for us, than the works, if any there were, published before the catastrophe that annihilated that highly scientific nation, which, according to Bailly, existed in remote ages in the centre of Tartary, or at the roots of Caucasus. In the sciences we should have had all to create ; and at the moment when the human mind should have emerged from its stupor and shaken off its slumbers, we should have been no more advanced than the Greeks were after the taking of Troy." He adds, that this consideration inspires feelings towards the religious orders very different from those which, when he wrote, were prevalent among his countrymen.

Except so far as their religious opinions interfered, it was natural that men who lived a life of quiet and study, and were necessarily in a great measure removed from the absorbing and blinding interests with which practical life occupies the thoughts, should cultivate science more successfully than others, precisely because their ideas on speculative subjects had time and opportunity to become clear and steady. The studies which were cultivated under the name of the Seven Liberal Arts, necessarily tended to favor this effect. The *Trivium*,[18] indeed, which consisted of Grammar, Logic, and Rhetoric, had no direct bearing upon those ideas with which physical science is concerned; but the *Quadrivium*, Music, Arithmetic, Geometry, Astronomy, could not be pursued with any attention, without a corresponding improvement of the mind for the purposes of sound knowledge.[19]

9. *Popular Opinions.*—That, even in the best intellects, something was wanting to fit them for scientific progress and discovery, is obvious from the fact that science was so long absolutely stationary. And I have endeavored to show that one part of this deficiency was the want of the requisite clearness and vigor of the fundamental scientific ideas. If these were wanting, even in the most powerful and most cultivated minds, we may easily conceive that still greater confusion and obscurity prevailed in the common class of mankind. They actually adopted the belief, however crude and inconsistent, that the form of the earth and heavens really is what at any place it appears to be ; that the earth is flat, and the waters of the sky sustained above a material floor, through which in showers they descend. Yet the true doctrines of

---

[18] Bruck. iii. 597.

[19] Roger Bacon, in his *Specula Mathematica*, cap. i. says, "Harum scientiarum porta et clavis est mathematica, quam sancti a principio mundi invenerunt, etc. Cujus negligentia *jam per triginta vel quadraginta annos* destruxit totum studium Latinorum." I do not know on what occasion this neglect took place.

astronomy appear to have had some popular circulation. For instance, a French poem of the time of Edward the Second, called *Ymage du Monde*, contains a metrical account of the earth and heavens, according to the Ptolemaic views; and in a manuscript of this poem, preserved in the library of the University of Cambridge, there are representations, in accordance with the text, of a spherical earth, with men standing upright upon it on every side; and by way of illustrating the tendency of all things to the centre, perforations of the earth, entirely through its mass, are described and depicted; and figures are exhibited dropping balls down each of these holes, so as to meet in the interior. And, as bearing upon the perplexity which attends the motions of *up* and *down*, when applied to the globular earth, and the change of the direction of gravity which would occur in passing the centre, the readers of Dante will recollect the extraordinary manner in which the poet and his guide emerge from the bottom of the abyss; and the explanation which Virgil imparts to him of what he there sees. After they have crept through the aperture in which Lucifer is placed, the poet says,

> "Io levai gli occhi e credetti vedere
> Lucifero com' io l' avea lasciato,
> E vidile le gambe in su tenere."
> . . . . . "Questi come è fitto
> Si sottasopra?" . . . . .
> "Quando mi volsi, tu passast' il punto
> Al qual si traggon d' ogni parte i pesi."
>
> *Inferno*, xxxiv.
>
> . . . . . "I raised mine eyes,
> Believing that I Lucifer should see
> Where he was lately left, but saw him now
> With legs held upward." . . . .
> "How standeth he in posture thus reversed?"
> . . . .
> "Thou wast on the other side so long as I
> Descended; when I turned, thou didst o'erpass
> That point to which from every part is dragged
> All heavy substance." CARY.

This is more philosophical than Milton's representation, in a more scientific age, of Uriel sliding to the earth on a sunbeam, and sliding back again, when the sun had sunk below the horizon.

> . . . . . "Uriel to his charge
> Returned on that bright beam whose point now raised,
> Bore him slope downward to the sun, now fallen
> Beneath the Azores." *Par. Lost*, B. iv.

The philosophical notions of up and down are too much at variance with the obvious suggestions of our senses, to be held steadily and justly by minds undisciplined in science. Perhaps it was some misunderstood statement of the curved surface of the ocean, which gave rise to the tradition of there being a part of the sea directly over the earth, from which at times an object has been known to fall or an anchor to be let down. Even such whimsical fancies are not without instruction, and may serve to show the reader what that vagueness and obscurity of ideas is, of which I have been endeavoring to trace the prevalence in the dark ages.

We now proceed to another of the features which appears to me to mark, in a very prominent manner, the character of the stationary period.

CHAPTER II.

The Commentatorial Spirit of the Middle Ages.

WE have already noticed, that, after the first great achievements of the founders of sound speculation, in the different departments of human knowledge, had attracted the interest and admiration which those who became acquainted with them could not but give to them, there appeared a disposition among men to lean on the authority of some of these teachers;—to study the opinions of others as the only mode of forming their own;—to read nature through books;—to attend to what had been already thought and said, rather than to what really is and happens. This tendency of men's minds requires our particular consideration. Its manifestations were very important, and highly characteristic of the stationary period; it gave, in a great degree, a peculiar bias and direction to the intellectual activity of many centuries; and the kind of labor with which speculative men were occupied in consequence of this bias, took the place of that examination of realities which must be their employment, in order that real knowledge may make any decided progress.

In some subjects, indeed, as, for instance, in the domains of morals, poetry, and the arts, whose aim is the production of beauty, this opposition between the study of former opinion and present reality, may not be so distinct; inasmuch as it may be said by some, that, in these subjects, opinions are realities; that the thoughts and feelings which prevail in men's minds are the material upon which we must work, the particulars from which we are to generalize, the instruments which we are to use; and that, therefore, to reject the study of antiquity, or even its authority, would be to show ourselves ignorant of the extent and mutual bearing of the elements with which we have to deal;—would be to cut asunder that which we ought to unite into a vital whole. Yet even in the provinces of history and poetry, the poverty and servility of men's minds during the middle ages, are shown by indications so strong as to be truly remarkable; for instance, in the efforts of the antiquarians of almost every European country to assimilate the early history of their own state to the poet's account of the foundation of Rome, by bringing from the sack of Troy, Brutus to England, Bavo to Flanders, and so on. But however this may be, our business at present is, to trace the varying spirit of the *physical* philosophy of different ages; trusting that, hereafter, this prefatory study will enable us to throw some light upon the other parts of philosophy. And in physics the case undoubtedly was, that the labor of observation, which is one of the two great elements of the progress of knowledge, was in a great measure superseded by the collection, the analysis, the explanation, of previous authors and opinions; experimenters were replaced by commentators; criticism took the place of induction; and instead of great discoverers we had learned men.

1. *Natural Bias to Authority.*—It is very evident that, in such a bias of men's studies, there is something very natural; however strained and technical this erudition may have been, the propensities on which it depends are very general, and are easily seen. Deference to the authority of thoughtful and sagacious men, a disposition which men in general neither reject nor think they ought to reject in practical matters, naturally clings to them, even in speculation. It is a satisfaction to us to suppose that there are, or have been, minds of transcendent powers, of wide and wise views, superior to the common errors and blindness of our nature. The pleasure of admiration, and the repose of confidence, are inducements to such a belief. There are also other reasons why we willingly believe that there are in philosophy great teachers, so profound and sagacious, that, in order to arrive at truth, we have only to learn their thoughts, to understand their writings. There is a peculiar interest which men feel in dealing with the thoughts of their fellow-men, rather than with brute matter. Matter feels and excites no sympathies: in seeking for mere laws of nature, there is nothing of mental intercourse with the great spirits of the past, as there is in stu-

dying Aristotle or Plato. Moreover, a large portion of this employment is of a kind the most agreeable to most speculative minds; it consists in tracing the consequences of assumed principles: it is deductive like geometry: and the principles of the teachers being known, and being undisputed, the deduction and application of their results is an obvious, self-satisfying, and inexhaustible exercise of ingenuity.

These causes, and probably others, make criticism and commentation flourish, when invention begins to fail, oppressed and bewildered by the acquisitions it has already made; and when the vigor and hope of men's minds are enfeebled by civil and political changes. Accordingly,[1] the Alexandrian school was eminently characterized by a spirit of erudition, of literary criticism, of interpretation, of imitation. These practices, which reigned first in their full vigor in "the Museum," are likely to be, at all times, the leading propensities of similar academical institutions.

How natural it is to select a great writer as a paramount authority, and to ascribe to him extraordinary profundity and sagacity, we may see, in the manner in which the Greeks looked upon Homer; and the fancy which detected in his poems traces of the origin of all arts and sciences, has, as we know, found favor even in modern times. To pass over earlier instances of this feeling, we may observe, that Strabo begins his Geography by saying that he agrees with Hipparchus, who had declared Homer to be the first author of our geographical knowledge; and he does not confine the application of this assertion to the various and curious topographical information which the Iliad and Odyssey contain, concerning the countries surrounding the Mediterranean; but in phrases which, to most persons, might appear the mere play of a poetical fancy, or a casual selection of circumstances, he finds unquestionable evidence of a correct knowledge of general geographical truths. Thus,[2] when Homer speaks of the sun "rising from the soft and deep-flowing ocean," of his "splendid blaze plunging in the ocean;" of the northern constellation

"Alone unwashen by the ocean wave;"

and of Jupiter, "who goes to the ocean to feast with the blameless Ethiopians;" Strabo is satisfied from these passages that Homer knew the dry land to be surrounded with water: and he reasons in like manner with respect to other points of geography.

---

[1] Degerando, *Hist. des Syst. de Philos.* iii. p. 134.

[2] Strabo, i. p. 5.

2. *Character of Commentators.*—The spirit of commentation, as has already been suggested, turns to questions of taste, of metaphysics, of morals, with far more avidity than to physics. Accordingly, critics and grammarians were peculiarly the growth of this school; and, though the commentators sometimes chose works of mathematical or physical science for their subject (as Proclus, who commented on Euclid's Geometry, and Simplicius, on Aristotle's Physics), these commentaries were, in fact, rather metaphysical than mathematical. It does not appear that the commentators have, in any instance, illustrated the author by bringing his assertions of facts to the test of experiment. Thus, when Simplicius comments on the passage concerning a vacuum, which we formerly adduced, he notices the argument which went upon the assertion, that a vessel full of ashes would contain as much water as an empty vessel; and he mentions various opinions of different authors, but no trial of the fact. Eudemus had said, that the ashes contained something hot, as quicklime does, and that by means of this, a part of the water was evaporated; others supposed the water to be condensed, and so on.[3]

The Commentator's professed object is to explain, to enforce, to illustrate doctrines assumed as true. He endeavors to adapt the work on which he employs himself to the state of information and of opinion in his own time; to elucidate obscurities and technicalities; to supply steps omitted in the reasoning; but he does not seek to obtain additional truths or new generalizations. He undertakes only to give what is virtually contained in his author; to develop, but not to create. He is a cultivator of the thoughts of others: his labor is not spent on a field of his own; he ploughs but to enrich the granary of another man. Thus he does not work as a freeman, but as one in a servile condition; or rather, his is a menial, and not a productive service: his office is to adorn the appearance of his master, not to increase his wealth.

Yet though the Commentator's employment is thus subordinate and dependent, he is easily led to attribute to it the greatest importance and dignity. To elucidate good books is, indeed, a useful task; and when those who undertake this work execute it well, it would be most unreasonable to find fault with them for not doing more. But the critic, long and earnestly employed on one author, may easily underrate the relative value of other kinds of mental exertion. He may

[3] Simplicius, p. 170.

ascribe too large dimensions to that which occupies the whole of his own field of vision. Thus he may come to consider such study as the highest aim, and best evidence of human genius. To understand Aristotle, or Plato, may appear to him to comprise all that is possible of profundity and acuteness. And when he has travelled over a portion of their domain, and satisfied himself that of this he too is master, he may look with complacency at the circuit he has made, and speak of it as a labor of vast effort and difficulty. We may quote, as an expression of this temper, the language of Sir Henry Savile, in concluding a course of lectures on Euclid, delivered at Oxford.[4] "By the grace of God, gentlemen hearers, I have performed my promise; I have redeemed my pledge. I have explained, according to my ability, the definitions, postulates, axioms, and *first eight propositions* of the Elements of Euclid. Here, sinking under the weight of years, I lay down my art and my instruments."

We here speak of the peculiar province of the Commentator; for undoubtedly, in many instances, a commentary on a received author has been made the vehicle of conveying systems and doctrines entirely different from those of the author himself; as, for instance, when the New Platonists wrote, taking Plato for their text. The labors of learned men in the stationary period, which came under this description, belong to another class.

3. *Greek Commentators on Aristotle.*—The commentators or disciples of the great philosophers did not assume at once their servile character. At first their object was to supply and correct, as well as to explain their teacher. Thus among the earlier commentators of Aristotle, Theophrastus invented five moods of syllogism in the first figure, in addition to the four invented by Aristotle, and stated with additional accuracy the rules of hypothetical syllogisms. He also not only collected much information concerning animals, and natural events, which Aristotle had omitted, but often differed with his master; as, for instance, concerning the saltness of the sea: this, which the Stagirite attributed to the effect of the evaporation produced by the sun's rays, was ascribed by Theophrastus to beds of salt at the bottom. Porphyry,[5] who flourished in the third century, wrote a book on the *Predicables*, which was found to be so suitable a complement

---

[4] Exolvi per Dei gratiam, Domini auditores, promissum; liberavi fidem meam; explicavi pro meo modulo, definitiones, petitiones, communes sententias, et *octo priores propositiones* Elementorum Euclidis. Hic, annis fessus, cyclos artemque repono.

[5] Buhle, Arist. i. 284.

to the *Predicaments* or Categories of Aristotle, that it was usually prefixed to that treatise; and the two have been used as an elementary work together, up to modern times. The Predicables are the five steps which the gradations of generality and particularity introduce;—*genus, species, difference, individual, accident*:—the Categories are the ten heads under which assertions or predications may be arranged; —*substance, quantity, relation, quality, place, time, position, habit, action, passion.*

At a later period, the Aristotelian commentators became more servile, and followed the author step by step, explaining, according to their views, his expressions and doctrines; often, indeed, with extreme prolixity, expanding his clauses into sentences, and his sentences into paragraphs. Alexander Aphrodisiensis, who lived at the end of the second century, is of this class; "sometimes useful," as one of the recent editors of Aristotle says;[6] "but by the prolixity of his interpretation, by his perverse itch for himself discussing the argument expounded by Aristotle, for defending his opinions, and for refuting or reconciling those of others, he rather obscures than enlightens." At various times, also, some of the commentators, and especially those of the Alexandrian school, endeavored to reconcile, or combined without reconciling, opposing doctrines of the great philosophers of the earlier times. Simplicius, for instance, and, indeed, a great number of the Alexandrian Philosophers,[7] as Alexander, Ammonius, and others, employed themselves in the futile task of reconciling the doctrines of the Pythagoreans, of the Eleatics, of Plato, and of the Stoics, with those of Aristotle. Boethius[8] entertained the design of translating into Latin the whole of Aristotle's and Plato's works, and of showing their agreement; a gigantic plan, which he never executed. Others employed themselves in disentangling the confusion which such attempts produced, as John the Grammarian, surnamed Philoponus, "the Labor-loving;" who, towards the end of the seventh century, maintained that Aristotle was entirely misunderstood by Porphyry and Proclus,[9] who had pretended to incorporate his doctrines into those of the New Platonic school, or even to reconcile him with Plato himself on the subject of *ideas*. Others, again, wrote Epitomes, Compounds, Abstracts; and endeavored to throw the works of the philosopher into some simpler and more obviously regular form, as John of Damascus, in

---

[6] *Ib.* i. 288.

[7] Ib. i. 311.

[8] Degerando, *Hist. des Syst.* iv. 100.

[9] Ib. iv. 155.

the middle of the eighth century, who made abstracts of some of Aristotle's works, and introduced the study of the author into theological education. These two writers lived under the patronage of the Arabs; the former was favored by Amrou, the conqueror of Egypt; the latter was at first secretary to the Caliph, but afterwards withdrew to a monastery.[10]

At this period the Arabians became the fosterers and patrons of philosophy, rather than the Greeks. Justinian had, by an edict, closed the school of Athens, the last of the schools of heathen philosophy. Leo, the Isaurian, who was a zealous Iconoclast, abolished also the schools where general knowledge had been taught, in combination with Christianity,[11] yet the line of the Aristotelian commentators was continued, though feebly, to the later ages of the Greek empire. Anna Comnena[12] mentions a Eustratus who employed himself upon the dialectic and moral treatises, and whom she does not hesitate to elevate above the Stoics and Platonists, for his talent in philosophical discussions. Nicephorus Blemmydes wrote logical and physical epitomes for the use of John Ducas; George Pachymerus composed an epitome of the philosophy of Aristotle, and a compend of his logic; Theodore Metochytes, who was famous in his time alike for his eloquence and his learning, has left a paraphrase of the books of Aristotle on Physics, on the Soul, the Heavens,[13] &c. Fabricius states that this writer has a chapter, the object of which is to prove, that all philosophers, and Aristotle and Plato in particular, have disdained the authority of their predecessors. He could hardly help remarking in how different a spirit philosophy had been pursued since their time.

4. *Greek Commentators of Plato and others.*—I have spoken principally of the commentators of Aristotle, for he was the great subject of the commentators proper; and though the name of his rival, Plato, was graced by a list of attendants, hardly less numerous, these, the Neoplatonists, as they are called, had introduced new elements into the doctrines of their nominal master, to such an extent that they must be placed in a different class. We may observe here, however, how, in this school as in the Peripatetic, the race of commentators multiplied itself. Porphyry, who commented on Aristotle, was commented on by Ammonius; Plotinus's Enneads were commented on by Proclus and Dexippus. Psellus[14] the elder was a paraphrast of Aris-

[10] Deg. iv. 150. [11] Ib. iv. 163. [12] Ib. iv. 167. [13] Ib. iv. 168. [14] Deg. iv. 169.

totle; Psellus the younger, in the eleventh century, attempted to restore the New Platonic school. The former of these two writers had for his pupils two men, the emperor Leo, surnamed the Philosopher, and Photius the patriarch, who exerted themselves to restore the study of literature at Constantinople. We still possess the Collection of Extracts of Photius, which, like that of Stobæus and others, shows the tendency of the age to compilations, abstracts, and epitomes, —the extinction of philosophical vitality.

5. *Arabian Commentators of Aristotle.*—The reader might perhaps have expected, that when the philosophy of the Greeks was carried among a new race of intellects, of a different national character and condition, the train of this servile tradition would have been broken; that some new thoughts would have started forth; that some new direction, some new impulse, would have been given to the search for truth. It might have been anticipated that we should have had schools among the Arabians which should rival the Peripatetic, Academic, and Stoic among the Greeks;—that they would preoccupy the ground on which Copernicus and Galileo, Lavoisier and Linnæus, won their fame;—that they would make the next great steps in the progressive sciences. Nothing of this, however, happened. The Arabians cannot claim, in science or philosophy, any really great names; they produced no men and no discoveries which have materially influenced the course and destinies of human knowledge; they tamely adopted the intellectual servitude of the nation which they conquered by their arms; they joined themselves at once to the string of slaves who were dragging the car of Aristotle and Plotinus. Nor, perhaps, on a little further reflection, shall we be surprised at this want of vigor and productive power, in this period of apparent national youth. The Arabians had not been duly prepared rightly to enjoy and use the treasures of which they became possessed. They had, like most uncivilized nations, been passionately fond of their indigenous poetry; their imagination had been awakened, but their rational powers and speculative tendencies were still torpid. They received the Greek philosophy without having passed through those gradations of ardent curiosity and keen research, of obscurity brightening into clearness, of doubt succeeded by the joy of discovery, by which the Greek mind had been enlarged and exercised. Nor had the Arabians ever enjoyed, as the Greeks had, the individual consciousness, the independent volition, the intellectual freedom, arising from the freedom of political institutions. They had not felt the contagious mental activity of a small city,—the elation arising from the general

sympathy in speculative pursuits diffused through an intelligent and acute audience; in short, they had not had a national education such as fitted the Greeks to be disciples of Plato and Hipparchus. Hence, their new literary wealth rather encumbered and enslaved, than enriched and strengthened them: in their want of taste for intellectual freedom, they were glad to give themselves up to the guidance of Aristotle and other dogmatists. Their military habits had accustomed them to look to a leader; their reverence for the book of their law had prepared them to accept a philosophical Koran also. Thus the Arabians, though they never translated the Greek poetry, translated, and merely translated, the Greek philosophy; they followed the Greek philosophers without deviation, or, at least, without any philosophical deviations. They became for the most part Aristotelians;—studied not only Aristotle, but the commentators of Aristotle; and themselves swelled the vast and unprofitable herd.

The philosophical works of Aristotle had, in some measure, made their way in the East, before the growth of the Saracen power. In the sixth century, a Syrian, Uranus,[15] encouraged by the love of philosophy manifested by Cosroes, had translated some of the writings of the Stagirite; about the same time, Sergius had given some translations in Syriac. In the seventh century, Jacob of Edessa translated into this language the Dialectics, and added Notes to the work. Such labors became numerous; and the first Arabic translations of Aristotle were formed upon these Persian or Syriac texts. In this succession of transfusions, some mistakes must inevitably have been introduced.

The Arabian interpreters of Aristotle, like a large portion of the Alexandrian ones, gave to the philosopher a tinge of opinions borrowed from another source, of which I shall have to speak under the head of *Mysticism.* But they are, for the most part, sufficiently strong examples of the peculiar spirit of commentation, to make it fitting to notice them here. At the head of them stands[16] Alkindi, who appears to have lived at the court of Almamon, and who wrote commentaries on the Organon of Aristotle. But Alfarabi was the glory of the school of Bagdad; his knowledge included mathematics, astronomy, medicine, and philosophy. Born in an elevated rank, and possessed of a rich patrimony, he led an austere life, and devoted himself altogether to study and meditation. He employed himself particularly in unfolding the import of Aristotle's treatise On the Soul.[17] Avicenna (Ebn Sina)

[15] Deg. iv. 196. [16] Ib. iv. 187. [17] Ib. iv. 205.

was at once the Hippocrates and the Aristotle of the Arabians; and certainly the most extraordinary man that the nation produced. In the course of an unfortunate and stormy life, occupied by politics and by pleasures, he produced works which were long revered as a sort of code of science. In particular, his writings on medicine, though they contain little besides a compilation of Hippocrates and Galen, took the place of both, even in the universities of Europe; and were studied as models at Paris and Montpelier, till the end of the seventeenth century, at which period they fell into an almost complete oblivion. Avicenna is conceived, by some modern writers,[18] to have shown some power of original thinking in his representations of the Aristotelian Logic and Metaphysics. Averroes (Ebn Roshd) of Cordova, was the most illustrious of the Spanish Aristotelians, and became the guide of the schoolmen,[19] being placed by them on a level with Aristotle himself, or above him. He translated Aristotle from the first Syriac version, not being able to read the Greek text. He aspired to, and retained for centuries, the title of the *Commentator;* and he deserves this title by the servility with which he maintains that Aristotle[20] carried the sciences to the highest possible degree, measured their whole extent, and fixed their ultimate and permanent boundaries; although his works are conceived to exhibit a trace of the New Platonism. Some of his writings are directed against an Arabian skeptic, of the name of Algazel, whom we have already noticed.

When the schoolmen had adopted the supremacy of Aristotle to the extent in which Averroes maintained it, their philosophy went further than a system of mere commentation, and became a system of dogmatism; we must, therefore, in another chapter, say a few words more of the Aristotelians in this point of view, before we proceed to the revival of science; but we must previously consider some other features in the character of the Stationary Period.

---

[18] Deg. iv. 206. [19] Ib. iv. 247. Averroes died A. D. 1206. [20] Ib. iv. 248.

CHAPTER III.

Of the Mysticism of the Middle Ages.

IT has been already several times hinted, that a new and peculiar element was introduced into the Greek philosophy which occupied the attention of the Alexandrian school; and that this element tinged a large portion of the speculations of succeeding ages. We may speak of this peculiar element as *Mysticism;* for, from the notion usually conveyed by this term, the reader will easily apprehend the general character of the tendency now spoken of; and especially when he sees its effect pointed out in various subjects. Thus, instead of referring the events of the external world to space and time, to sensible connection and causation, men attempted to reduce such occurrences under spiritual and supersensual relations and dependencies; they referred them to superior intelligences, to theological conditions, to past and future events in the moral world, to states of mind and feelings, to the creatures of an imaginary mythology or demonology. And thus their physical Science became Magic, their Astronomy became Astrology, the study of the Composition of bodies became Alchemy, Mathematics became the contemplation of the Spiritual Relations of number and figure, and Philosophy became Theosophy.

The examination of this feature in the history of the human mind is important for us, in consequence of its influence upon the employments and the thoughts of the times now under our notice. This tendency materially affected both men's speculations and their labors in the pursuit of knowledge. By its direct operation, it gave rise to the newer Platonic philosophy among the Greeks, and to corresponding doctrines among the Arabians; and by calling into a prominent place astrology, alchemy, and magic, it long occupied most of the real observers of the material world. In this manner it delayed and impeded the progress of true science; for we shall see reason to believe that human knowledge lost more by the perversion of men's minds and the misdirection of their efforts, than it gained by any increase of zeal arising from the peculiar hopes and objects of the mystics.

It is not our purpose to attempt any general view of the progress and fortunes of the various forms of Mystical Philosophy; but only to exhibit some of its characters, in so far as they illustrate those tendencies of thought which accompanied the retrogradation of inductive science. And of these, the leading feature which demands our notice is that already alluded to; namely, the practice of referring things and events, not to clear and distinct relations, obviously applicable to such cases;—not to general rules capable of direct verification; but to notions vague, distant, and vast, which we cannot bring into contact with facts, because they belong to a different region from the facts; as when we connect natural events with moral or historical causes, or seek spiritual meanings in the properties of number and figure. Thus the character of Mysticism is, that it refers particulars, not to generalizations homogeneous and immediate, but to such as are heterogeneous and remote; to which we must add, that the process of this reference is not a calm act of the intellect, but is accompanied with a glow of enthusiastic feeling.

1. *Neoplatonic Theosophy.*—The *Newer Platonism* is the first example of this Mystical Philosophy which I shall consider. The main points which here require our notice are, the doctrine of an Intellectual World resulting from the act of the Divine Mind, as the only reality; and the aspiration after the union of the human soul with this Divine Mind, as the object of human existence. The "Ideas" of Plato were Forms of our *knowledge*; but among the Neoplatonists they became really existing, indeed the only really existing, Objects; and the inaccessible scheme of the universe which these ideas constitute, was offered as the great subject of philosophical contemplation. The desire of the human mind to approach towards its Creator and Preserver, and to obtain a spiritual access to Him, leads to an employment of the thoughts which is well worth the notice of the religious philosopher; but such an effort, even when founded on revelation and well regulated, is not a means of advance in physics; and when it is the mere result of natural enthusiasm, it may easily obtain such a place in men's minds as to unfit them for the successful prosecution of natural philosophy. The temper, therefore, which introduces such supernatural communion into the general course of its speculations, may be properly treated as mystical, and as one of the causes of the decline of science in the Stationary Period. The Neoplatonic philosophy requires our notice as one of the most remarkable forms of this Mysticism.

Though Ammonius Saccas, who flourished at the end of the second century, is looked upon as the beginner of the Neoplatonists, his disciple Plotinus is, in reality, the great founder of the school, both by his works, which still remain to us, and by the enthusiasm which his character and manners inspired among his followers. He lived a life of meditation, gentleness, and self-denial, and died in the second year of the reign of Claudius (A. D. 270). His disciple, Porphyry, has given us a Life of him, from which we may see how well his habitual manners were suited to make his doctrines impressive. "Plotinus, the philosopher of our time," Porphyry thus begins his biography, "appeared like a person ashamed that he was in the body. In consequence of this disposition, he could not bear to talk concerning his family, or his parents, or his country. He would not allow himself to be represented by a painter or statuary; and once, when Aurelius entreated him to permit a likeness of him to be taken, he said, 'Is it not enough for us to carry this image in which nature has inclosed us, but we must also try to leave a more durable image of this image, as if it were so great a sight?' And he retained the same temper to the last. When he was dying, he said, 'I am trying to bring the divinity which is in us to the divinity which is in the universe.'" He was looked upon by his successors with extraordinary admiration and reverence; and his disciple Porphyry collected from his lips, or from fragmental notes, the six *Enneads* of his doctrines (that is, parts each consisting of *nine* Books), which he arranged and annotated.

We have no difficulty in finding in this remarkable work examples of mystical speculation. The Intelligible World of realities or essences corresponds to the world of sense[1] in the classes of things which it includes. To the Intelligible World, man's mind ascends, by a triple road which Plotinus figuratively calls that of the Musician, the Lover, the Philosopher.[2] The activity of the human soul is identified by analogy with the motion of the heavens. "This activity is about a middle point, and thus it is circular; but a middle point is not the same in body and in the soul: in that, the middle point is local; in this, it is that on which the rest depends. There is, however, an analogy; for as in one case, so in the other, there must be a middle point, and as the sphere revolves about its centre, the soul revolves about God through its affections."

The conclusion of the work is,[3] as might be supposed, upon the approach to, union with, and fruition of God. The author refers again to the analogy between the movements of the soul and those of the heavens. "We move round him like a choral dance; even when we

---

[1] vi. Ennead, iii. 1. [2] ii. E. ii. 2. [3] vi. Enn. ix. 8.

look from him we revolve about him : we do not always look at him, but when we do, we have satisfaction and rest, and the harmony which belongs to that divine movement. In this movement, the mind beholds the fountain of life, the fountain of mind, the origin of being, the cause of good, the root of the soul."[4] "There will be a time when this vision shall be continual; the mind being no more interrupted, nor suffering any perturbation from the body. Yet that which beholds is not that which is disturbed; and when this vision becomes dim, it does not obscure the knowledge which resides in demonstration, and faith, and reasoning; but the vision itself is not reason, but greater than reason, and before reason."[5]

The fifth book of the third Ennead has for its subject the Dæmon which belongs to each man. It is entitled "Concerning Love;" and the doctrine appears to be, that the Love, or common source of the passions which is in each man's mind, is "the Dæmon which they say accompanies each man."[6] These dæmons were, however (at least by later writers), invested with a visible aspect and with a personal character, including a resemblance of human passions and motives. It is curious thus to see an untenable and visionary generalization falling back into the domain of the senses and the fancy, after a vain attempt to support itself in the region of the reason. This imagination soon produced pretensions to the power of making these dæmons or genii visible; and the Treatise on the Mysteries of the Egyptians, which is attributed to Iamblichus, gives an account of the secret ceremonies, the mysterious words, the sacrifices and expiations, by which this was to be done.

It is unnecessary for us to dwell on the progress of this school; to point out the growth of the Theurgy which thus arose; or to describe the attempts to claim a high antiquity for this system, and to make Orpheus, the poet, the first promulgator of its doctrines. The system, like all mystical systems, assumed the character rather of religion than of a theory. The opinions of its disciples materially influenced their lives. It gave the world the spectacle of an austere morality, a devotional exaltation, combined with the grossest superstitions of Paganism. The successors of Iamblichus appeared rather to hold a priesthood, than the chair of a philosophical school.[7] They were persecuted by Constantine and Constantius, as opponents of Christianity. Sopater, a

---

[4] vi. Enn. ix. 9. [5] vi. Enn. ix. 10. [6] Ficinus, *Comm.* in. v. Enn. iii.
[7] Deg. iii. 407.

Syrian philosopher of this school, was beheaded by the former emperor on a charge that he had bound the winds by the power of magic.[8] But Julian, who shortly after succeeded to the purple, embraced with ardor the opinions of Iamblichus. Proclus (who died A. D. 487) was one of the greatest of the teachers of this school;[9] and was, both in his life and doctrines, a worthy successor of Plotinus, Porphyry, and Iamblichus. We possess a biography, or rather a panegyric of him, by his disciple Marinus, in which he is exhibited as a representation of the ideal perfection of the philosophic character, according to the views of the Neoplatonists. His virtues are arranged as physical, moral, purificatory, theoretic, and theurgic. Even in his boyhood, Apollo and Minerva visited him in his dreams: he studied oratory at Alexandria, but it was at Athens that Plutarch and Lysianus initiated him in the mysteries of the New Platonists. He received a kind of consecration at the hands of the daughter of Plutarch, the celebrated Asclepigenia, who introduced him to the traditions of the Chaldeans, and the practices of theurgy; he was also admitted to the mysteries of Eleusis. He became celebrated for his knowledge and eloquence; but especially for his skill in the supernatural arts which were connected with the doctrines of his sect. He appears before us rather as a hierophant than a philosopher. A large portion of his life was spent in evocations, purifications, fastings, prayers, hymns, intercourse with apparitions, and with the gods, and in the celebration of the festivals of Paganism, especially those which were held in honor of the Mother of the Gods. His religious admiration extended to all forms of mythology. The philosopher, said he, is not the priest of a single religion, but of all the religions of the world. Accordingly, he composed hymns in honor of all the divinities of Greece, Rome, Egypt, Arabia;—Christianity alone was excluded from his favor.

The reader will find an interesting view of the *School of Alexandria*, in M. Barthelemy Saint-Hilaire's *Rapport* on the *Mémoires* sent to the Academy of Moral and Political Sciences at Paris, in consequence of its having, in 1841, proposed this as the subject of a prize, which was awarded in 1844. M. Saint-Hilaire has prefixed to this *Rapport* a dissertation on the Mysticism of that school. He, however, uses the term *Mysticism* in a wider sense than my purpose, which regarded mainly the bearing of the doctrines of this school upon the progress of the Inductive Sciences, has led me to do. Although he finds much to ad-

---

[8] Gibbon, iii. 352.

[9] Deg. iii. 419.

mire in the Alexandrian philosophy, he declares that they were incapable of treating scientific questions. The extent to which this is true is well illustrated by the extract which he gives from Plotinus, on the question, "Why objects appear smaller in proportion as they are more distant." Plotinus denies that the reason of this is that the angles of vision become smaller. His reason for this denial is curious enough. If it were so, he says, how could the heaven appear smaller than it is, since it occupies the whole of the visual angle?

2. *Mystical Arithmetic.*—It is unnecessary further to exemplify, from Proclus, the general mystical character of the school and time to which he belonged; but we may notice more specially one of the forms of this mysticism, which very frequently offers itself to our notice, especially in him; and which we may call *Mystical Arithmetic.* Like all the kinds of Mysticism, this consists in the attempt to connect our conceptions of external objects by general and inappropriate notions of goodness, perfection, and relation to the divine essence and government; instead of referring such conceptions to those appropriate ideas, which, by due attention, become perfectly distinct, and capable of being positively applied and verified. The subject which is thus dealt with, in the doctrines of which we now speak, is Number; a notion which tempts men into these visionary speculations more naturally than any other. For number is really applicable to moral notions—to emotions and feelings, and to their objects—as well as to the things of the material world. Moreover, by the discovery of the principle of musical concords, it had been found, probably most unexpectedly, that numerical relations were closely connected with sounds which could hardly be distinguished from the expression of thought and feeling; and a suspicion might easily arise, that the universe, both of matter and of thought, might contain many general and abstract truths of some analogous kind. The relations of number have so wide a bearing, that the ramifications of such a suspicion could not easily be exhausted, supposing men willing to follow them into darkness and vagueness; which it is precisely the mystical tendency to do. Accordingly, this kind of speculation appeared very early, and showed itself first among the Pythagoreans, as we might have expected, from the attention which they gave to the theory of harmony: and this, as well as some other of the doctrines of the Pythagorean philosophy, was adopted by the later Platonists, and, indeed, by Plato himself, whose speculations concerning number have decidedly a mystical character. The mere mathematical relations of numbers,—as odd and even, perfect and imperfect,

abundant and defective,—were, by a willing submission to an enthusiastic bias, connected with the notions of good and beauty, which were suggested by the terms expressing their relations; and principles resulting from such a connection were woven into a wide and complex system. It is not necessary to dwell long on this subject; the mere titles of the works which treated of it show its nature. Archytas[10] is said to have written a treatise on the number *ten:* Telaugé, the daughter of Pythagoras, wrote on the number *four*. This number, indeed, which was known by the name of the *Tetractys*, was very celebrated in the school of Pythagoras. It is mentioned in the "Golden Verses," which are ascribed to him: the pupil is conjured to be virtuous,

> Ναὶ μὰ τὸν ἁμετέρᾳ ψυχᾶ παραδόντα τετρακτὺν
> Παγὰν ἀεννάου φύσεως . . . .

> By him who stampt *The Four* upon the mind,—
> *The Four*, the fount of nature's endless stream.

In Plato's works, we have evidence of a similar belief in religious relations of Number; and in the new Platonists, this doctrine was established as a system. Proclus, of whom we have been speaking, founds his philosophy, in a great measure, on the relation of Unity and Multiple; from this, he is led to represent the causality of the Divine Mind by three Triads of abstractions; and in the development of one part of this system, the number seven is introduced.[11] "The intelligible and intellectual gods produce all things triadically; for the monads in these latter are divided according to number; and what the monad was in the former, the number is in these latter. And the intellectual gods produce all things hebdomically; for they evolve the intelligible, and at the same time intellectual triads, into intellectual hebdomads, and expand their contracted powers into intellectual variety." Seven is what is called by arithmeticians a *prime* number, that is, it cannot be produced by the multiplication of other numbers. In the language of the New Platonists, the number seven is said to be a virgin, and without a mother, and it is therefore sacred to Minerva. The number six is a perfect number, and is consecrated to Venus.

The relations of space were dealt with in like manner, the Geometrical properties being associated with such physical and metaphysical notions as vague thought and lively feeling could anyhow connect with them. We may consider, as an example of this,[12] Plato's opinion

---

[10] Mont. ii. 123. [11] Procl. v. 3, Taylor's translation. [12] Stanley, *Hist. Phil.*

concerning the particles of the four elements. He gave to each kind of particle one of the five regular solids, about which the geometrical speculations of himself and his pupils had been employed. The particles of fire were pyramids, because they are sharp, and tend upwards; those of earth are cubes, because they are stable, and fill space; the particles of air are octahedral, as most nearly resembling those of fire; those of water are the icositetrahedron, as most nearly spherical. The dodecahedron is the figure of the element of the heavens, and shows its influence in other things, as in the twelve signs of the zodiac. In such examples we see how loosely space and number are combined or confounded by these mystical visionaries.

These numerical dreams of ancient philosophers have been imitated by modern writers; for instance, by Peter Bungo and Kircher, who have written De Mysteriis Numerorum. Bungo treats of the mystical properties of each of the numbers in order, at great length. And such speculations have influenced astronomical theories. In the first edition of the Alphonsine Tables,[13] the precession was represented by making the first point of Aries move, in a period of 7000 years, through a circle of which the radius was 18 degrees, while the circle moved round the ecliptic in 49,000 years; and these numbers, 7000 and 49,000, were chosen probably by Jewish calculators, or with reference to Jewish Sabbatarian notions.

3. *Astrology.*—Of all the forms which mysticism assumed, none was cultivated more assiduously than astrology. Although this art prevailed most universally and powerfully during the stationary period, its existence, even as a detailed technical system, goes back to a very early age. It probably had its origin in the East; it is universally ascribed to the Babylonians and Chaldeans; the name *Chaldean* was, at Rome, synonymous with *mathematicus*, or astrologer; and we read repeatedly that this class of persons were expelled from Italy by a decree of the senate, both during the times of the republic and of the empire.[14] The recurrence of this act of legislation shows that it was not effectual: "It is a class of men," says Tacitus, "which, in our city, will always be prohibited, and will always exist." In Greece, it does not appear that the state showed any hostility to the professors of this art. They undertook, it would seem, then, as at a later period, to determine the course of a man's character and life from the configuration of the stars at the moment of his birth. We do not possess any of the

[13] Montucla, i. 511. [14] Tacit. *Ann.* ii. 82. xii. 52. *Hist.* I. 22, II. 62.

speculations of the early astrologers; and we cannot therefore be certain that the notions which operated in men's minds when the art had its birth, agreed with the views on which it was afterwards defended, when it became a matter of controversy. But it appears probable, that, though it was at later periods supported by physical analogies, it was originally suggested by mythological belief. The Greeks spoke of the *influences* or *effluxes* (ἀπόῤῥοιας) which proceeded from the stars; but the Chaldeans had probably thought rather of the powers which they exercised as *deities*. In whatever manner the sun, moon, and planets came to be identified with gods and goddesses, it is clear that the characters ascribed to these gods and goddesses regulate the virtues and powers of the stars which bear their names. This association, so manifestly visionary, was retained, amplified, and pursued, in an enthusiastic spirit, instead of being rejected for more distinct and substantial connections; and a pretended science was thus formed, which bears the obvious stamp of mysticism.

That common sense of mankind which teaches them that theoretical opinions are to be calmly tried by their consequences and their accordance with facts, appears to have counteracted the prevalence of astrology in the better times of the human mind. Eudoxus, as we are informed by Cicero,[15] rejected the pretensions of the Chaldeans; and Cicero himself reasons against them with arguments as sensible and intelligent as could be adduced by a writer of the present day; such as the different fortunes and characters of persons born at the same time; and the failure of the predictions, in the case of Pompey, Crassus, Cæsar, to whom the astrologers had foretold glorious old age and peaceful death. He also employs an argument which the reader would perhaps not expect from him,—the very great remoteness of the planets as compared with the distance of the moon. "What contagion can reach us," he asks, "from a distance almost infinite?"

Pliny argues on the same side, and with some of the same arguments.[16] "Homer," he says, "tells us that Hector and Polydamus were born the same night;—men of such different fortune. And every hour, in every part of the world, are born lords and slaves, kings and beggars."

The impression made by these arguments is marked in an anecdote told concerning Publius Nigidius Figulus, a Roman of the time of Julius Cæsar, whom Lucan mentions as a celebrated astrologer. It is

---

15 Cic. *de Div.* ii. 42.

16 *Hist. Nat.* vii. 49.

said, that when an opponent of the art urged as an objection the different fates of persons born in two successive instants, Nigidius bade him make two contiguous marks on a potter's wheel, which was revolving rapidly near them. On stopping the wheel, the two marks were found to be really far removed from each other; and Nigidius is said to have received the name of Figulus (the potter), in remembrance of this story. His argument, says St. Augustine, who gives us the narrative, was as fragile as the ware which the wheel manufactured.

As the darkening times of the Roman empire advanced, even the stronger minds seem to have lost the *clear energy* which was requisite to throw off this delusion. Seneca appears to take the influence of the planets for granted; and even Tacitus[17] seems to hesitate. "For my own part," says he, "I doubt; but certainly the majority of mankind cannot be weaned from the opinion, that, at the birth of each man, his future destiny is fixed; though some things may fall out differently from the predictions, by the ignorance of those who profess the art; and that thus the art is unjustly blamed, confirmed as it is by noted examples in all ages." The occasion which gives rise to these reflections of the historian is the mention of Thrasyllus, the favorite astrologer of the Emperor Tiberius, whose skill is exemplified in the following narrative. Those who were brought to Tiberius on any important matter, were admitted to an interview in an apartment situated on a lofty cliff in the island of Capreæ. They reached this place by a narrow path, accompanied by a single freedman of great bodily strength; and on their return, if the emperor had conceived any doubts of their trustworthiness, a single blow buried the secret and its victim in the ocean below. After Thrasyllus had, in this retreat, stated the results of his art as they concerned the emperor, Tiberius asked him whether he had calculated how long he himself had to live. The astrologer examined the aspect of the stars, and while he did this, as the narrative states, showed hesitation, alarm, increasing terror, and at last declared that, "the present hour was for him critical, perhaps fatal." Tiberius embraced him, and told him "he was right in supposing he had been in danger, but that he should escape it;" and made him thenceforth his confidential counsellor.

The belief in the power of astrological prediction which thus obtained dominion over the minds of men of literary cultivation and practical energy, naturally had a more complete sway among the speculative

---

[17] *Ann.* vi. 22.

but unstable minds of the later philosophical schools of Alexandria, Athens, and Rome. We have a treatise on astrology by Proclus, which will serve to exemplify the mystical principle in this form. It appears as a commentary on a work on the same subject called "Tetrabiblos," ascribed to Ptolemy; though we may reasonably doubt whether the author of the "Megale Syntaxis" was also the writer of the astrological work. A few notices of the commentary of Proclus will suffice.[18] The science is defended by urging how powerful we know the physical effects of the heavenly bodies to be. "The sun regulates all things on earth;—the birth of animals, the growth of fruits, the flowing of waters, the change of health, according to the seasons: he produces heat, moisture, dryness, cold, according to his approach to our zenith. The moon, which is the nearest of all bodies to the earth, gives out much *influence;* and all things, animate and inanimate, sympathize with her: rivers increase and diminish according to her light; the advance of the sea, and its recess, are regulated by her rising and setting; and along with her, fruits and animals wax and wane, either wholly or in part." It is easy to see that by pursuing this train of associations (some real and some imaginary) very vaguely and very enthusiastically, the connections which astrology supposes would receive a kind of countenance. Proclus then proceeds to state[19] the doctrines of the science. "The sun," he says, "is productive of heat and dryness; this power is moderate in its nature, but is more perceived than that of the other luminaries, from his magnitude, and from the change of seasons. The nature of the moon is for the most part moist; for being the nearest to the earth, she receives the vapors which rise from moist bodies, and thus she causes bodies to soften and rot. But by the illumination she receives from the sun, she partakes in a moderate degree of heat. Saturn is cold and dry, being most distant both from the heating power of the sun, and the moist vapors of the earth. His cold, however, is most prevalent, his dryness is more moderate. Both he and the rest receive additional powers from the configurations which they make with respect to the sun and moon." In the same manner it is remarked that Mars is dry and caustic, from his fiery nature, which, indeed, his color shows. Jupiter is well compounded of warm and moist, as is Venus. Mercury is variable in his character. From these notions were derived others concerning the beneficial or hurtful effect of these stars. Heat and

---

18 I. 2.

19 I. 4.

moisture are generative and creative elements; hence the ancients, says Proclus, deemed Jupiter, and Venus, and the Moon to have a good power; Saturn and Mercury, on the other hand, had an evil nature.

Other distinctions of the character of the stars are enumerated, equally visionary, and suggested by the most fanciful connections. Some are masculine, and some feminine: the Moon and Venus are of the latter kind. This appears to be merely a mythological or etymological association. Some are diurnal, some nocturnal: the Moon and Venus are of the latter kind, the Sun and Jupiter of the former; Saturn and Mars are both.

The fixed stars, also, and especially those of the zodiac, had especial influences and subjects assigned to them. In particular, each sign was supposed to preside over a particular part of the body; thus Aries had the head assigned to it, Taurus the neck, and so on.

The most important part of the sky in the astrologer's consideration, was that sign of the zodiac which rose at the moment of the child's birth; this was, properly speaking, the *horoscope*, the *ascendant*, or the *first house;* the whole circuit of the heavens being divided into twelve *houses*, in which life and death, marriage and children, riches and honors, friends and enemies, were distributed.

We need not attempt to trace the progress of this science. It prevailed extensively among the Arabians, as we might expect from the character of that nation. Albumasar, of Balkh in Khorasan, who flourished in the ninth century, who was one of their greatest astronomers, was also a great astrologer; and his work on the latter subject, "De Magnis Conjunctionibus, Annorum Revolutionibus ac eorum Perfectionibus," was long celebrated in Europe. Aboazen Haly (the writer of a treatise "De Judiciis Astrorum"), who lived in Spain in the thirteenth century, was one of the classical authors on this subject.

It will easily be supposed that when this *apotelesmatic* or *judicial* astrology obtained firm possession of men's minds, it would be pursued into innumerable subtle distinctions and extravagant conceits; and the more so, as experience could offer little or no check to such exercises of fancy and subtlety. For the correction of rules of astrological divination by comparison with known events, though pretended to by many professors of the art, was far too vague and fallible a guidance to be of any real advantage. Even in what has been called Natural Astrology, the dependence of the weather on the heavenly bodies, it is easy to see what a vast accumulation of well-observed facts is requisite to establish any true rule; and it is well known how long, in spite of facts, false and groundless rules (as the dependence of the weather on the moon) may keep their hold on men's minds. When the facts are such loose and many-sided things as human characters, passions, and happiness, it was hardly to be expected that even the most powerful minds should be able to find a footing sufficiently firm, to enable them to resist the impression of a theory constructed of sweeping and bold assertions, and filled out into a complete system of details. Accordingly, the connection of the stars with human persons and actions was, for a long period, undisputed. The vague, obscure, and heterogeneous character of such a connection, and its unfitness for any really scientific reasoning, could, of course, never be got rid of; and the bewildering feeling of earnestness and solemnity, with which the connection of the heavens with man was contemplated, never died away. In other respects, however, the astrologers fell into a servile commentatorial spirit; and employed themselves in annotating and illustrating the works of their predecessors to a considerable extent, before the revival of true science.

It may be mentioned, that astrology has long been, and probably is, an art held in great esteem and admiration among other eastern nations besides the Mohammedans: for instance, the Jews, the Indians, the Siamese, and the Chinese. The prevalence of vague, visionary, and barren notions among these nations, cannot surprise us; for with regard to them we have no evidence, as with regard to Europeans we have, that they are capable, on subjects of physical speculation, of originating sound and rational general principles. The Arts may have had their birth in all parts of the globe; but it is only Europe, at particular favored periods of its history, which has ever produced Sciences.

We are, however, now speaking of a long period, during which this productive energy was interrupted and suspended. During this period Europe descended, in intellectual character, to the level at which the other parts of the world have always stood. Her Science was then a mixture of Art and Mysticism; we have considered several forms of this Mysticism, but there are two others which must not pass unnoticed, Alchemy and Magic.

We may observe, before we proceed, that the deep and settled influence which Astrology had obtained among them, appears perhaps most strongly in the circumstance, that the most vigorous and clear-sighted minds which were concerned in the revival of science, did not, for a long period, shake off the persuasion that there was, in this art, some element of truth. Roger Bacon, Cardan, Kepler, Tycho Brahe,

Francis Bacon, are examples of this. These, or most of them, rejected all the more obvious and extravagant absurdities with which the subject had been loaded; but still conceived that some real and valuable truth remained when all these were removed. Thus Campanella,[20] whom we shall have to speak of as one of the first opponents of Aristotle, wrote an "Astrology purified from all the Superstitions of the Jews and Arabians, and treated physiologically."

4. *Alchemy.*—Like other kinds of Mysticism, Alchemy seems to have grown out of the notions of moral, personal, and mythological qualities, which men associated with terms, of which the primary application was to physical properties. This is the form in which the subject is presented to us in the earliest writings which we possess on the subject of chemistry;—those of Geber[21] of Seville, who is supposed to have lived in the eighth or ninth century. The very titles of Geber's works show the notions on which this pretended science proceeds. They are, "Of the Search of Perfection;" "Of the Sum of Perfection, or of the Perfect Magistery;" "Of the Invention of Verity, or Perfection." The basis of this phraseology is the distinction of metals into more or less *perfect;* gold being the most perfect, as being the most valuable, most beautiful, most pure, most durable; silver the next; and so on. The "Search of Perfection" was, therefore, the attempt to convert other metals into gold; and doctrines were adopted which represented the metals as all compounded of the same elements, so that this was theoretically possible. But the mystical trains of association were pursued much further than this; gold and silver were held to be the most noble of metals; gold was their King, and silver their Queen. Mythological associations were called in aid of these fancies, as had been done in astrology. Gold was Sol, silver was Luna, the moon; copper, iron, tin, lead, were assigned to Venus, Mars, Jupiter, Saturn. The processes of mixture and heat were spoken of as personal actions and relations, struggles and victories. Some elements were conquerors, some conquered; there existed preparations which possessed the power of changing the whole of a body into a substance of another kind: these were called *magisteries.*[22] When gold and quicksilver are combined, the king and the queen are married, to produce children of their own kind. It will easily be conceived, that when chemical operations were described in phraseology of this sort, the enthusiasm of the

---

[20] Bacon, *De Aug.* iii. 4.

[21] Thomson's *Hist. of Chem.* i. 117.

[22] Boyle, Thomson's *Hist. Ch.* i. 25. Carolus Musitanus.

fancy would be added to that of the hopes, and observation would not be permitted to correct the delusion, or to suggest sounder and more rational views.

The exaggeration of the vague notion of perfection and power in the object of the alchemist's search, was carried further still. The same preparation which possessed the faculty of turning baser metals into gold, was imagined to be also a universal medicine, to have the gift of curing or preventing diseases, prolonging life, producing bodily strength and beauty: the *philosophers' stone* was finally invested with every desirable efficacy which the fancy of the "philosophers" could devise.

It has been usual to say that Alchemy was the mother of Chemistry; and that men would never have made the experiments on which the real science is founded, if they had not been animated by the hopes and the energy which the delusive art inspired. To judge whether this is truly said, we must be able to estimate the degree of interest which men feel in purely speculative truth, and in the real and substantial improvement of art to which it leads. Since the fall of Alchemy, and the progress of real Chemistry, these motives have been powerful enough to engage in the study of the science, a body far larger than the Alchemists ever were, and no less zealous. There is no apparent reason why the result should not have been the same, if the progress of true science had begun sooner. Astronomy was long cultivated without the bribe of Astrology. But, perhaps, we may justly say this;—that, in the stationary period, men's minds were so far enfeebled and degraded, that pure speculative truth had not its full effect upon them; and the mystical pursuits in which some dim and disfigured images of truth were sought with avidity, were among the provisions by which the human soul, even when sunk below its best condition, is perpetually directed to something above the mere objects of sense and appetite;—a contrivance of compensation, as it were, in the intellectual and spiritual constitution of man.

5. *Magic.*—Magical Arts, so far as they were believed in by those who professed to practise them, and so far as they have a bearing in science, stand on the same footing as astrology; and, indeed, a close alliance has generally been maintained between the two pursuits. Incapacity and indisposition to perceive natural and philosophical causation, an enthusiastic imagination, and such a faith as can devise and maintain supernatural and spiritual connections, are the elements of this, as of other forms of Mysticism. And thus, that temper which led men to aim at the magician's supposed authority over the elements,

is an additional exemplification of those habits of thought which prevented the progress of real science, and the acquisition of that command over nature which is founded on science, during the interval now before us.

But there is another aspect under which the opinions connected with this pursuit may serve to illustrate the mental character of the Stationary Period.

The tendency, during the middle ages, to attribute the character of Magician to almost all persons eminent for great speculative or practical knowledge, is a feature of those times, which shows how extensive and complete was the inability to apprehend the nature of real science. In cultivated and enlightened periods, such as those of ancient Greece, or modern Europe, knowledge is wished for and admired, even by those who least possess it: but in dark and degraded periods, superior knowledge is a butt for hatred and fear. In the one case, men's eyes are open; their thoughts are clear; and, however high the philosopher may be raised above the multitude, they can catch glimpses of the intervening path, and see that it is free to all, and that elevation is the reward of energy and labor. In the other case, the crowd are not only ignorant, but spiritless; they have lost the pleasure in knowledge, the appetite for it, and the feeling of dignity which it gives: there is no sympathy which connects them with the learned man: they see him above them, but know not how he is raised or supported: he becomes an object of aversion and envy, of vague suspicion and terror; and these emotions are embodied and confirmed by association with the fancies and dogmas of superstition. To consider superior knowledge as Magic, and Magic as a detestable and criminal employment, was the form which these feelings of dislike assumed; and at one period in the history of Europe, almost every one who had gained any eminent literary fame, was spoken of as a magician. Naudæus, a learned Frenchman, in the seventeenth century, wrote "An Apology for all the Wise Men who have been unjustly reported Magicians, from the Creation to the present Age." The list of persons whom he thus thinks it necessary to protect, are of various classes and ages. Alkindi, Geber, Artephius, Thebit, Raymund Lully, Arnold de Villâ Novâ, Peter of Apono, and Paracelsus, had incurred the black suspicion as physicians or alchemists. Thomas Aquinas, Roger Bacon, Michael Scott, Picus of Mirandula, and Trithemius, had not escaped it, though ministers of religion. Even dignitaries, such as Robert Grosteste, Bishop of Lincoln, Albertus Magnus, Bishop of Ratisbon,

Popes Sylvester the Second and Gregory the Seventh, had been involved in the wide calumny. In the same way in which the vulgar confounded the eminent learning and knowledge which had appeared in recent times, with skill in dark supernatural arts, they converted into wizards all the best-known names in the rolls of fame; as Aristotle, Solomon, Joseph, Pythagoras; and, finally, the poet Virgil was a powerful and skilful necromancer, and this fancy was exemplified by many strange stories of his achievements and practices.

The various results of the tendency of the human mind to mysticism, which we have here noticed, form prominent features in the intellectual character of the world, for a long course of centuries. The theosophy and theurgy of the Neoplatonists, the mystical arithmetic of the Pythagoreans and their successors, the predictions of the astrologers, the pretences of alchemy and magic, represent, not unfairly, the general character and disposition of men's thoughts, with reference to philosophy and science. That there were stronger minds, which threw off in a greater or less degree this train of delusive and unsubstantial ideas, is true; as, on the other hand, Mysticism, among the vulgar or the foolish, often went to an extent of extravagance and superstition, of which I have not attempted to convey any conception. The lesson which the preceding survey teaches us is, that during the Stationary Period, Mysticism, in its various forms, was a leading character, both of the common mind, and of the speculations of the most intelligent and profound reasoners; and that this Mysticism was the opposite of that habit of thought which we have stated Science to require; namely, clear Ideas, distinctly employed to connect well-ascertained Facts; inasmuch as the Ideas in which it dealt were vague and unstable, and the temper in which they were contemplated was an urgent and aspiring enthusiasm, which could not submit to a calm conference with experience upon even terms. The fervor of thought in some degree supplied the place of reason in producing belief; but opinions so obtained had no enduring value; they did not exhibit a permanent record of old truths, nor a firm foundation for new. Experience collected her stores in vain, or ceased to collect them, when she had only to pour them into the flimsy folds of the lap of Mysticism; who was, in truth, so much absorbed in looking for the treasures which were to fall from the skies, that she heeded little how scantily she obtained, or how loosely she held, such riches as might be found near her.

CHAPTER IV.

Of the Dogmatism of the Stationary Period.

IN speaking of the character of the age of commentators, we noticed *principally the ingenious servility* which it displays;—the acuteness with which it finds ground for speculation in the expression of other men's thoughts;—the want of all vigor and fertility in acquiring any real and new truths. Such was the character of the reasoners of the stationary period from the first; but, at a later day, this character, from various causes, was modified by new features. The servility which had yielded itself to the yoke, insisted upon forcing it on the necks of others: the subtlety which found all the truth it needed in certain accredited writings, resolved that no one should find there, or in any other region, any other truths; speculative men became tyrants without ceasing to be slaves; to their character of Commentators they added that of Dogmatists.

1. *Origin of the Scholastic Philosophy.*—The causes of this change have been very happily analyzed and described by several modern writers.[1] The general nature of the process may be briefly stated to have been the following.

The tendencies of the later times of the Roman empire to a commenting literature, and a second-hand philosophy, have already been noticed. The loss of the dignity of political freedom, the want of the cheerfulness of advancing prosperity, and the substitution of the less philosophical structure of the Latin language for the delicate intellectual mechanism of the Greek, fixed and augmented the prevalent feebleness and barrenness of intellect. Men forgot, or feared, to consult nature, to seek for new truths, to do what the great discoverers of other times had done; they were content to consult libraries, to study and defend old opinions, to talk of what great geniuses had said. They sought their philosophy in accredited treatises, and dared not question such doctrines as they there found.

The character of the philosophy to which they were thus led, was determined by this want of courage and originality. There are various

---

[1] Dr. Hampden, in the Life of Thomas Aquinas, in the *Encyc. Metrop.* Degerando, *Hist. Comparée*, vol. iv. Also Tennemann, *Hist. of Phil.* vol. viii. Introduction.

antagonist principles of opinion, which seem alike to have their root in the intellectual constitution of man, and which are maintained and developed by opposing sects, when the intellect is in vigorous action. Such principles are, for instance—the claims of Authority and of Reason to our assent;—the source of our knowledge in Experience or in Ideas;—the superiority of a Mystical or of a Skeptical turn of thought. Such oppositions of doctrine were found in writers of the greatest fame; and two of those, who most occupied the attention of students, Plato and Aristotle, were, on several points of this nature, very diverse from each other in their tendency. The attempt to reconcile these philosophers by Boëthius and others, we have already noticed; and the attempt was so far successful, that it left on men's minds the belief in the possibility of a great philosophical system which should be based on both these writers and have a claim to the assent of all sober speculators.

But, in the mean time, the Christian Religion had become the leading subject of men's thoughts; and divines had put forward its claims to be, not merely the guide of men's lives, and the means of reconciling them to their heavenly Master, but also to be a Philosophy in the widest sense in which the term had been used;—a consistent speculative view of man's condition and nature, and of the world in which he is placed.

These claims had been acknowledged; and, unfortunately, from the intellectual condition of the times, with no due apprehension of the necessary ministry of Observation, and Reason dealing with observation, by which alone such a system can be embodied. It was held without any regulating principle, that the philosophy which had been bequeathed to the world by the great geniuses of heathen antiquity, and the Philosophy which was deduced from, and implied by, the Revelations made by God to man, must be identical; and, therefore, that Theology is the only true philosophy. Indeed, the Neoplatonists had already arrived, by other roads, at the same conviction. John Scot Erigena, in the reign of Alfred, and consequently before the existence of the Scholastic Philosophy, properly so called, had reasserted this doctrine.[2] Anselm, in the eleventh century, again brought it forward;[3] and Bernard de Chartres, in the thirteenth.[4]

This view was confirmed by the opinion which prevailed, concerning the nature of philosophical truth; a view supported by the theory

[2] Deg. iv. 351. [3] Ib. iv. 388. [4] Ib. iv. 418.

of Plato, the practice of Aristotle, and the general propensities of the human mind: I mean the opinion that all science may be obtained by *the use of reasoning alone*;—*that by analyzing* and combining the notions which common language brings before us, we may learn all that we can know. Thus Logic came to include the whole of Science; and accordingly this Abelard expressly maintained.[5] I have already explained, in some measure, the fallacy of this belief, which consists, as has been well said,[6] "in mistaking the universality of the theory of language for the generalization of facts." But on all accounts this opinion is readily accepted; and it led at once to the conclusion, that the Theological Philosophy which we have described, is complete as well as true.

Thus a Universal Science was established, with the authority of a Religious Creed. Its universality rested on erroneous views of the relation of words and truths; its pretensions as a science were admitted by the servile temper of men's intellects; and its religious authority was assigned it, by making all truth part of religion. And as Religion claimed assent within her own jurisdiction under the most solemn and imperative sanctions, Philosophy shared in her imperial power, and dissent from their doctrines was no longer blameless or allowable. Error became wicked, dissent became heresy; to reject the received human doctrines, was nearly the same as to doubt the Divine declarations. The *Scholastic Philosophy* claimed the assent of all believers.

The external form, the details, and the text of this philosophy, were taken, in a great measure, from Aristotle; though, in the spirit, the general notions, and the style of interpretation, Plato and the Platonists had no inconsiderable share. Various causes contributed to the elevation of Aristotle to this distinction. His Logic had early been adopted as an instrument of theological disputation; and his spirit of systematization, of subtle distinction, and of analysis of words, as well as his disposition to argumentation, afforded the most natural and grateful employment to the commentating propensities. Those principles which we before noted as the leading points of his physical philosophy, were selected and adopted; and these, presented in a most technical form, and applied in a systematic manner, constitute a large portion of the philosophy of which we now speak, so far as it pretends to deal with physics.

2. *Scholastic Dogmas.*—But before the complete ascendency of Aristotle was thus established, when something of an intellectual waking

---

[5] Deg. iv. 407. [6] *Enc. Met.* 807.

took place after the darkness and sleep of the ninth and tenth centuries, the Platonic doctrines seem to have had, at first, a strong attraction for men's minds, as better falling in with the mystical speculations and contemplative piety which belonged to the times. John Scot Erigena[7] may be looked upon as the reviver of the New Platonism in the tenth century. Towards the end of the eleventh, Peter Damien,[8] in Italy, reproduced, involved in a theological discussion, some Neoplatonic ideas. Godefroy[9] also, censor of St. Victor, has left a treatise, entitled *Microcosmus;* this is founded on a mystical analogy, often afterwards again brought forward, between Man and the Universe. "Philosophers and theologians," says the writer, "agree in considering man as a little world; and as the world is composed of four elements, man is endowed with four faculties, the senses, the imagination, reason, and understanding." Bernard of Chartres,[10] in his *Megascosmus* and *Microcosmus*, took up the same notions. Hugo, abbot of St. Victor, made a contemplative life the main point and crown of his philosophy; and is said to have been the first of the scholastic writers who made psychology his special study.[11] He says the faculties of the mind are "the senses, the imagination, the reason, the memory, the understanding, and the intelligence."

Physics does not originally and properly form any prominent part of the Scholastic Philosophy, which consists mainly of a series of questions and determinations upon the various points of a certain technical divinity. Of this kind is the *Book of Sentences* of Peter the Lombard (bishop of Paris), who is, on that account, usually called "Magister Sententiarum;" a work which was published in the twelfth century, and was long the text and standard of such discussions. The questions are decided by the authority of Scripture and of the Fathers of the Church, and are divided into four Books, of which the first contains questions concerning God and the doctrine of the Trinity in particular; the second is concerning the Creation; the third, concerning Christ and the Christian Religion; and the fourth treats of Religious and Moral Duties. In the second book, as in many of the writers of this time, the nature of Angels is considered in detail, and the Orders of their Hierarchy, of which there were held to be nine. The physical discussions enter only as bearing upon the scriptural history of the creation, and cannot be taken as a specimen of the work; but I may observe, that in speaking of the division of the waters above the fir-

---

[7] Deg. iv. 85. [8] Ib. iv. 367. [9] Ib. iv. 413. [10] Ib. iv. 419. [11] Ib. iv. 415.

mament, from the waters under the firmament, he gives one opinion, that of Bede, that the former waters are the solid crystalline heavens in which the stars are fixed,[12] "for crystal, which is so hard and transparent, is made of water." But he mentions also the opinion of St. Augustine, that the waters above the heavens are in a state of vapor, (*vaporaliter*) and in minute drops; "if, then, water can, as we see in clouds, be so minutely divided that it may be thus supported as vapor on air, which is naturally lighter than water; why may we not believe that it floats above that lighter celestial element in still minuter drops and still lighter vapors? But in whatever manner the waters are there, we do not doubt that they are there."

The celebrated *Summa Theologiæ* of Thomas Aquinas is a work of the same kind; and any thing which has a physical bearing forms an equally small part of it. Thus, of the 512 Questions of the *Summa*, there is only one (Part. I., Quest. 115), "on Corporeal Action," or on any part of the material world; though there are several concerning the celestial Hierarchies, as "on the Act of Angels," "on the Speaking of Angels," "on the Subordination of Angels," "on Guardian Angels," and the like. This, of course, would not be remarkable in a treatise on Theology, except this Theology were intended to constitute the whole of Philosophy.

We may observe, that in this work, though Plato, Avecibron, and many other heathen as well as Christian philosophers, are adduced as authority, Aristotle is referred to in a peculiar manner as "the philosopher." This is noticed by John of Salisbury, as attracting attention in his time (he died A.D. 1182). "The various Masters of Dialectic," says he,[13] "shine each with his peculiar merit; but all are proud to worship the footsteps of Aristotle; so much so, indeed, that the name of *philosopher*, which belongs to them all, has been pre-eminently appropriated to him. He is called the philosopher *autonomatice*, that is, by excellence."

The Question concerning Corporeal Action, in Aquinas, is divided into six Articles; and the conclusion delivered upon the first is,[14] that "Body being compounded of power and act, is active as well as passive." Against this it is urged, that quantity is an attribute of body, and that quantity prevents action; that this appears in fact, since a larger body is more difficult to move. The author replies, that "quan-

[12] Lib. ii. Distinct. xiv. *De opere secundæ diei.*

[13] *Metalogicus*, lib. ii. cap. 16.

[14] *Summæ*, P. i. Q. 115. Art. 1.

tity does not prevent corporeal form from action altogether, but prevents it from being a universal agent, inasmuch as the form is individualized, which, in matter subject to quantity, it is. Moreover, the illustration deduced from the ponderousness of bodies is not to the purpose; first, because the addition of quantity is not the cause of gravity, as is proved in the fourth book, De Cœlo and De Mundo" (we see that he quotes familiarly the physical treatises of Aristotle); "second, because it is false that ponderousness makes motion slower; on the contrary, in proportion as any thing is heavier, the more does it move with its proper motion; thirdly, because action does not take place by local motion, as Democritus asserted; but by this, that something is drawn from power into act."

It does not belong to our purpose to consider either the theological or the metaphysical doctrines which form so large a portion of the treatises of the schoolmen. Perhaps it may hereafter appear, that some light is thrown on some of the questions which have occupied metaphysicians in all ages, by that examination of the history of the Progressive Sciences in which we are now engaged; but till we are able to analyze the leading controversies of this kind, it would be of little service to speak of them in detail. It may be noticed, however, that many of the most prominent of them refer to the great question, "What is the relation between actual things and general terms?" Perhaps in modern times, the actual things would be more commonly taken as the point to start from; and men would begin by considering how classes and universals are obtained from individuals. But the schoolmen, founding their speculations on the received modes of considering such subjects, to which both Aristotle and Plato had contributed, travelled in the opposite direction, and endeavored to discover how individuals were deduced from genera and species;—what was "the Principle of Individuation." This was variously stated by different reasoners. Thus Bonaventura[15] solves the difficulty by the aid of the Aristotelian distinction of Matter and Form. The individual derives from the Form the property of *being something*, and from the Matter the property of being that *particular thing*. Duns Scotus,[16] the great adversary of Thomas Aquinas in theology, placed the principle of Individuation in "a certain determining positive entity," which his school called *Hæcceity* or *thisness*. "Thus an individual man is Peter, because his *humanity* is combined with *Petreity*." The force

---

[15] Deg. iv. 573. [16] Ib. iv. 523.

of abstract terms is a curious question, and some remarkable experiments in their use had been made by the Latin Aristotelians before this time. In the same way in which we talk of the *quantity* and *quality* of a thing, they spoke of its *quiddity*.[17]

We may consider the reign of mere disputation as fully established at the time of which we are now speaking; and the only kind of philosophy henceforth studied was one in which no sound physical science had or could have a place. The wavering abstractions, indistinct generalizations, and loose classifications of common language, which we have already noted as the fountain of the physics of the Greek Schools of philosophy, were also the only source from which the Schoolmen of the middle ages drew their views, or rather their arguments: and though these notional and verbal relations were invested with a most complex and pedantic technicality, they did not, on that account, become at all more precise as notions, or more likely to lead to a single real truth. Instead of acquiring distinct ideas, they multiplied abstract terms; instead of real generalizations, they had recourse to verbal distinctions. The whole course of their employments tended to make them, not only ignorant of physical truth, but incapable of conceiving its nature.

Having thus taken upon themselves the task of raising and discussing questions by means of abstract terms, verbal distinctions, and logical rules alone, there was no tendency in their activity to come to an end, as there was no progress. The same questions, the same answers, the same difficulties, the same solutions, the same verbal subtleties,—sought for, admired, cavilled at, abandoned, reproduced, and again admired,—might recur without limit. John of Salisbury[18] observes of the Parisian teachers, that, after several years' absence, he found them not a step advanced, and still employed in urging and parrying the same arguments; and this, as Mr. Hallam remarks,[19] "was equally applicable to the period of centuries." The same knots were tied and

[17] Deg. iv. 494.

[18] He studied logic at Paris, at St. Geneviève, and then left them. "Duodecennium mihi elapsum est diversis studiis occupatum. Jucundum itaque visum est veteres quos reliqueram, et quos adhuc Dialectica detinebat in monte, (Sanctæ Genovefæ) revisere socios, conferre cum eis super ambiguitatibus pristinis; ut nostrûm invicem collatione mutuâ commetiremur profectum. Inventi sunt, qui fuerant, et ubi; neque enim ad palmam visi sunt processisse ad quæstiones pristinis dirimendas, neque propositiunculam unam adjecerant. Quibus urgebant stimulis eisdem et ipsi urgebantur," &c. *Metalogicus*, lib. ii. cap. 10.

[19] *Middle Ages*, iii. 537.

untied; the same clouds were formed and dissipated. The poet's censure of "the Sons of Aristotle," is just as happily expressed:

> They stand
> Locked up together hand in hand
> Every one leads as he is led,
> The same bare path they tread,
> And dance like Fairies a fantastic round,
> But neither change their motion nor their ground.

It will therefore be unnecessary to go into any detail respecting the history of the School Philosophy of the thirteenth, fourteenth, and fifteenth centuries. We may suppose it to have been, during the intermediate time, such as it was at first and at last. An occasion to consider its later days will be brought before us by the course of our subject. But, even during the most entire ascendency of the scholastic doctrines, the elements of change were at work. While the doctors and the philosophers received all the ostensible homage of men, a doctrine and a philosophy of another kind were gradually forming: the practical instincts of man, their impatience of tyranny, the progress of the useful arts, the promises of alchemy, were all disposing men to reject the authority and deny the pretensions of the received philosophical creed. Two antagonist forms of opinion were in existence, which for some time went on detached, and almost independent of each other; but, finally, these came into conflict, at the time of Galileo; and the war speedily extended to every part of civilized Europe.

3. *Scholastic Physics.*—It is difficult to give briefly any appropriate examples of the nature of the Aristotelian physics which are to be found in the works of this time. As the gravity of bodies was one of the first subjects of dispute when the struggle of the rival methods began, we may notice the mode in which it was treated.[20] "Zabarella maintains that the proximate cause of the motion of elements is the *form*, in the Aristotelian sense of the term: but to this sentence we," says Keckerman, "cannot agree; for in all other things the *form* is the proximate cause, not of the *act*, but of the power or faculty from which the act flows. Thus in man, the rational soul is not the cause of the act of laughing, but of the risible faculty or power." Keckerman's system was at one time a work of considerable authority: it was published in 1614. By comparing and systematizing what he finds in Aristotle, he is led to state his results in the form of definitions

---

[20] Keckerman, p. 1428.

and theorems. Thus, "gravity is a motive quality, arising from cold, density, and bulk, by which the elements are carried downwards." "Water is the lower, intermediate element, cold and moist." The first theorem concerning water is, "The moistness of the water is controlled by its coldness, so that it is less than the moistness of the air; though, according to the sense of the vulgar, water appears to moisten more than air." It is obvious that the two properties of fluids, to have their parts easily moved, and to wet other bodies, are here confounded. I may, as a concluding specimen of this kind, mention those propositions or maxims concerning fluids, which were so firmly established, that, when Boyle propounded the true mechanical principles of fluid action, he was obliged to state his opinions as "hydrostatical *paradoxes*." These were,—that fluids do not gravitate *in proprio loco;* that is, that water has no gravity in or on water, since it is in its own place;—that air has no gravity on water, since it is above water, which is its proper place;—that earth in water tends to descend, since its place is below water;—that the water rises in a pump or siphon, because nature abhors a vacuum;—that some bodies have a positive levity in others, as oil in water; and the like.

4. *Authority of Aristotle among the Schoolmen.*—The authority of Aristotle, and the practice of making him the text and basis of the system, especially as it regarded physics, prevailed during the period of which we speak. This authority was not, however, without its fluctuations. Launoy has traced one part of its history in a book *On the various Fortune of Aristotle in the University of Paris.* The most material turns of this fortune depend on the bearing which the works of Aristotle were supposed to have upon theology. Several of Aristotle's works, and more especially his metaphysical writings, had been translated into Latin, and were explained in the schools of the University of Paris, as early as the beginning of the thirteenth century.[21] At a council held at Paris in 1209, they were prohibited, as having given occasion to the heresy of Almeric (or Amauri), and because "they might give occasion to other heresies not yet invented." The Logic of Aristotle recovered its credit some years after this, and was publicly taught in the University of Paris in the year 1215; but the Natural Philosophy and Metaphysics were prohibited by a decree of Gregory the Ninth, in 1231. The Emperor Frederic the Second employed a number of learned men to translate into Latin, from the Greek and

21 Mosheim, iii. 157.

Arabic, certain books of Aristotle, and of other ancient sages; and we have a letter of Peter de Vineis, in which they are recommended to the attention of the University of Bologna: probably the same recommendation was addressed to other universities. Both Albertus Magnus and Thomas Aquinas wrote commentaries on Aristotle's works; and as this was done soon after the decree of Gregory the Ninth, Launoy is much perplexed to reconcile the fact with the orthodoxy of the two doctors. Campanella, who was one of the first to cast off the authority of Aristotle, says, "We are by no means to think that St. Thomas *aristotleized;* he only expounded Aristotle, that he might correct his errors; and I should conceive he did this with the license of the Pope." This statement, however, by no means gives a just view of the nature of Albertus's and Aquinas's commentaries. Both have followed their authors with profound deference.[22] For instance, Aquinas[23] attempts to defend Aristotle's assertion, that if there were no resistance, a body would move through a space in no time; and the same defence is given by Scotus.

We may imagine the extent of authority and admiration which Aristotle would attain when thus countenanced, both by the powerful and the learned. In universities, no degree could be taken without a knowledge of the philosopher. In 1452, Cardinal Totaril established this rule in the University of Paris.[24] When Ramus, in 1543, published an attack upon Aristotle, it was repelled by the power of the court and the severity of the law. Francis the First published an edict, in which he states that he had appointed certain judges, who had been of opinion,[25] "que le dit Ramus avoit été téméraire, arrogant et impudent; et que parcequ'en son livre des animadversions il reprenait Aristotle, estait évidemment connue et manifeste son ignorance." The books are then declared to be suppressed. It was often a complaint of pious men, that theology was corrupted by the influence of Aristotle and his commentators. Petrarch says,[26] that one of the Italian learned men conversing with him, after expressing much contempt for the Apostles and Fathers, exclaimed, "Utinam tu Averroen pati posses, ut videres quanto ille tuis his nugatoribus major sit!"

When the revival of letters began to take place, and a number of men of ardent and elegant minds, susceptible to the impressions of beauty of style and dignity of thought, were brought into contact with Greek literature, Plato had naturally greater charms for them. A

---

22 Deg. N. 475. 23 F. Piccolomini, ii. 835. 24 Launoy, pp. 108, 128.
25 Launoy, p. 132. 26 Hallam, *M. A.* iii. 536.

powerful school of Platonists (not Neoplatonists) was formed in Italy, including some of the principal scholars and men of genius of the time; as Picus of Mirandula in the middle, Marsilius Ficinus at the end, of the fifteenth century. At one time, it appeared as if the ascendency of Aristotle was about to be overturned; but, in physics at least, his authority passed unshaken through this trial. It was not by disputation that Aristotle could be overthrown; and the Platonists were not persons whose doctrines led them to use the only decisive method in such cases, the observation and unfettered interpretation of facts.

The history of their controversies, therefore, does not belong to our design. For like reasons we do not here speak of other authors, who opposed the scholastic philosophy on general theoretical grounds of various kinds. Such examples of insurrection against the dogmatism which we have been reviewing, are extremely interesting events in the history of the philosophy of science. But, in the present work, we are to confine ourselves to the history of science itself; in the hope that we may thus be able, hereafter, to throw a steadier light upon that philosophy by which the succession of stationary and progressive periods, which we are here tracing, may be in some measure explained. We are now to close our account of the stationary period, and to enter upon the great subject of the progress of physical science in modern times.

5. *Subjects omitted. Civil Law. Medicine.*—My object has been to make my way, as rapidly as possible, to this period of progress; and in doing this, I have had to pass over a long and barren track, where almost all traces of the right road disappear. In exploring this region, it is not without some difficulty that he who is travelling with objects such as mine, continues a steady progress in the proper direction; for many curious and attractive subjects of research come in his way: he crosses the track of many a controversy, which in its time divided the world of speculators, and of which the results may be traced, even now, in the conduct of moral, or political, or metaphysical discussions; or in the common associations of thought, and forms of language. The wars of the Nominalists and Realists; the disputes concerning the foundations of morals, and the motives of human actions; the controversies concerning predestination, free will, grace, and the many other points of metaphysical divinity; the influence of theology and metaphysics upon each other, and upon other subjects of human curiosity; the effects of opinion upon politics, and of political condition upon opinion; the influence of literature and philosophy

upon each other, and upon society; and many other subjects;—might be well worth examination, if our hope of success did not reside in pursuing, steadily and directly, those inquiries in which we can look for a definite and certain reply. We must even neglect two of the leading studies of those times, which occupied much of men's time and thoughts, and had a very great influence on society; the one dealing with Notions, the other with Things; the one employed about moral rules, the other about material causes, but both for practical ends; I mean the study of the *Civil Law*, and of *Medicine*. The second of these studies will hereafter come before us, as one of the principal occasions which led to the cultivation of chemistry; but, in itself, its progress is of too complex and indefinite a nature to be advantageously compared with that of the more exact sciences. The Roman Law is held, by its admirers, to be a system of deductive science, as exact as the mathematical sciences themselves; and it may, therefore, be useful to consider it, if we should, in the sequel, have to examine how far there can exist an analogy between moral and physical science. But after a few more words on the middle ages, we must return to our task of tracing the progress of the latter.

CHAPTER V.

Progress of the Arts in the Middle Ages.

ART AND SCIENCE.—I shall, before I resume the history of science, say a few words on the subject described in the title of this chapter, both because I might otherwise be accused of doing injustice to the period now treated of; and also, because we shall by this means bring under our notice some circumstances which were important as being the harbingers of the revival of progressive knowledge.

The accusation of injustice towards the state of science in the middle ages, if we were to terminate our survey of them with what has hitherto been said, might be urged from obvious topics. How do we recognize, it might be asked, in a picture of mere confusion and mysticism of thought, of servility and dogmatism of character, the powers and acquirements to which we owe so many of the most important inventions which we now enjoy? Parchment and paper, printing and engraving, improved glass and steel, gunpowder, clocks, telescopes, the mariner's compass, the reformed calendar, the decimal notation, algebra, trigonometry, chemistry, counterpoint, an invention equivalent to a new creation of music;—these are all possessions which we inherit from that which has been so disparagingly termed the Stationary Period. Above all, let us look at the monuments of architecture of this period;—the admiration and the despair of modern architects, not only for their beauty, but for the skill disclosed in their construction. With all these evidences before us, how can we avoid allowing that the masters of the middle ages not only made some small progress in Astronomy, which has, grudgingly as it would seem, been admitted in a former Book; but also that they were no small proficients in other sciences, in Optics, in Harmonics, in Physics, and, above all, in Mechanics?

If, it may be added, we are allowed, in the present day, to refer to the perfection of our arts as evidence of the advanced state of our physical philosophy;—if our steam-engines, our gas-illumination, our buildings, our navigation, our manufactures, are cited as triumphs of science;—shall not prior inventions, made under far heavier disadvantages,—shall not greater works, produced in an earlier state of knowledge, also be admitted as witnesses that the middle ages had their share, and that not a small or doubtful one, of science?

To these questions I answer, by distinguishing between Art, and Science in that sense of general Inductive Systematic Truth, which it bears in this work. To separate and compare, with precision, these two processes, belongs to the Philosophy of Induction; and the attempt must be reserved for another place: but the leading differences are sufficiently obvious. Art is practical, Science is speculative: the former is seen in doing; the latter rests in the contemplation of what is known. The art of the builder appears in his edifice, though he may never have meditated on the abstract propositions on which its stability and strength depends. The Science of the mathematical mechanician consists in his seeing that, under certain conditions, bodies must sustain each other's pressure, though he may never have applied his knowledge in a single case.

Now the remark which I have to make is this:—in all cases the Arts are prior to the related Sciences. Art is the parent, not the progeny, of Science; the realization of principles in practice forms part of the prelude, as well as of the sequel, of theoretical discovery. And thus the inventions of the middle ages, which have been above enumerated, though at the present day they may be portions of our sciences, are no evidence that the sciences then existed; but only that those powers of practical observation and practical skill were at work, which prepare the way for theoretical views and scientific discoveries.

It may be urged, that the great works of art do virtually take for granted principles of science; and that, therefore, it is unreasonable to deny science to great artists. It may be said, that the grand structures of Cologne, or Amiens, or Canterbury, could not have been erected without a profound knowledge of mechanical principles.

To this we reply, that *such* knowledge is manifestly not of the nature of that which we call *science*. If the beautiful and skilful structures of the middle ages prove that mechanics then existed as a science, mechanics must have existed as a science also among the builders of the Cyclopean walls of Greece and Italy, or of our own Stonehenge; for the masses which are there piled on each other, could not be raised without considerable mechanical skill. But we may go much further. The actions of every man who raises and balances weights, or walks along a pole, take for granted the laws of equilibrium; and even animals constantly avail themselves of such principles. Are these, then, acquainted with mechanics as a science? Again, if actions which are performed by taking advantage of mechanical properties prove a knowledge of the science of mechanics, they must also be allowed to prove a knowledge of the science of geometry, when they proceed on geometrical properties. But the most familiar actions of men and animals proceed upon geometrical truths. The Epicureans held, as Proclus informs us, that even asses knew that two sides of a triangle are greater than the third. And animals may truly be said to have a practical knowledge of this truth; but they have not, therefore, a science of geometry. And in like manner among men, if we consider the matter strictly, a practical assumption of a principle does not imply a speculative knowledge of it.

We may, in another way also, show how inadmissible are the works of the Master Artists of the middle ages into the series of events which mark the advance of Science. The following maxim is applicable to a history, such as we are here endeavoring to write. We are employed in tracing the progress of such general principles as constitute each of the sciences which we are reviewing; and no facts or subordinate truths belong to our scheme, except so far as they lead to or are included in these higher principles; nor are they important to us, any further than as they prove such principles. Now with regard to processes of art like those which we have referred to, namely, the inventions of the middle ages, let us ask, *what* principle each of them illustrates? What chemical doctrine rests for its support on the phenomena of gunpowder, or glass, or steel? What new harmonical truth was illustrated in the Gregorian chant? What mechanical principle unknown to Archimedes was displayed in the printing-press? The practical value and use, the ingenuity and skill of these inventions is not questioned; but what is their place in the history of speculative knowledge? Even in those cases in which they enter into such a history, how minute a figure do they make! how great is the contrast between their practical and theoretical importance! They may in their operation have changed the face of the world; but in the history of the principles of the sciences to which they belong, they may be omitted without being missed.

As to that part of the objection which was stated by asking, why, if the arts of our age prove its scientific eminence, the arts of the middle ages should not be received as proof of theirs; we must reply to it, by giving up some of the pretensions which are often put forwards on behalf of the science of our times. The perfection of the mechanical and other arts among us proves the advanced condition of our sciences, only in so far as these arts have been perfected by the application of some great scientific truth, with a clear insight into its nature. The greatest improvement of the steam-engine was due to the steady apprehension of an atmological doctrine by Watt; but what distinct theoretical principle is illustrated by the beautiful manufactures of porcelain, or steel, or glass? A chemical view of these compounds, which would explain the conditions of success and failure in their manufacture, would be of great value in art; and it would also be a novelty in chemical theory; so little is the present condition of those processes a triumph of science, shedding intellectual glory on our age. And the same might be said of many, or of most, of the processes of the arts as now practised.

2. *Arabian Science.*—Having, I trust, established the view I have stated, respecting the relation of Art and Science, we shall be able very rapidly to dispose of a number of subjects which otherwise might seem to require a detailed notice. Though this distinction has been recognized by others, it has hardly been rigorously adhered to, in consequence of the indistinct notion of *science* which has commonly prevailed. Thus Gibbon, in speaking of the knowledge of the period now under our notice, says,[1] "Much useful experience had been acquired in

[1] *Decline and Fall*, vol. x. p. 43.

the practice of arts and manufactures; but the *science* of chemistry owes its origin and improvement to the industry of the Saracens. They," he adds, "first invented and named the alembic for the purposes of distillation, analyzed the substances of the three kingdoms of nature, tried the distinction and affinities of alkalies and acids, and converted the poisonous minerals into soft and salutary medicines." The formation and realization of the notions of *analysis* and of *affinity*, were important steps in chemical science, which, as I shall hereafter endeavor to show, it remained for the chemists of Europe to make at a much later period. If the Arabians had done this, they might with justice have been called the authors of the science of chemistry; but no doctrines can be adduced from their works which give them any title to this eminent distinction. Their claims are dissipated at once by the application of the maxim above stated. *What* analysis of theirs tended to establish any received principle of chemistry? *What* true doctrine concerning the differences and affinities of acids and alkalies did they teach? We need not wonder if Gibbon, whose views of the boundaries of scientific chemistry were probably very wide and indistinct, could include the arts of the Arabians within its domain; but they cannot pass the frontier of science if philosophically defined, and steadily guarded.

The judgment which we are thus led to form respecting the chemical knowledge of the middle ages, and of the Arabians in particular, may serve to measure the condition of science in other departments; for chemistry has justly been considered one of their strongest points. In botony, anatomy, zoology, optics, acoustics, we have still the same observations to make, that the steps in science which, in the order of progress, next followed what the Greeks had done, were left for the Europeans of the sixteenth and seventeenth centuries. The merits and advances of the Arabian philosophers in astronomy and pure mathematics, we have already described.

3. *Experimental Philosophy of the Arabians.*—The estimate to which we have thus been led, of the scientific merits of the learned men of the middle ages, is much less exalted than that which has been formed by many writers; and, among the rest, by some of our own time. But I am persuaded that any attempt to answer the questions just asked, will expose the untenable nature of the higher claims which have been advanced in favor of the Arabians. We can deliver no just decision, except we will consent to use the terms of science in a strict and precise sense: and if we do this, we shall find little, either in the particu-

lar discoveries or general processes of the Arabians, which is important in the history of the Inductive Sciences.[2]

The credit due to the Arabians for improvements in the general methods of philosophizing, is a more difficult question; and cannot be discussed at length by us, till we examine the history of such methods in the abstract, which, in the present work, it is not our intention to do. But we may observe, that we cannot agree with those who rank their merits high in this respect. We have already seen, that their minds were completely devoured by the worst habits of the stationary period,—Mysticism and Commentation. They followed their Greek leaders, for the most part, with abject servility, and with only that kind of acuteness and independent speculation which the Commentator's vocation implies. And in their choice of the standard subjects of their studies, they fixed upon those works, the Physical Books of Aristotle, which have never promoted the progress of science, except in so far as they incited men to refute them; an effect which they never produced on the Arabians. That the Arabian astronomers made some advances beyond the Greeks, we have already stated: the two great instances are, the discovery of the Motion of the Sun's Apogee by Albategnius, and the discovery (recently brought to light) of the existence of the Moon's Second Inequality, by Aboul Wefa. But we cannot but observe in how different a manner they treated these discoveries, from that with which Hipparchus or Ptolemy would have done. The Variation of the Moon, in particular, instead of being incorporated into the system by means of an Epicycle, as Ptolemy had done with the Evection, was allowed, almost immediately, so far as we can judge, to fall into neglect and oblivion: so little were the learned Arabians prepared to take their lessons from observation as well as from books. That in many subjects they made experiments, may easily be allowed: there never was a period of the earth's history, and least of all a period of commerce

[2] If I might take the liberty of criticising an author who has given a very interesting view of the period in question (*Mahometanism Unveiled*, by the Rev. Charles Forster, 1829), I would remark, that in his work this caution is perhaps too little observed. Thus, he says, in speaking of Alhazen (vol. ii. p. 270), "the theory of the telescope may be found in the work of this astronomer;" and of another, "the uses of magnifying glasses and telescopes, and the principle of their construction, are explained in the Great Work of (Roger) Bacon, with a truth and clearness which have commanded universal admiration." Such phrases would be much too strong, even if used respecting the optical doctrines of Kepler, which were yet incomparably more true and clear than those of Bacon. To employ such language, in such cases, is to deprive such terms as *theory* and *principle* of all meaning.

and manufactures, luxury and art, medicine and engineering, in which there were not going on innumerable processes, which may be termed Experiments; and, in addition to these, the Arabians adopted the pursuit of alchemy, and the love of exotic plants and animals. But so far from their being, as has been maintained,[3] a people whose "experimental intellect" fitted them to form sciences which the "abstract intellect" of the Greeks failed in producing, it rather appears, that several of the sciences which the Greeks had founded, were never even comprehended by the Arabians. I do not know any evidence that these pupils ever attained to understand the real principles of Mechanics, Hydrostatics, and Harmonics, which their masters had established. At any rate, when these sciences again became progressive, Europe had to start where Europe had stopped. There is no Arabian name which any one has thought of interposing between Archimedes the ancient, and Stevinus and Galileo the moderns.

4. *Roger Bacon.*—There is one writer of the middle ages, on whom much stress has been laid, and who was certainly a most remarkable person. Roger Bacon's works are not only so far beyond his age in the knowledge which they contain, but so different from the temper of the times, in his assertion of the supremacy of experiment, and in his contemplation of the future progress of knowledge, that it is difficult to conceive how such a character could then exist. That he received much of his knowledge from Arabic writers, there can be no doubt; for they were in his time the repositories of all traditionary knowledge. But that he derived from them his disposition to shake off the authority of Aristotle, to maintain the importance of experiment, and to look upon knowledge as in its infancy, I cannot believe, because I have not myself hit upon, nor seen quoted by others, any passages in which Arabian writers express such a disposition. On the other hand, we do find in European writers, in the authors of Greece and Rome, the solid sense, the bold and hopeful spirit, which suggest such tendencies. We have already seen that Aristotle asserts, as distinctly as words can express, that all knowledge must depend on observation, and that science must be collected from facts by induction. We have seen, too, that the Roman writers, and Seneca in particular, speak with an enthusiastic confidence of the progress which science must make in the course of ages. When Roger Bacon holds similar language in the thirteenth century, the resemblance is probably rather a sympathy of character, than a matter of direct derivation; but I know of nothing

[3] *Mahometanism Unveiled*, ii. 271.

which proves even so much as this sympathy in the case of Arabian philosophers.

A good deal has been said of late of the coincidences between his views, and those of his great namesake in later times, Francis Bacon.[4] The resemblances consist mainly in such points as I have just noticed; and we cannot but acknowledge, that many of the expressions of the Franciscan Friar remind us of the large thoughts and lofty phrases of the Philosophical Chancellor. How far the one can be considered as having anticipated the method of the other, we shall examine more advantageously, when we come to consider what the character and effect of Francis Bacon's works really are.[5]

5. *Architecture of the Middle Ages.*—But though we are thus compelled to disallow several of the claims which have been put forwards in support of the scientific character of the middle ages, there are two points in which we may, I conceive, really trace the progress of scientific ideas among them; and which, therefore, may be considered as the prelude to the period of discovery. I mean their practical architecture, and their architectural treatises.

In a previous chapter of this book, we have endeavored to explain how the indistinctness of ideas, which attended the decline of the Roman empire, appears in the forms of their architecture;—in the disregard, which the decorative construction exhibits, of the necessary mechanical conditions of support. The original scheme of Greek ornamental architecture had been horizontal masses resting on vertical columns: when the arch was introduced by the Romans, it was concealed, or kept in a state of subordination: and the lateral support which it required was supplied latently, marked by some artifice. But the struggle between the *mechanical* and the *decorative construction*[6] ended in the complete disorganization of the classical style. The

---

[4] Hallam's *Middle Ages*, iii. 549. Forster's *Mahom. U.* ii. 313.

[5] In the *Philosophy of the Inductive Sciences*, I have given an account at considerable length of Roger Bacon's mode of treating Arts and Sciences; and have also compared more fully his philosophy with that of Francis Bacon; and I have given a view of the bearing of this latter upon the progress of Science in modern times. See *Phil. Ind. Sc.* book xii. chaps. 7 and 11. See also the Appendix to this volume.

[6] See Mr. Willis's admirable *Remarks on the Architecture of the Middle Ages*, chap. ii. Since the publication of my first edition, Mr. Willis has shown that much of the "mason-craft" of the middle ages consisted in the geometrical methods by which the artists wrought out of the blocks the complex forms of their decorative system.

To the general indistinctness of speculative notions on mechanical subjects prevalent in the middle ages, there may have been some exceptions, and especially so long as there were readers of Archimedes. Boëthius had translated the

inconsistencies and extravagances, of which we have noticed the occurrence, were results and indications of the fall of good architecture. The elements of the ancient system had lost all principle of connection and regard to rule. Building became not only a mere art, but an art exercised by masters without skill, and without feeling for real beauty.

When, after this deep decline, architecture rose again, as it did in the twelfth and succeeding centuries, in the exquisitely beautiful and skilful forms of the Gothic style, what was the nature of the change which had taken place, so far as it bears upon the progress of science? It was this:—the idea of true mechanical relations in an edifice had been revived in men's minds, as far as was requisite for the purposes of art and beauty: and this, though a very different thing from the possession of the idea as an element of speculative science, was the proper preparation for that acquisition. The notion of support and stability again became conspicuous in the decorative construction, and universal in the forms of building. The eye which, looking for beauty in definite and significant relations of parts, is never satisfied except the weights appear to be duly supported,[7] was again gratified. Architecture threw off its barbarous characters: a new decorative construction was matured, not thwarting and controlling, but assisting and harmonizing with the mechanical construction. All the ornamental parts were made to enter into the apparent construction. Every member, almost every moulding, became a sustainer of weight; and by the multiplicity of props assisting each other, and the consequent subdivision of weight, the eye was satisfied of the stability of the structure, notwithstanding the curiously slender forms of the separate parts. The arch and the vault, no longer trammelled by an incompatible system of decoration, but favored by more tractable forms, were only limited by the skill of the builders. Every thing showed that, practically at least, men possessed and applied, with steadiness and pleasure, the idea of mechanical pressure and support.

The possession of this idea, as a principle of art, led, in the course of time, to its speculative development as the foundation of a science;

---

mechanical works of Archimedes into Latin, as we learn from the enumeration of his work by his friend Cassiodorus (*Variar.* lib. i. cap. 45), "*Mechanicum* etiam Archimedem latialem siculis reddidisti." But *Mechanicus* was used in those times rather for one skilled in the art of constructing wonderful machines than in the speculative theory of them. The letter from which the quotation is taken is sent by King Theodoric to Boëthius, to urge him to send the king a water-clock.

7 Willis, pp. 15–21. I have throughout this description of the formation of the Gothic style availed myself of Mr. Willis's well-chosen expressions.

and thus Architecture prepared the way for Mechanics. But this advance required several centuries. The interval between the admirable cathedrals of Salisbury, Amiens, Cologne, and the mechanical treatises of Stevinus, is not less than three hundred years. During this time, men were advancing towards science; but in the mean time, and perhaps from the very beginning of the time, art had begun to decline. The buildings of the fifteenth century, erected when the principles of mechanical support were just on the verge of being enunciated in general terms, exhibit those principles with a far less impressive simplicity and elegance than those of the thirteenth. We may hereafter inquire whether we find any other examples to countenance the belief, that the formation of Science is commonly accompanied by the decline of Art.

The leading principle of the style of the Gothic edifices was, not merely that the weights were supported, but that they were seen to be so; and that not only the mechanical relations of the larger masses, but of the smaller members also, were displayed. Hence we cannot admit, as an origin or anticipation of the Gothic, a style in which this principle is not manifested. I do not see, in any of the representations of the early Arabic buildings, that distribution of weights to supports, and that mechanical consistency of parts, which would elevate them above the character of barbarous architecture. Their masses are broken into innumerable members, without subordination or meaning, in a manner suggested apparently by caprice and the love of the marvellous. "In the construction of their mosques, it was a favorite artifice of the Arabs to sustain immense and ponderous masses of stone by the support of pillars so slender, that the incumbent weight seemed, as it were, suspended in the air by an invisible hand."[8] This pleasure in the contemplation of apparent impossibilities is a very general disposition among mankind; but it appears to belong to the infancy, rather than the maturity of intellect. On the other hand, the pleasure in the contemplation of what is clear, the craving for a thorough insight into the reasons of things, which marks the European mind, is the temper which leads to science.

6. *Treatises on Architecture.*—No one who has attended to the architecture which prevailed in England, France, and Germany, from the twelfth to the fifteenth century, so far as to comprehend its beauty, harmony, consistency, and uniformity, even in the minutest parts and most obscure relations, can look upon it otherwise than as a remark-

---

[8] *Mahometanism Unveiled*, ii. 255.

ably connected and definite artificial system. Nor can we doubt that it was exercised by a class of artists who formed themselves by laborious study and practice, and by communication with each other. There must have been bodies of masters and of scholars, discipline, traditions, precepts of art. How these associated artists diffused themselves over Europe, and whether history enables us to trace them in a distinct form, I shall not here discuss. But the existence of a course of instruction, and of a body of rules of practice, is proved beyond dispute by the great series of European cathedrals and churches, so nearly identical in their general arrangements, and in their particular details. The question then occurs, have these rules and this system of instruction anywhere been committed to writing? Can we, by such evidence, trace the progress of the scientific idea, of which we see the working in these buildings?

We are not to be surprised, if, during the most flourishing and vigorous period of the art of the middle ages, we find none of its precepts in books. Art has, in all ages and countries, been taught and transmitted by practice and verbal tradition, not by writing. It is only in our own times, that the thought occurs as familiar, of committing to books all that we wish to preserve and convey. And, even in our own times, most of the Arts are learned far more by practice, and by intercourse with practitioners, than by reading. Such is the case, not only with Manufactures and Handicrafts, but with the Fine Arts, with Engineering, and even yet, with that art, Building, of which we are now speaking.

We are not, therefore, to wonder, if we have no treatises on Architecture belonging to the great period of the Gothic masters;—or if it appears to have required some other incitement and some other help, besides their own possession of their practical skill, to lead them to shape into a literary form the precepts of the art which they knew so well how to exercise:—or if, when they did write on such subjects, they seem, instead of delivering their own sound practical principles, to satisfy themselves with pursuing some of the frivolous notions and speculations which were then current in the world of letters.

Such appears to be the case. The earliest treatises on Architecture come before us under the form which the commentatorial spirit of the middle ages inspired. They are Translations of Vitruvius, with Annotations. In some of these, particularly that of Cesare Cesariano, published at Como, in 1521, we see, in a very curious manner, how the habit of assuming that, in every department of literature, the ancients must needs be their masters, led these writers to subordinate the members of their own architecture to the precepts of the Roman author. We have Gothic shafts, mouldings, and arrangements, given as parallelisms to others, which profess to represent the Roman style, but which are, in fact, examples of that mixed manner which is called the style of the *Cinque cento* by the Italians, of the *Renaissance* by the French, and which is commonly included in our *Elizabethan*. But in the early architectural works, besides the superstitions and mistaken erudition which thus choked the growth of real architectural doctrines, another of the peculiar elements of the middle ages comes into view; —its mysticism. The dimensions and positions of the various parts of edifices and of their members, are determined by drawing triangles, squares, circles, and other figures, in such a manner as to bound them; and to these geometrical figures were assigned many abstruse significations. The plan and the front of the Cathedral at Milan are thus represented in Cesariano's work, bounded and subdivided by various equilateral triangles; and it is easy to see, in the earnestness with which he points out these relations, the evidence of a fanciful and mystical turn of thought.[9]

We thus find erudition and mysticism take the place of much of that development of the architectural principles of the middle ages which would be so interesting to us. Still, however, these works are by no means without their value. Indeed many of the arts appear to flourish not at all the worse, for being treated in a manner somewhat mystical; and it may easily be, that the relations of geometrical figures, for which fantastical reasons are given, may really involve principles of beauty or stability. But independently of this, we find, in the best works of the architects of all ages (including engineers), evidence that the true idea of mechanical pressure exists among them more distinctly than among men in general, although it may not be developed in a scientific form. This is true up to our own time, and the arts which such persons cultivate could not be successfully exer-

[9] The plan which he has given, fol. 14, he has entitled "Ichnographia Fundamenti sacræ Ædis baricephalæ, Germanico more, à Trigono ac Pariquadrato perstructa, uti etiam ea quæ nunc Milani videtur."

The work of Cesariano was translated into German by Gualter Rivius, and published at Nuremberg, in 1548, under the title of *Vitruvius Teutsch*, with copies of the Italian diagrams. A few years ago, in an article in the *Wiener Jahrbücher* (Oct.—Dec., 1821), the reviewer maintained, on the authority of the diagrams in Rivius's book, that Gothic architecture had its origin in Germany and not in England.

cised if it were not so. Hence the writings of architects and engineers during the middle ages do really form a prelude to the works on scientific mechanics. Vitruvius, in his *Architecture*, and Julius Frontinus, who, under Vespasian, wrote *On Aqueducts*, of which he was superintendent, have transmitted to us the principal part of what we know respecting the practical mechanics and hydraulics of the Romans. In modern times the series is resumed. The early writers on architecture are also writers on engineering, and often on hydrostatics: for example, Leonardo da Vinci wrote on the equilibrium of water. And thus we are led up to Stevinus of Bruges, who was engineer to Prince Maurice of Nassau, and inspector of the dykes in Holland; and in whose work, on the processes of his art, is contained the first clear modern statement of the scientific principles of hydrostatics.

Having thus explained both the obstacles and the prospects which the middle ages offered to the progress of science, I now proceed to the history of the progress, when that progress was once again resumed.

BOOK V.

HISTORY

OF

FORMAL ASTRONOMY

AFTER THE STATIONARY PERIOD.

. . . Cyclopum educta caminis
Mœnia conspicio, atque adverso fornice portas.

. . . . . .

His demum exactis, perfecto munere Divæ,
Devenere locos lætos et amœna vireta
Fortunatorum nemorum sedesque beatas.
Largior hic campos æther et lumine vestit
Purpureo : solemque suum, sua sidera norunt.

VIRGIL, *Æn.* vi. 630.

They leave at length the nether gloom, and stand
Before the portals of a better land :
To happier plains they come, and fairer groves,
The seats of those whom heaven, benignant, loves ;
A brighter day, a bluer ether, spreads
Its lucid depths above their favored heads ;
And, purged from mists that veil our earthly skies,
Shine suns and stars unseen by mortal eyes.

INTRODUCTION.

Of Formal and Physical Astronomy.

WE have thus rapidly traced the causes of the almost complete blank which the history of physical science offers, from the decline of the Roman empire, for a thousand years. Along with the breaking up of the ancient forms of society, were broken up the ancient energy of thinking, the clearness of idea, and steadiness of intellectual action. This mental declension produced a servile admiration for the genius of the better periods, and thus, the spirit of Commentation: Christianity established the claim of truth to govern the world; and this principle, misinterpreted and combined with the ignorance and servility of the times, gave rise to the Dogmatic System: and the love of speculation, finding no secure and permitted path on solid ground, went off into the regions of Mysticism.

The causes which produced the inertness and blindness of the stationary period of human knowledge, began at last to yield to the influence of the principles which tended to progression. The indistinctness of thought, which was the original feature in the decline of sound knowledge, was in a measure remedied by the steady cultivation of Pure Mathematics and Astronomy, and by the progress of inventions in the Arts, which call out and fix the distinctness of our conceptions of the relations of natural phenomena. As men's minds became clear, they became less servile: the perception of the nature of truth drew men away from controversies about mere opinion; when they saw distinctly the relations of *things*, they ceased to give their whole attention to what had been *said* concerning them; and thus, as science rose into view, the spirit of commentation lost its way. And when men came to feel what it was to think for themselves on subjects of science, they soon rebelled against the right of others to impose opinions upon them. When they threw off their blind admiration for the ancients, they were disposed to cast away also their passive obedience to the ancient system of doctrines. When they were no longer inspired by the spirit of commentation, they were no longer submissive to the dogmatism of the schools. When they began to feel that they could discover truths, they felt also a persuasion of a right and a growing will so to do.

Thus the revived clearness of ideas, which made its appearance at the revival of letters, brought on a struggle with the authority, intellectual and civil, of the established schools of philosophy. This clearness of idea showed itself, in the first instance, in Astronomy, and was embodied in the system of Copernicus; but the contest did not come to a crisis till a century later, in the time of Galileo and other disciples of the new doctrine. It is our present business to trace the principles of this series of events in the history of philosophy.

I do not profess to write a history of Astronomy, any further than is necessary in order to exhibit the principles on which the progression of science proceeds; and, therefore, I neglect subordinate persons and occurrences, in order to bring into view the leading features of great changes. Now in the introduction of the Copernican system into general acceptation, two leading views operated upon men's minds; the consideration of the system as exhibiting the apparent motions of the universe, and the consideration of this system with reference to its causes;—the *formal* and the *physical* aspect of the Theory;—the relations of Space and Time, and the relations of Force and Matter. These two divisions of the subject were at first not clearly separated; the second was long mixed, in a manner very dim and obscure, with the first, without appearing as a distinct subject of attention; but at last it was extricated and treated in a manner suitable to its nature. The views of Copernicus rested mainly on the formal condition of the universe, the relations of space and time; but Kepler, Galileo, and others, were led, by controversies and other causes, to give a gradually increasing attention to the physical relations of the heavenly bodies; an impulse was given to the study of Mechanics (the Doctrine of Motion), which became very soon an important and extensive science; and in no long period, the discoveries of Kepler, suggested by a vague but intense belief in the physical connection of the parts of the universe, led to the decisive and sublime generalizations of Newton.

The distinction of *formal* and *physical* Astronomy thus becomes necessary, in order to treat clearly of the discussions which the propounding of the Copernican theory occasioned. But it may be observed that, besides this great change, Astronomy made very great advances in the same path which we have already been tracing, namely, the determination of the quantities and laws of the celestial motions, in so far as they were exhibited by the ancient theories, or might be represented by obvious modifications of those theories. I speak of new Inequalities, new Phenomena, such as Copernicus, Galileo, and Tycho Brahe discovered. As, however, these were very soon referred to the Copernican rather than the Ptolemaic hypothesis, they may be considered as developments rather of the new than of the old Theory; and I shall, therefore, treat of them, agreeably to the plan of the former part, as the sequel of the Copernican Induction.

CHAPTER I.

Prelude to the Inductive Epoch of Copernicus.

THE Doctrine of Copernicus, that the Sun is the true centre of the celestial motions, depends primarily upon the consideration that such a supposition explains very simply and completely all the obvious appearances of the heavens. In order to see that it does this, nothing more is requisite than a distinct conception of the nature of Relative Motion, and a knowledge of the principal Astronomical Phenomena. There was, therefore, no reason why such a doctrine might not be *discovered*, that is, suggested as a theory plausible at first sight, long before the time of Copernicus; or rather, it was impossible that this guess, among others, should not be propounded as a solution of the appearances of the heavens. We are not, therefore, to be surprised if we find, in the earliest times of Astronomy, and at various succeeding periods, such a system spoken of by astronomers, and maintained by some as true, though rejected by the majority, and by the principal writers.

When we look back at such a difference of opinion, having in our minds, as we unavoidably have, the clear and irresistible considerations by which the Copernican Doctrine is established *for us*, it is difficult for us not to attribute superior sagacity and candor to those who held that side of the question, and to imagine those who clung to the Ptolemaic Hypothesis to have been blind and prejudiced; incapable of seeing the beauty of simplicity and symmetry, or indisposed to resign established errors, and to accept novel and comprehensive truths. Yet in judging thus, we are probably ourselves influenced by prejudices arising from the knowledge and received opinions of our own times. For is it, in reality, clear that, before the time of Copernicus, the *Helio-*

centric Theory (that which places the centre of the celestial motions in the Sun) had a claim to assent so decidedly superior to the Geocentric Theory, which places the Earth in the centre? What is the basis of the heliocentric theory?—That the *relative* motions are *the same*, on that and on the other supposition. So far, therefore, the two hypotheses are exactly on the same footing. But, it is urged, on the heliocentric side we have the advantage of simplicity:—true; but we have, on the other side, the testimony of our senses; that is, the geocentric doctrine (which asserts that the Earth rests and the heavenly bodies move) is the obvious and spontaneous interpretation of the appearances. Both these arguments, *simplicity* on the one side, and *obviousness* on the other, are vague, and we may venture to say, both indecisive. We cannot establish any strong preponderance of probability in favor of the former doctrine, without going much further into the arguments of the question.

Nor, when we speak of the superior *simplicity* of the Copernican theory, must we forget, that though this theory has undoubtedly, in this respect, a great advantage over the Ptolemaic, yet that the Copernican system itself is very complex, when it undertakes to account, as the Ptolemaic did, for the *Inequalities* of the Motions of the sun, moon, and planets; and, that in the hands of Copernicus, it retained a large share of the eccentrics and epicycles of its predecessor, and, in some parts, with increased machinery. The heliocentric theory, without these appendages, would not approach the Ptolemaic, in the accurate explanation of facts; and as those who had placed the sun in the centre had never, till the time of Copernicus, shown how the inequalities were to be explained on that supposition, we may assert that after the promulgation of the theory of eccentrics and epicycles on the geocentric hypothesis, there was no *published* heliocentric theory which could bear a comparison with that hypothesis.

It is true, that all the contrivances of epicycles, and the like, by which the geocentric hypothesis was made to represent the phenomena, were susceptible of an easy adaptation to a heliocentric method, *when a good mathematician had once proposed to himself the problem:* and this was precisely what Copernicus undertook and executed. But, till the appearance of his work, the heliocentric system had never come before the world except as a hasty and imperfect hypothesis; which bore a favorable comparison with the phenomena, so long as their general features only were known; but which had been completely thrown into the shade by the labor and intelligence bestowed upon the Hipparchian or Ptolemaic theories by a long series of great astronomers of all civilized countries.

But, though the astronomers who, before Copernicus, held the heliocentric opinion, cannot, on any good grounds, be considered as much more enlightened than their opponents, it is curious to trace the early and repeated manifestations of this view of the universe. The distinct assertion of the heliocentric theory among the Greeks is an evidence of the clearness of their thoughts, and the vigor of their minds; and it is a proof of the feebleness and servility of intellect in the stationary period, that, till the period of Copernicus, no one was found to try the fortune of this hypothesis, modified according to the improved astronomical knowledge of the time.

The most ancient of the Greek philosophers to whom the ancients ascribe the heliocentric doctrine, is Pythagoras; but Diogenes Laertius makes Philolaus, one of the followers of Pythagoras, the first author of this doctrine. We learn from Archimedes, that it was held by his contemporary, Aristarchus. "Aristarchus of Samos," says he,[1] "makes this supposition,—that the fixed stars and the sun remain at rest, and that the earth revolves round the sun in a circle." Plutarch[2] asserts that this, which was only a hypothesis in the hands of Aristarchus, was *proved* by Seleucus; but we may venture to say that, at that time, no such proof was possible. Aristotle had recognized the existence of this doctrine by arguing against it. "All things," says he,[3] "tend to the centre of the earth and rest there, and therefore the whole mass of the earth cannot rest except there." Ptolemy had in like manner argued against the diurnal motion of the earth: such a revolution would, he urged, disperse into surrounding space all the loose parts of the earth. Yet he allowed that such a supposition would facilitate the explanation of some phenomena. Cicero appears to make Mercury and Venus revolve about the sun, as does Martianus Capella at a later period; and Seneca says,[4] it is a worthy subject of contemplation, whether the earth be at rest or in motion: but at this period, as we may see from Seneca himself, that habit of intellect which was requisite for the solution of such a question, had been succeeded by indistinct views, and rhetorical forms of speech. If there were any good mathematicians and good observers at this period, they were employed in cultivating and verifying the Hipparchian theory.

Next to the Greeks, the Indians appear to have possessed that

---

1 Archim. *Arenarius*.

2 *Quest. Plat.* Delamb. *A. A.* vi.

3 Quoted by Copernic. i. 7.

4 *Quest. Nat.* vii. 2.

original vigor and clearness of thought, from which true science springs. It is remarkable that the Indians, also, had their heliocentric theorists. Aryabatta[5] (A. D. 1322), and other astronomers of that country, are said to have advocated the doctrine of the earth's revolution on its axis; which opinion, however, was rejected by subsequent philosophers among the Hindoos.

Some writers have thought that the heliocentric doctrine was *derived* by Pythagoras and other European philosophers, from some of the oriental nations. This opinion, however, will appear to have little weight, if we consider that the heliocentric hypothesis, in the only shape in which the ancients knew it, was too obvious to require much teaching; that it did not and could not, so far as we know, receive any additional strength from any thing which the oriental nations could teach; and that each astronomer was induced to adopt or reject it, not by any information which a master could give him, but by his love of geometrical simplicity on the one hand, or the prejudices of sense on the other. Real science, depending on a clear view of the relation of phenomena to general theoretical ideas, cannot be communicated in the way of secret and exclusive traditions, like the mysteries of certain arts and crafts. If the philosopher do not *see* that the theory is true, he is little the better for having heard or read the words which assert its truth.

It is impossible, therefore, for us to assent to those views which would discover in the heliocentric doctrines of the ancients, traces of a more profound astronomy than any which they have transmitted to us. Those doctrines were merely the plausible conjectures of men with sound geometrical notions; but they were never extended so as to embrace the details of the existing astronomical knowledge; and perhaps we may say, that the analysis of the phenomena into the arrangements of the Ptolemaic system, was so much more obvious than any other, that it must necessarily come first, in order to form an introduction to the Copernican.

The true foundation of the heliocentric theory for the ancients was, as we have intimated, its perfect geometrical consistency with the general features of the phenomena, and its simplicity. But it was unlikely that the human mind would be content to consider the subject under this strict and limited aspect alone. In its eagerness for wide speculative views, it naturally looked out for other and vaguer principles of connection and relation. Thus, as it had been urged in

[5] Lib. U. K. *Hist. Ast.* p. 11.

favor of the geocentric doctrine, that the heaviest body must be in the centre, it was maintained, as a leading recommendation of the opposite opinion, that it placed the Fire, the noblest element, in the Centre of the Universe. The authority of mythological ideas was called in on both sides to support these views. Numa, as Plutarch[6] informs us, built a circular temple over the ever-burning Fire of Vesta; typifying, not the earth, but the Universe, which, according to the Pythagoreans, has the Fire seated at its Centre. The same writer, in another of his works, makes one of his interlocutors say, "Only, my friend, do not bring me before a court of law on a charge of impiety; as Cleanthes said, that Aristarchus the Samian ought to be tried for impiety, because he removed the Hearth of the Universe." This, however, seems to have been intended as a pleasantry.

The prevalent physical views, and the opinions concerning the causes of the motions of the parts of the universe, were scarcely more definite than the ancient opinions concerning the relations of the four elements, till Galileo had founded the true Doctrine of Motion. Though, therefore, arguments on this part of the subject were the most important part of the controversy after Copernicus, the force of such arguments was at his time almost balanced. Even if more had been known on such subjects, the arguments would not have been conclusive: for instance, the vast mass of the heavens, which is commonly urged as a reason why the heavens do not move round the earth, would not make such a motion impossible; and, on the other hand, the motions of bodies at the earth's surface, which were alleged as inconsistent with its motion, did not really disprove such an opinion. But according to the state of the science of motion before Copernicus, all reasonings from such principles were utterly vague and obscure.

We must not omit to mention a modern who preceded Copernicus, in the assertion at least of the heliocentric doctrine. This was Nicholas of Cusa (a village near Treves), a cardinal and bishop, who, in the first half of the fifteenth century, was very eminent as a divine and mathematician; and who in a work, *De Doctâ Ignorantiâ*, propounded the doctrine of the motion of the earth; more, however, as a paradox than as a reality. We cannot consider this as any distinct anticipation of a profound and consistent view of the truth.

We shall now examine further the promulgation of the Heliocentric System by Copernicus, and its consequences.

---

[6] *De Facie in Orbe Lunæ*, 6.

CHAPTER II.

Induction of Copernicus.—The Heliocentric Theory asserted on formal grounds.

IT will be recollected that the *formal* are opposed to the *physical* grounds of a theory; the former term indicating that it gives a satisfactory account of the relations of the phenomena in Space and Time, that is, of the Motions themselves; while the latter expression implies further that we include in our explanation the Causes of the motions, the laws of Force and Matter. The strongest of the considerations by which Copernicus was led to invent and adopt his system of the universe were of the former kind. He was dissatisfied, he says, in his Preface addressed to the Pope, with the want of symmetry in the Eccentric Theory, as it prevailed in his days; and weary of the uncertainty of the mathematical traditions. He then sought through all the works of philosophers, whether any had held opinions concerning the motions of the world, different from those received in the established mathematical schools. He found, in ancient authors, accounts of Philolaus and others, who had asserted the motion of the earth. "Then," he adds, "I, too, began to meditate concerning the motion of the earth; and though it appeared an absurd opinion, yet since I knew that, in previous times, others had been allowed the privilege of feigning what circles they chose, in order to explain the phenomena, I conceived that I also might take the liberty of trying whether, on the supposition of the earth's motion, it was possible to find better explanations than the ancient ones, of the revolutions of the celestial orbs.

"Having then assumed the motions of the earth, which are hereafter explained, by laborious and long observation I at length found, that if the motions of the other planets be compared with the revolution of the earth, not only their phenomena follow from the suppositions, but also that the several orbs, and the whole system, are so connected in order and magnitude, that no one part can be transposed without disturbing the rest, and introducing confusion into the whole universe."

Thus the satisfactory explanation of the apparent motions of the planets, and the simplicity and symmetry of the system, were the grounds on which Copernicus adopted his theory; as the craving for these qualities was the feeling which led him to seek for a new theory. It is manifest that in this, as in other cases of discovery, a clear and steady possession of abstract Ideas, and an aptitude in comprehending real Facts under these general conceptions, must have been leading characters in the discoverer's mind. He must have had a good geometrical head, and great astronomical knowledge. He must have seen, with peculiar distinctness, the consequences which flowed from his suppositions as to the relations of space and time,—the apparent motions which resulted from the assumed real ones; and he must also have known well all the irregularities of the apparent motions for which he had to account. We find indications of these qualities in his expressions. A steady and calm contemplation of the theory is what he asks for, as the main requisite to its reception. If you suppose the earth to revolve and the heaven to be at rest, you will find, he says, "*si serio animadvertas*," if you think steadily, that the apparent diurnal motion will follow. And after alleging his reasons for his system, he says,[1] "We are, therefore, not ashamed to confess, that the whole of the space within the orbit of the moon, along with the centre of the earth, moves round the sun in a year among the other planets; the magnitude of the world being so great, that the distance of the earth from the sun has no apparent magnitude when compared with the sphere of the fixed stars." "All which things, though they be difficult and almost inconceivable, and against the opinion of the majority, yet, in the sequel, by God's favor, we will make clearer than the sun, at least to those who are not ignorant of mathematics."

It will easily be understood, that since the ancient geocentric hypothesis ascribed to the planets those motions which were apparent only, and which really arose from the motion of the earth round the sun in the new hypothesis, the latter scheme must much simplify the planetary theory. Kepler[2] enumerates eleven motions of the Ptolemaic system, which are at once exterminated and rendered unnecessary by the new system. Still, as the real motions, both of the earth and the planets, are unequable, it was requisite to have some mode of representing their inequalities; and, accordingly, the ancient theory of eccentrics and epicycles was retained, so far as was requisite for this purpose. The planets revolved round the sun by means of a Deferent, and a

[1] Nicolai Copernici Torinensis *de Revolutionibus Orbium Cœlestium Libri VI.* Norimbergæ, M.D.XLIII. p. 9.

[2] *Myst. Cosm.* cap. 1.

great and small Epicycle; or else by means of an Eccentric and Epicycle, modified from Ptolemy's, for reasons which we shall shortly mention. This mode of representing the motions of the planets continued in use, until it was expelled by the discoveries of Kepler.

Besides the daily rotation of the earth on its axis, and its annual circuit about the sun, Copernicus attributed to the axis a "motion of declination," by which, during the whole annual revolution, the pole was constantly directed towards the same part of the heavens. This constancy in the absolute direction of the axis, or its moving parallel to itself, may be more correctly viewed as not indicating any separate motion. The axis continues in the same direction, because there is nothing to make it change its direction; just as a straw, lying on the surface of a cup of water, continues to point nearly in the same direction when the cup is carried round a room. And this was noticed by Copernicus's adherent, Rothman,[3] a few years after the publication of the work *De Revolutionibus.* "There is no occasion," he says, in a letter to Tycho Brahe, "for the triple motion of the earth: the annual and diurnal motions suffice." This error of Copernicus, if it be looked upon as an error, arose from his referring the position of the axis to a limited space, which he conceived to be carried round the sun along with the earth, instead of referring it to fixed or absolute space. When, in a Planetarium (a machine in which the motions of the planets are imitated), the earth is carried round the sun by being fastened to a material radius, it is requisite to give a motion to the axis by *additional* machinery, in order to enable it to *preserve* its parallelism. A similar confusion of geometrical conception, produced by a double reference to absolute space and to the centre of revolution, often leads persons to dispute whether the moon, which revolves about the earth, always turning to it the same face, revolves about her axis or not.

It is also to be noticed that the precession of the equinoxes made it necessary to suppose the axis of the earth to be not *exactly* parallel to itself, but to deviate from that position by a slight annual difference. Copernicus erroneously supposes the precession to be unequable; and his method of explaining this change, which is simpler than that of the ancients, becomes more simple still, when applied to the true state of the facts.

The tendencies of our speculative nature, which carry us onwards in

---

[3] Tycho. Epist. i. p. 184, A. D. 1590.

pursuit of symmetry and rule, and which thus produced the theory of Copernicus, as they produce all theories, perpetually show their vigor by overshooting their mark. They obtain something by aiming at much more. They detect the order and connection which exist, by imagining relations of order and connection which have no existence. Real discoveries are thus mixed with baseless assumptions; profound sagacity is combined with fanciful conjecture; not rarely, or in peculiar instances, but commonly, and in most cases; probably in all, if we could read the thoughts of the discoverers as we read the books of Kepler. To try wrong guesses is apparently the only way to hit upon right ones. The character of the true philosopher is, not that he never conjectures hazardously, but that his conjectures are clearly conceived and brought into rigid contact with facts. He sees and compares distinctly the ideas and the things,—the relations of his notions to each other and to phenomena. Under these conditions it is not only excusable, but necessary for him, to snatch at every semblance of general rule;—to try all promising forms of simplicity and symmetry.

Copernicus is not exempt from giving us, in his work, an example of this character of the inventive spirit. The axiom that the celestial motions must be *circular* and *uniform*, appeared to him to have strong claims to acceptation; and his theory of the inequalities of the planetary motions is fashioned upon it. His great desire was to apply it more rigidly than Ptolemy had done. The time did not come for rejecting this axiom, till the observations of Tycho Brahe and the calculations of Kepler had been made.

I shall not attempt to explain, in detail, Copernicus's system of the planetary inequalities. He retained epicycles and eccentrics, altering their centres of motion; that is, he retained what was *true* in the old system, *translating* it into his own. The peculiarities of his method consisted in making such a combination of epicycles as to supply the place of the *equant*,[4] and to make all the motions equable about the centres of motion. This device was admired for a time, till Kepler's elliptic theory expelled it, with all other forms of the theory of epicycles: but we must observe that Copernicus was aware of some of the discrepancies which belonged to that theory as it had, up to that time, been propounded. In the case of Mercury's orbit, which is more eccentric than that of the other planets, he makes suppositions which are complex indeed, but which show his perception of the imperfection of

---

[4] See B. iii. Chap. iii. Sect. 7.

the common theory; and he proposes a new theory of the moon, for the very reason which did at last overturn the doctrine of epicycles, namely, that the ratio of their distances from the earth at different times was inconsistent with the circular hypothesis.[5]

It is obvious, that, along with his mathematical clearness of view, and his astronomical knowledge, Copernicus must have had great intellectual boldness and vigor, to conceive and fully develop a theory so different as his was from all received doctrines. His pupil and expositor, Rheticus, says to Schener, "I beg you to have this opinion concerning that learned man, my Preceptor; that he was an ardent admirer and follower of Ptolemy; but when he was compelled by phenomena and demonstration, he thought he did well to aim at the same mark at which Ptolemy had aimed, though with a bow and shafts of a very different material from his. We must recollect what Ptolemy says, *Δεῖ δ' ἐλευθέρον εἶναι τῇ γνώμῃ τὸν μέλλοντα φιλοσοφεῖν.* 'He who is to follow philosophy must be a freeman in mind.'" Rheticus then goes on to defend his master from the charge of disrespect to the ancients: "That temper," he says, "is alien from the disposition of every good man, and most especially from the spirit of philosophy, and from no one more utterly than from my Preceptor. He was very far from rashly rejecting the opinions of ancient philosophers, except for weighty reasons and irresistible facts, through any love of novelty. His years, his gravity of character, his excellent learning, his magnanimity and nobleness of spirit, are very far from having any liability to such a temper, which belongs either to youth, or to ardent and light minds, or to those *τῶν μέγα φρονούντων ἐπὶ θεωρίᾳ μικρῇ*, 'who think much of themselves and know little,' as Aristotle says." Undoubtedly this deference for the great men of the past, joined with the talent of seizing the spirit of their methods when the letter of their theories is no longer tenable, *is* the true mental constitution of discoverers.

Besides the intellectual energy which was requisite in order to construct a system of doctrines so novel as those of Copernicus, some courage was necessary to the publication of such opinions; certain, as they were, to be met, to a great extent, by rejection and dispute, and perhaps by charges of heresy and mischievous tendency. This last danger, however, must not be judged so great as we might infer from the angry controversies and acts of authority which occurred in Gali-

---

[5] *De Rev.* iv. c. 2.

leo's time. The Dogmatism of the stationary period, which identified the cause of philosophical and religious truth, had not yet distinctly felt itself attacked by the advance of physical knowledge; and therefore had not begun to look with alarm on such movements. Still, the claims of Scripture and of ecclesiastical authority were asserted as paramount on all subjects; and it was obvious that many persons would be disquieted or offended with the new interpretation of many scriptural expressions, which the true theory would make necessary. This evil Copernicus appears to have foreseen; and this and other causes long withheld him from publication. He was himself an ecclesiastic; and, by the patronage of his maternal uncle, was prebendary of the church of St. John at Thorn, and a canon of the church of Frauenburg, in the diocese of Ermeland.[6] He had been a student at Bologna, and had taught mathematics at Rome in the year 1500; and he afterwards pursued his studies and observations at his residence near the mouth of the Vistula.[7] His discovery of his system must have occurred before 1507, for in 1543 he informs Pope Paulus the Third, in his dedication, that he had kept his book by him for four times the nine years recommended by Horace, and then only published it at the earnest entreaty of his friend Cardinal Schomberg, whose letter is prefixed to the work. "Though I know," he says, "that the thoughts of a philosopher do not depend on the judgment of the many, his study being to seek out truth in all things as far as that is permitted by God to human reason: yet when I considered," he adds, "how absurd my doctrine would appear, I long hesitated whether I should publish my book, or whether it were not better to follow the example of the Pythagoreans and others, who delivered their doctrines only by tradition and to friends." It will be observed that he speaks here of the opposition of the established school of Astronomers, not of Divines. The latter, indeed, he appears to consider as a less formidable danger. "If perchance," he says at the end of his preface, "there be *ματαιολόγοι*, vain babblers, who knowing nothing of mathematics, yet assume the right of judging on account of some place of Scripture perversely wrested to their purpose, and who blame and attack my undertaking; I heed them not, and look upon their judgments as rash and contemptible." He then goes on to show that the globular figure of the earth (which was, of course, at that time, an undisputed point among astronomers), had been opposed on similar grounds by Lactantius, who,

---

[6] Rheticus, *Nar.* p. 94. [7] Riccioli.

though a writer of credit in other respects, had spoken very childishly in that matter. In another epistle prefixed to the work (by Andreas Osiander), the reader is reminded that the hypotheses of astronomers are not necessarily asserted to be true, by those who propose them, but only to be a way of *representing* facts. We may observe that, in the time of Copernicus, when the motion of the earth had not been connected with the physical laws of matter and motion, it could not be considered so distinctly real as it necessarily was held to be in after times.

The delay of the publication of Copernicus's work brought it to the end of his life; he died in the year 1543, in which it was published. It was entitled *De Revolutionibus Orbium Cœlestium Libri VI.* He received the only copy he ever saw on the day of his death, and never opened it: he had then, says Gassendi, his biographer, other cares. His system was, however, to a certain extent, promulgated, and his fame diffused before that time. Cardinal Schomberg, in his letter of 1536, which has been already mentioned, says, "Some years ago, when I heard tidings of your merit by the constant report of all persons, my affection for you was augmented, and I congratulated the men of our time, among whom you flourish in so much honor. For I had understood that you were not only acquainted with the discoveries of ancient mathematicians, but also had formed a new system of the world, in which you teach that the Earth moves, the Sun occupies the lowest, and consequently, the middle place, the sphere of the fixed stars remains immovable and fixed, and the Moon, along with the elements included in her sphere, placed between the orbits (*cœlum*) of Mars and Venus, travels round the sun in a yearly revolution."[8] The writer goes on to say that he has heard that Copernicus has written a book (*Commentarios*), in which this system is applied to the construction of Tables of the Planetary Motions (*erraticarum stellarum*). He then proceeds to entreat him earnestly to publish his lucubrations.

---

[8] This passage has so important a place in the history, that I will give it in the original:—"Intellexeram te non modo veterum mathematicorum inventa egregie callere sed etiam novam mundi rationem constituisse: Qua doceas terram moveri: solem imum mundi, atque medium locum obtinere: cœlum octavum immotum atque fixum perpetuo manere: Lunam se una cum inclusis suæ spheræ elementis, inter Martis et Veneris cœlum sitam, anniversario cursu circum solem convertere. Atque de hac tota astronomiæ ratione commentarios a te confectos esse, ac erraticarum stellarum motus calculis subductos tabulis te contulisse, maxima omnium cum admiratione. Quamobrem vir doctissime, nisi tibi molestus sum, te etiam atque etiam oro vehementer ut hoc tuum inventum studiosis communices, et tuas de mundi sphæra lucubrationes, una cum Tabulis et si quid habes præterea quod ad eandem rem pertineat primo quoque tempore ad me mittas."

This letter is dated 1536, and implies that the work of Copernicus was then written, and known to persons who studied astronomy. Delambre says that Achilles Gassarus of Lindau, in a letter dated 1540, sends to his friend George Vogelin of Constance, the book *De Revolutionibus*. But Mr. De Morgan[9] has pointed out that the printed work which Gassarus sent to Vogelin was the *Narratio* by Rheticus of Feldkirch, a eulogium of Copernicus and his system prefixed to the second edition of the *De Revolutionibus*, which appeared in 1566. In this Narration, Rheticus speaks of the work of Copernicus as a Palingenesia, or New Birth of astronomy. Rheticus, it appears, had gone to Copernicus for the purpose of getting knowledge about triangles and trigonometrical tables, and had had his attention called to the heliocentric theory, of which he became an ardent admirer. He speaks of his "Preceptor" with strong admiration, as we have seen. "He appears to me," says he, "more to resemble Ptolemy than any other astronomers." This, it must be recollected, was selecting the highest known subject of comparison.

CHAPTER III.

SEQUEL TO COPERNICUS.—THE RECEPTION AND DEVELOPMENT OF THE COPERNICAN THEORY.

Sect. 1.—*First Reception of the Copernican Theory.*

THE theories of Copernicus made their way among astronomers, in the manner in which true astronomical theories always obtain the assent of competent judges. They led to the construction of Tables of the motion of the sun, moon, and planets, as the theories of Hipparchus and Ptolemy had done; and the verification of the doctrines was to be looked for, from the agreement of these Tables with observation, through a sufficient course of time. The work *De Revolutionibus* contains such Tables. In 1551 Reinhold improved and republished Tables founded on the principles of Copernicus. "We owe," he says in his preface, "great obligations to Copernicus, both for his laborious

---

[9] *Ast. Mod.* i. p. 138. I owe this and many other corrections to the personal kindness of Mr. De Morgan.

observations, and for restoring the doctrine of the Motions. But though his geometry is perfect, the good old man appears to have been, at times, careless in his numerical calculations. I have, therefore, recalculated the whole, from a comparison of his observations with those of Ptolemy and others, following nothing but the general plan of Copernicus's demonstrations." These "Prutenic Tables" were republished in 1571 and 1585, and continued in repute for some time; till superseded by the Rudolphine Tables of Kepler in 1627. The name *Prutenic*, or Prussian, was employed by the author as a mark of gratitude to his benefactor Albert, Markgrave of Brandenbourg. The discoveries of Copernicus had inspired neighboring nations with the ambition of claiming a place in the literary community of Europe. In something of the same spirit, Rheticus wrote an *Encomium Borussiæ*, which was published along with his *Narratio*.

The Tables founded upon the Copernican system were, at first, much more generally adopted than the heliocentric doctrine on which they were founded. Thus Magini published at Venice, in 1587, *New Theories of the Celestial Orbits, agreeing with the Observations of Nicholas Copernicus*. But in the preface, after praising Copernicus, he says, "Since, however, he, either for the sake of showing his talents, or induced by his own reasons, has revived the opinion of Nicetas, Aristarchus, and others, concerning the motion of the earth, and has disturbed the established constitution of the world, which was a reason why many rejected, or received with dislike, his hypothesis, I have thought it worth while, that, rejecting the suppositions of Copernicus, I should accommodate other causes to his observations, and to the Prutenic Tables."

This doctrine, however, was, as we have shown, received with favor by many persons, even before its general publication. The doctrine of the motion of the earth was first publicly maintained at Rome by Widmanstadt,[1] who professed to have received it from Copernicus, and explained the System before the Pope and the Cardinals, but did not teach it to the public.

Leonardo da Vinci, who was an eminent mathematician, as well as painter, about 1510, explained how a body, by describing a kind of spiral, might descend towards a revolving globe, so that its apparent motion relative to a point in the surface of the globe, might be in a

---

[1] See Venturi, *Essai sur les Ouvrages Physico-Mathématiques de Leonard da Vinci, avec des Fragmens tirés de ses Manuscrits apportés d'Italie.* Paris, 1797; and, as there quoted, *Marini Archiatri Pontificii*, tom. ii. p. 251.

straight line leading to the centre. He thus showed that he had entertained in his thoughts the hypothesis of the earth's rotation, and was employed in removing the difficulties which accompanied this supposition, by means of the consideration of the composition of motions.

In like manner we find the question stirred by other eminent men. Thus John Muller of Konigsberg, a celebrated astronomer who died in 1476, better known by the name of Regiomontanus, wrote a dissertation on the subject "Whether the earth be in motion or at rest," in which he decides *ex professo*[2] against the motion. Yet such discussions must have made generally known the arguments for the heliocentric theory.

We have already seen the enthusiasm with which Rheticus, who was Copernicus's pupil in the latter years of his life, speaks of him. "Thus," says he, "God has given to my excellent preceptor a reign without end; which may He vouchsafe to guide, govern, and increase, to the restoration of astronomical truth. Amen."

Of the immediate converts of the Copernican system, who adopted it before the controversy on the subject had attracted attention, I shall only add Mastlin, and his pupil, Kepler. Mastlin published in 1588 an *Epitome Astronomiæ*, in which the immobility of the earth is asserted; but in 1596 he edited Kepler's *Mysterium Cosmographicum*, and the *Narratio* of Rheticus: and in an epistle of his own, which he inserts, he defends the Copernican system by those physical reasonings which we shall shortly have to mention, as the usual arguments in this dispute. Kepler himself, in the outset of the work just named, says, "When I was at Tübingen, attending to Michael Mæstlin, being disturbed by the manifold inconveniences of the usual opinion concerning the world, I was so delighted with Copernicus, of whom he made great mention in his lectures, that I not only defended his opinions in our disputations of the candidates, but wrote a thesis concerning the First Motion which is produced by the revolution of the earth." This must have been in 1590.

The differences of opinion respecting the Copernican system, of which we thus see traces, led to a controversy of some length and extent. This controversy turned principally upon physical considerations, which were much more distinctly dealt with by Kepler, and others of the followers of Copernicus, than they had been by the dis-

---

[2] Schoneri *Opera*, part ii. p. 129.

coverer himself. I shall, therefore, give a separate consideration to this part of the subject. It may be proper, however, in the first place, to make a few observations on the progress of the doctrine, independently of these physical speculations.

Sect. 2.—*Diffusion of the Copernican Theory.*

THE diffusion of the Copernican opinions in the world did not take place rapidly at first. Indeed, it was necessarily some time before the progress of observation and of theoretical mechanics gave the heliocentric doctrine that superiority in argument, which now makes us wonder that men should have hesitated when it was presented to them. Yet there were some speculators of this kind, who were attracted at once by the enlarged views of the universe which it opened to them. Among these was the unfortunate Giordano Bruno of Nola, who was burnt as a heretic at Rome in 1600. The heresies which led to his unhappy fate were, however, not his astronomical opinions, but a work which he published in England, and dedicated to Sir Philip Sydney, under the title of *Spaccio della Bestia Trionfante*, and which is understood to contain a bitter satire of religion and the papal government. Montucla conceives that, by his rashness in visiting Italy after putting forth such a work, he compelled the government to act against him. Bruno embraced the Copernican opinions at an early period, and connected with them the belief in innumerable worlds besides that which we inhabit; as also certain metaphysical or theological doctrines, which he called the Nolan philosophy. In 1591 he published *De innumerabilibus, immenso, et infigurabili, seu de Universo et Mundis*, in which he maintains that each star is a sun, about which revolve planets like our earth; but this opinion is mixed up with a large mass of baseless verbal speculations.

Giordano Bruno is a disciple of Copernicus on whom we may look with peculiar interest, since he probably had a considerable share in introducing the new opinions into England;[3] although other persons, as Recorde, Field, Dee, had adopted it nearly thirty years earlier; and Thomas Digges ten years before, much more expressly. Bruno visited this country in the reign of Queen Elizabeth, and speaks of her and of her councillors in terms of praise, which appear to show that

[3] See Burton's *Anat. Mel.* Pref. "Some prodigious tenet or paradox of the earth's motion," &c. "Bruno," &c.

his book was intended for English readers; though he describes the mob which was usually to be met with in the streets of London with expressions of great disgust: "Una plebe la quale in essere irrespettevole, incivile, rozza, rustica, selvatica, et male allevata, non cede ad altra che pascer possa la terra nel suo seno."[4] The work to which I refer is *La Cena de le Cenere*, and narrates what took place at a supper held on the evening of Ash Wednesday (about 1583, see p. 145 of the book), at the house of Sir Fulk Greville, in order to give "Il Nolano" an opportunity of defending his peculiar opinions. His principal antagonists are two "Dottori d' Oxonia," whom Bruno calls Nundinio and Torquato. The subject is not treated in any very masterly manner on either side; but the author makes himself have greatly the advantage not only in argument, but in temper and courtesy: and in support of his representations of "pedantesca, ostinatissima ignoranza et presunzione, mista con una rustica incivilità, che farebbe prevaricar la pazienza di Giobbe," in his opponents, he refers to a public disputation which he had held at Oxford with these doctors of theology, in presence of Prince Alasco, and many of the English nobility.[5]

Among the evidences of the difficulties which still lay in the way of the reception of the Copernican system, we may notice Bacon, who, as is well known, never gave a full assent to it. It is to be observed, however, that he does not reject the opinion of the earth's motion in so peremptory and dogmatical a manner as he is sometimes accused of doing: thus in the *Thema Cœli* he says, "The earth, then, being supposed to be at rest (for that now appears to us the *more true* opinion)." And in his tract *On the Cause of the Tides*, he says, "If the tide of the sea be the extreme and diminished limit of the diurnal motion of the heavens, it will follow that the earth is immovable; or at least that it moves with a much slower motion than the water." In the *Descriptio Globi Intellectualis* he gives his reasons for not accepting the heliocentric theory. "In the system of Copernicus there are many and grave difficulties: for the threefold motion with which he encumbers the earth is a serious inconvenience; and the separation of the sun from the planets, with which he has so many affections in common, is likewise a harsh step; and the introduction of so many immovable bodies into nature, as when he makes the sun and the stars immovable, the bodies which are peculiarly lucid and radiant; and his making the moon adhere to the earth in a sort of epicycle; and some

[4] *Opere di Giordano Bruno*, vol. i. p. 146.

[5] Ib. vol. i. p. 179.

other things which he assumes, are proceedings which mark a man who thinks nothing of introducing fictions of any kind into nature, provided his calculations turn out well." We have already explained that, in attributing *three* motions to the earth, Copernicus had presented his system encumbered with a complexity not really belonging to it. But it will be seen shortly, that Bacon's fundamental objection to this system was his wish for a system which could be supported by sound physical considerations; and it must be allowed, that at the period of which we are speaking, this had not yet been done in favor of the Copernican hypothesis. We may add, however, that it is not quite clear that Bacon was in full possession of the details of the astronomical systems which that of Copernicus was intended to supersede; and that thus he, perhaps, did not see how much less harsh were these fictions, as he called them, than those which were the inevitable alternatives. Perhaps he might even be liable to a little of that indistinctness, with respect to strictly geometrical conceptions, which we have remarked in Aristotle. We can hardly otherwise account for his not seeing any use in resolving the apparently irregular motion of a planet into separate regular motions. Yet he speaks slightingly of this important step.[6] "The motion of planets, which is constantly talked of as the motion of regression, or renitency, from west to east, and which is ascribed to the planets as a proper motion, is not true; but only arises from appearance, from the greater advance of the starry heavens towards the west, by which the planets are left behind to the east." Undoubtedly those who spoke of such a motion of *regression*, were aware of this; but they saw how the motion was simplified by this way of conceiving it, which Bacon seems not to have seen. Though, therefore, we may admire Bacon for the steadfastness with which he looked forward to physical astronomy as the great and proper object of philosophical interest, we cannot give him credit for seeing the full value and meaning of what had been done, up to his time, in Formal Astronomy.

Bacon's contemporary, Gilbert, whom he frequently praises as a philosopher, was much more disposed to adopt the Copernican opinions, though even he does not appear to have made up his mind to assent to the whole of the system. In his work, *De Magnete* (printed 1600), he gives the principal arguments in favor of the Copernican system, and decides that the earth revolves on its axis.[7] He connects

---

[6] *Thema Cœli*, p. 246.

[7] Lib. vi. cap. 3, 4.

this opinion with his magnetic doctrines; and especially endeavors by that means to account for the precession of the equinoxes. But he does not seem to have been equally confident of its annual motion. In a posthumous work, published in 1651 (*De Mundo Nostro Sublunari Philosophia Nova*), he appears to hesitate between the systems of Tycho and Copernicus.[8] Indeed, it is probable that at this period many persons were in a state of doubt on such subjects. Milton, at a period somewhat later, appears to have been still undecided. In the opening of the eighth book of the *Paradise Lost*, he makes Adam state the difficulties of the Ptolemaic hypothesis, to which the archangel Raphael opposes the usual answers; but afterwards suggests to his pupil the newer system:

> . . . . What if seventh to these
> The planet earth, so steadfast though she seem,
> Insensibly three different motions move?
>
> *Par. Lost*, b. viii.

Milton's leaning, however, seems to have been for the new system; we can hardly believe that he would otherwise have conceived so distinctly, and described with such obvious pleasure, the motion of the earth:

> Or she from west her silent course advance
> With inoffensive pace, that spinning sleeps
> On her soft axle, while she paces even,
> And bears thee soft with the smooth air along.
>
> *Par. Lost*, b. viii.

Perhaps the works of the celebrated Bishop Wilkins tended more than any others to the diffusion of the Copernican system in England, since even their extravagances drew a stronger attention to them. In 1638, when he was only twenty-four years old, he published a book entitled *The Discovery of a New World; or, a Discourse tending to prove that it is probable there may be another habitable World in the Moon; with a Discourse concerning* the possibility of a passage thither. The latter part of his subject was, of course, an obvious mark for the sneers and witticisms of critics. Two years afterwards, in 1640, appeared his *Discourse concerning a new Planet; tending to prove that it is probable our Earth is one of the Planets:* in which he urged the reasons in favor of the heliocentric system; and explained away the opposite arguments, especially those drawn from the sup-

---

[8] Lib. ii. cap. 20.

posed declarations of Scripture. Probably a good deal was done for the establishment of those opinions by Thomas Salusbury, who was a warm admirer of Galileo, and published, in 1661, a translation of several of his works bearing upon this subject. The mathematicians of this country, in the seventeenth century, as Napier and Briggs, Horrox and Crabtree, Oughtred and Seth Ward, Wallis and Wren, were probably all decided Copernicans. Kepler dedicates one of his works to Napier, and Ward invented an approximate method of solving Kepler's problem, still known as "the simple elliptical hypothesis." Horrox wrote, and wrote well, in defence of the Copernican opinion, in his *Keplerian Astronomy defended and promoted*, composed (in Latin) probably about 1635, but not published till 1673, the author having died at the age of twenty-two, and his papers having been lost. But Salusbury's work was calculated for another circle of readers. "The book," he says in the introductory address, "being, for subject and design, intended chiefly for gentlemen, I have been as careless of using a studied pedantry in my style, as careful in contriving a pleasant and beautiful impression." In order, however, to judge of the advantage under which the Copernican system now came forward, we must consider the additional evidence for it which was brought to light by Galileo's astronomical discoveries.

Sect. 3.—The Heliocentric Theory confirmed by Facts.—Galileo's Astronomical Discoveries.

THE long interval which elapsed between the last great discoveries made by the ancients and the first made by the moderns, had afforded ample time for the development of all the important consequences of the ancient doctrines. But when the human mind had been thoroughly roused again into activity, this was no longer the course of events. Discoveries crowded on each other; one wide field of speculation was only just opened, when a richer promise tempted the laborers away into another quarter. Hence the history of this period contains the beginnings of many sciences, but exhibits none fully worked out into a complete or final form. Thus the science of Statics, soon after its revival, was eclipsed and overlaid by that of Dynamics; and the Copernican system, considered merely with reference to the views of its author, was absorbed in the commanding interest of Physical Astronomy.

Still, advances were made which had an important bearing on the heliocentric theory, in other ways than by throwing light upon its physical principles. I speak of the new views of the heavens which the Telescope gave; the visible inequalities of the moon's surface; the moon-like phases of the planet Venus; the discovery of the Satellites of Jupiter, and of the Ring of Saturn. These discoveries excited at the time the strongest interest; both from the novelty and beauty of the objects they presented to the sense; from the way in which they seemed to gratify man's curiosity with regard to the remote parts of the universe; and also from that of which we have here to speak, their bearing upon the conflict of the old and the new philosophy, the heliocentric and geocentric theories. It may be true, as Lagrange and Montucla say, that the laws which Galileo discovered in Mechanics implied a profounder genius than the novelties he detected in the sky: but the latter naturally attracted the greater share of the attention of the world, and were matter of keener discussion.

It is not to our purpose to speak here of the details and of the occasion of the invention of the Telescope; it is well known that Galileo constructed his about 1609, and proceeded immediately to apply it to the heavens. The discovery of the Satellites of Jupiter was almost immediately the reward of his activity; and these were announced in his *Nuncius Sidereus*, published at Venice in 1610. The title of this work will best convey an idea of the claim it made to public notice: "The *Sidereal Messenger*, announcing great and very wonderful spectacles, and offering them to the consideration of every one, but especially of philosophers and astronomers; which have been observed by *Galileo Galilei*, &c., &c., by the assistance of a perspective glass lately invented by him; namely, in the face of the moon, in innumerable fixed stars in the milky-way, in nebulous stars, but especially in four planets which revolve round Jupiter at different intervals and periods with a wonderful celerity; which, hitherto not known to any one, the author has recently been the first to detect, and has decreed to call the *Medicean stars*."

The interest this discovery excited was intense: and men were at this period so little habituated to accommodate their convictions on matters of science to newly observed facts, that several of the "paper-philosophers," as Galileo termed them, appear to have thought they could get rid of these new objects by writing books against them. The effect which the discovery had upon the reception of the Copernican system was immediately very considerable. It showed that the real universe was very different from that which ancient philosophers had imagined, and suggested at once the thought that it contained mechanism more various and more vast than had yet been conjectured. And when the system of the planet Jupiter thus offered to the bodily eye a model or image of the solar system according to the views of Copernicus, it supported the belief of such an arrangement of the planets, by an analogy all but irresistible. It thus, as a writer[9] of our own times has said, "gave the *holding turn* to the opinions of mankind respecting the Copernican system." We may trace this effect in Bacon, even though he does not assent to the motion of the earth. "We affirm," he says,[10] "the *sun-following arrangement* (solisequium) of Venus and Mercury; since it has been found by Galileo that Jupiter also has attendants."

The *Nuncius Sidereus* contained other discoveries which had the same tendency in other ways. The examination of the moon showed, or at least seemed to show, that she was a solid body, with a surface extremely rugged and irregular. This, though perhaps not bearing directly upon the question of the heliocentric theory, was yet a blow to the Aristotelians, who had, in their philosophy, made the moon a body of a kind altogether different from this, and had given an abundant quantity of reasons for the visible marks on her surface, all proceeding on these preconceived views. Others of his discoveries produced the same effect; for instance, the new stars invisible to the naked eye, and those extraordinary appearances called Nebulæ.

But before the end of the year, Galileo had new information to communicate, bearing more decidedly on the Copernican controversy. This intelligence was indeed decisive with regard to the motion of Venus about the sun; for he found that that planet, in the course of her revolution, assumes the same succession of phases which the moon exhibits in the course of a month. This he expressed by a Latin verse:

Cynthiæ figuras æmulatur mater amorum:
The Queen of Love like Cynthia shapes her forms:

transposing the letters of this line in the published account, according to the practice of the age; which thus showed the ancient love for combining verbal puzzles with scientific discoveries, while it betrayed the newer feeling, of jealousy respecting the priority of discovery of physical facts.

It had always been a formidable objection to the Copernican theory that this appearance of the planets had not been observed. The author

---

9 Sir J. Herschel.

10 *Thema Cœli*, ix. p. 253.

of that theory had endeavored to account for this, by supposing that the rays of the sun passed freely through the body of the planet; and Galileo takes occasion to praise him for not being deterred from adopting the system which, on the whole, appeared to agree best with the phenomena, by meeting with some appearances which it did not enable him to explain.[11] Yet while the fate of the theory was yet undecided, this could not but be looked upon as a weak point in its defences.

The objection, in another form also, was embarrassing alike to the Ptolemaic and Copernican systems. Why, it was asked, did not Venus appear four times as large when nearest to the earth, as when furthest from it? The author of the Epistle prefixed to Copernicus's work had taken refuge in this argument from the danger of being supposed to believe in the reality of the system; and Bruno had attempted to answer it by saying, that luminous bodies were not governed by the same laws of perspective as opake ones. But a more satisfactory answer now readily offered itself. Venus does not appear four times as large when she is four times as near, because her *bright part* is *not* four times as large, though her visible diameter is; and as she is too small for us to see her shape with the naked eye, we judge of her size only by the quantity of light.

The other great discoveries made in the heavens by means of telescopes, as that of Saturn's ring and his satellites, the spots in the sun, and others, belong to the further progress of astronomy. But we may here observe, that this doctrine of the motion of Mercury and Venus about the sun was further confirmed by Kepler's observation of the transit of the former planet over the sun in 1631. Our countryman Horrox was the first person who, in 1639, had the satisfaction of seeing a transit of Venus.

These events are a remarkable instance of the way in which a discovery in art (for at this period, the making of telescopes must be mainly so considered) may influence the progress of science. We shall soon have to notice a still more remarkable example of the way in which two sciences (Astronomy and Mechanics) may influence and promote the progress of each other.

---

[11] Drinkwater-Bethune, *Life of Galileo*, p. 35.

Sect. 4.—The Copernican System opposed on Theological Grounds.

The doctrine of the Earth's motion round the Sun, when it was asserted and promulgated by Copernicus, soon after 1500, excited no visible alarm among the theologians of his own time. Indeed, it was received with favor by the most intelligent ecclesiastics; and lectures in support of the heliocentric doctrine were delivered in the ecclesiastical colleges. But the assertion and confirmation of this doctrine by Galileo, about a century later, excited a storm of controversy, and was visited with severe condemnation. Galileo's own behavior appears to have provoked the interference of the ecclesiastical authorities; but there must have been a great change in the temper of the times to make it possible for his adversaries to bring down the sentence of the Inquisition upon opinions which had been so long current without giving any serious offence.

[2d Ed.] [It appears to me that the different degree of toleration accorded to the heliocentric theory in the time of Copernicus and of Galileo, must be ascribed in a great measure to the controversies and alarms which had in the mean time arisen out of the Reformation in religion, and which had rendered the Romish Church more jealous of innovations in received opinions than it had previously been. It appears too that the discussion of such novel doctrines was, at that time at least, less freely tolerated in Italy than in other countries. In 1597, Kepler writes to Galileo thus: "Confide Galilæe et progredere. Si bene conjecto, pauci de præcipuis Europæ Mathematicis a nobis secedere volent; tanta vis est veritatis. Si tibi Italia minus est idonea ad publicationem et si aliqua habitures es impedimenta, forsan Germania nobis hanc libertatem concedet."—Venturi, *Mem. di Galileo*, vol. i. p. 19.

I would not however be understood to assert the condemnation of new doctrines in science to be either a general or a characteristic practice of the Romish Church. Certainly the intelligent and cultivated minds of Italy, and many of the most eminent of her ecclesiastics among them, have always been the foremost in promoting and welcoming the progress of science: and, as I have stated, there were found among the Italian ecclesiastics of Galileo's time many of the earliest and most enlightened adherents of the Copernican system. The condemnation of the doctrine of the earth's motion, is, so far as I am aware, the only instance in which the Papal authority has pronounced a decree upon a point of science. And the most candid of the adherents of the Romish Church condemn the assumption of authority in such matters, which in this one instance, at least, was made by the ecclesiastical tribunals. The author of the *Ages of Faith* (book viii. p. 248) says, "A congregation, it is to be lamented, declared the new system to be opposed to Scripture, and therefore heretical." In more recent times, as I have elsewhere remarked,[12] the Church of Authority and the Church of Private Judgment have each its peculiar temptations and dangers, when there appears to be a discrepance between Scripture and Philosophy.

But though we may acquit the popes and cardinals in Galileo's time of stupidity and perverseness in rejecting manifest scientific truths, I do not see how we can acquit them of dissimulation and duplicity. Those persons appear to me to defend in a very strange manner the conduct of the ecclesiastical authorities of that period, who boast of the liberality with which Copernican professors were placed by them in important offices, at the very time when the motion of the earth had been declared by the same authorities contrary to Scripture. Such merits cannot make us approve of their conduct in demanding from Galileo a public recantation of the system which they thus favored in other ways, and which they had repeatedly told Galileo he might hold as much as he pleased. Nor can any one, reading the plain language of the Sentence passed upon Galileo, and of the Abjuration forced from him, find any value in the plea which has been urged, that the opinion was denominated a *heresy* only in a wide, improper, and technical sense.

But if we are thus unable to excuse the conduct of Galileo's judges, I do not see how we can give our unconditional admiration to the philosopher himself. Perhaps the conventional decorum which, as we have seen, was required in treating of the Copernican system, may excuse or explain the furtive mode of insinuating his doctrines which he often employs, and which some of his historians admire as subtle irony, while others blame it as insincerity. But I do not see with what propriety Galileo can be looked upon as a "Martyr of Science." Undoubtedly he was very desirous of promoting what he conceived to be the cause of philosophical truth; but it would seem that, while he was restless and eager in urging his opinions, he was always ready to make such submissions as the spiritual tribunals required. He would really have acted as a martyr, if he had uttered

[12] *Phil. Ind. Sci.* book x. chap. 4.

his "E pur si muove," in the place of his abjuration, not after it. But even in this case he would have been a martyr to a cause of which the merit was of a mingled scientific character; for his own special and favorite share in the reasonings by which the Copernican system was supported, was the argument drawn from the flux and reflux of the sea, which argument is altogether false. He considered this as supplying a mechanical ground of belief, without which the mere astronomical reasons were quite insufficient; but in this case he was deserted by the mechanical sagacity which appeared in his other speculations.]

The heliocentric doctrine had for a century been making its way into the minds of thoughtful men, on the general ground of its simplicity and symmetry. Galileo appears to have thought that now, when these original recommendations of the system had been reinforced by his own discoveries and reasonings, it ought to be universally acknowledged as a truth and a reality. And when arguments against the fixity of the sun and the motion of the earth were adduced from the expressions of Scripture, he could not be satisfied without maintaining his favorite opinion to be conformable to Scripture as well as to Philosophy; and he was very eager in his attempts to obtain from authority a declaration to this effect. The ecclesiastical authorities were naturally averse to express themselves in favor of a novel opinion, startling to the common mind, and contrary to the most obvious meaning of the words of the Bible; and when they were compelled to pronounce, they decided against Galileo and his doctrines. He was accused before the Inquisition in 1615; but at that period the result was that he was merely recommended to confine himself to the mathematical reasonings upon the system, and to abstain from meddling with the Scripture. Galileo's zeal for his opinions soon led him again to bring the question under the notice of the Pope, and the result was a declaration of the Inquisition that the doctrine of the earth's motion appeared to be contrary to the Sacred Scripture. Galileo was prohibited from defending and teaching this doctrine in any manner, and promised obedience to this injunction. But in 1632 he published his *Dialogo delli due Massimi Sistemi del Mondo, Tolemaico e Copernicano:*" and in this he defended the heliocentric system by all the strongest arguments which its admirers used. Not only so, but he introduced into this *Dialogue* a character under the name of Simplicius, in whose mouth was put the defence of all the ancient dogmas, and who was represented as defeated at all points in the discussion;

and he prefixed to the *Dialogue* a Notice, *To the Discreet Reader*, in which, in a vein of transparent irony, he assigned his reasons for the publication. "Some years ago," he says, "a wholesome edict was promulgated at Rome, which, in order to check the perilous scandals of the present age, imposed silence upon the Pythagorean opinion of the motion of the earth. There were not wanting," he adds, "persons who rashly asserted that this decree was the result, not of a judicious inquiry, but of a passion ill-informed; and complaints were heard that counsellors, utterly unacquainted with astronomical observations, ought not to be allowed, with their undue prohibitions, to clip the wings of speculative intellects. At the hearing of rash lamentations like these, my zeal could not keep silence." And he then goes on to say that he wishes, by the publication of his *Dialogue*, to show that the subject had been fully examined at Rome. The result of this was that Galileo was condemned for his infraction of the injunction laid upon him in 1616; his *Dialogue* was prohibited; he himself was commanded to abjure on his knees the doctrine which he had taught; and this abjuration he performed.

This celebrated event must be looked upon rather as a question of decorum than a struggle in which the interests of truth and free inquiry were deeply concerned. The general acceptance of the Copernican System was no longer a matter of doubt. Several persons in the highest positions, including the Pope himself, looked upon the doctrine with favorable eyes; and had shown their interest in Galileo and his discoveries. They had tried to prevent his involving himself in trouble by discussing the question on scriptural grounds. It is probable that his knowledge of those favorable dispositions towards himself and his opinions led him to suppose that the slightest color of professed submission to the Church in his belief, would enable his arguments in favor of the system to pass unvisited: the notice which I have quoted, in which the irony is quite transparent and the sarcasm glaringly obvious, was deemed too flimsy a veil for the purpose of decency, and indeed must have aggravated the offence. But it is not to be supposed that the inquisitors believed Galileo's abjuration to be sincere, or even that they wished it to be so. It is stated that when Galileo had made his renunciation of the earth's motion, he rose from his knees, and stamping on the earth with his foot, said, *E pur si muove*—"And yet it *does* move." This is sometimes represented as the heroic soliloquy of a mind cherishing its conviction of the truth in spite of persecution: I think we may more naturally conceive it uttered as a playful epigram in the ear of a cardinal's secretary, with a full knowledge that it would be immediately repeated to his master.

[2d Ed.] [Throughout the course of the proceedings against him, Galileo was treated with great courtesy and indulgence. He was condemned to a formal imprisonment and a very light discipline. "Te damnamus ad formalem carcerem hujus S. Officii ad tempus arbitrio nostro limitandum; et titulo pœnitentiæ salutaris præcipimus ut tribus annis futuris recites semil in hebdomadâ septem psalmos penitentiales." But this confinement was reduced to his being placed under some slight restrictions, first at the house of Nicolini, the ambassador of his own sovereign, and afterwards at the country seat of Archbishop Piccolomini, one of his own warmest friends.

It has sometimes been asserted or insinuated that Galileo was subjected to bodily torture. An argument has been drawn from the expressions used in his sentence: "Cum vero nobis videretur non esse a te integram veritatem pronunciatam circa tuam intentionem; judicavimus necesse esse venire ad rigorosum examen tui, in quo respondisti catholicè." It has been argued by M. Libri (*Hist. des Sciences Mathématiques en Italie*, vol. IV. p. 259), and M. Quinet (*L'Ultramontanisme*, IV. Leçon, p. 104), that the *rigorosum examen* necessarily implies bodily torture, notwithstanding that no such thing is mentioned by Galileo and his contemporaries, and notwithstanding the consideration with which he was treated in all other respects: but M. Biot more justly remarks (*Biogr. Univ.* Art. *Galileo*), that such a procedure is incredible.

To the opinion of M. Biot, we may add that of Delambre, who rejects the notion of Galileo's having been put to the torture, as inconsistent with the general conduct of the authorities towards him, and as irreconcilable with the accounts of the trial given by Galileo himself, and by a servant of his, who never quitted him for an instant. He adds also, that it is inconsistent with the words of his sentence, "ne tuus iste gravis et perniciosus error ac transgressio remaneat *omnino impunitus*;" for the error would have been already very far from impunity, if Galileo had been previously subjected to the rack. He adds, very reasonably, "il ne faut noircir personne sans preuve, pas même l'Inquisition;"—we must not calumniate even the Inquisition.]

The ecclesiastical authorities having once declared the doctrine of the earth's motion to be contrary to Scripture and heretical, long adhered in form to this declaration, and did not allow the Copernican system to be taught in any other way than as an "hypothesis." The

Padua edition of Galileo's works, published in 1744, contains the *Dialogue* which now, the editors say, "Esce finalmente a pubblico libero uso colle debite licenze," is now at last freely published with the requisite license; but they add, "quanto alla Quistione principale del moto della terra, anche noi ci conformiamo alla ritrazione et protesta dell' Autore, dichiarando nella piu solenne forma, che non può, nè dee ammetersi se non come pura Ipotesi Mathematice, che serve a spiegare piu agevolamento certi fenomeni;" "neither can nor ought to be admitted except as a convenient hypothesis." And in the edition of Newton's *Principia*, published in 1760, by Le Sueur and Jacquier, of the Order of Minims, the editors prefix to the Third Book their *Declaratio*, that though Newton assumes the hypothesis of the motion of the earth, and therefore they had used similar language, they were, in doing this, assuming a character which did not belong to them. "Hinc alienam coacti sumus gerere personam." They add, "Cæterum latis a summis Pontificibus contra telluris motum Decretis, nos obsequi profitemur."

By thus making decrees against a doctrine which in the course of time was established as an indisputable scientific truth, the See of Rome was guilty of an unwise and unfortunate stretch of ecclesiastical authority. But though we do not hesitate to pronounce such a judgment on this case, we may add that there is a question of no small real difficulty, which the progress of science often brings into notice, as it did then. The Revelation on which our religion is founded, seems to declare, or to take for granted, opinions on points on which Science also gives her decision; and we then come to this dilemma,—that doctrines, established by a scientific use of reason, may seem to contradict the declarations of Revelation, according to our view of its meaning;—and yet, that we cannot, in consistency with our religious views, make reason a judge of the truth of revealed doctrines. In the case of Astronomy, on which Galileo was called in question, the general sense of cultivated and sober-minded men has long ago drawn that distinction between religious and physical tenets, which is necessary to resolve this dilemma. On this point, it is reasonably held, that the phrases which are employed in Scripture respecting astronomical facts, are not to be made use of to guide our scientific opinions; they may be supposed to answer their end if they fall in with common notions, and are thus effectually subservient to the moral and religious import of Revelation. But the establishment of this distinction was not accomplished without long and distressing controversies. Nor, if we wish to include all cases in which the same dilemma may again come into play, is it easy to lay down an adequate canon for the purpose. For we can hardly foresee, beforehand, what part of the past history of the universe may eventually be found to come within the domain of science; or what bearing the tenets, which science establishes, may have upon our view of the providential and revealed government of the world. But without attempting here to generalize on this subject, there are two reflections which may be worth our notice: they are supported by what took place in reference to Astronomy on the occasion of which we are speaking; and may, at other periods, be applicable to other sciences.

In the first place, the meaning which any generation puts upon the phrases of Scripture, depends, more than is at first sight supposed, upon the received philosophy of the time. Hence, while men imagine that they are contending for Revelation, they are, in fact, contending for their own interpretation of Revelation, unconsciously adapted to what they believe to be rationally probable. And the new interpretation, which the new philosophy requires, and which appears to the older school to be a fatal violence done to the authority of religion, is accepted by their successors without the dangerous results which were apprehended. When the language of Scripture, invested with its new meaning, has become familiar to men, it is found that the ideas which it calls up, are quite as reconcilable as the former ones were, with the soundest religious views. And the world then looks back with surprise at the error of those who thought that the essence of Revelation was involved in their own arbitrary version of some collateral circumstance. At the present day we can hardly conceive how reasonable men should have imagined that religious reflections on the stability of the earth, and the beauty and use of the luminaries which revolve round it, would be interfered with by its being acknowledged that this rest and motion are apparent only.

In the next place, we may observe that those who thus adhere tenaciously to the traditionary or arbitrary mode of understanding Scriptural expressions of physical events, are always strongly condemned by succeeding generations. They are looked upon with contempt by the world at large, who cannot enter into the obsolete difficulties with which they encumbered themselves; and with pity by the more considerate and serious, who know how much sagacity and rightmindedness are requisite for the conduct of philosophers and religious men on such occasions; but who know also how weak and vain is the attempt to get rid of the difficulty by merely denouncing the new tenets as inconsistent with religious belief, and by visiting the promulgators of them with severity such as the state of opinions and institutions may allow. The prosecutors of Galileo are still up to the scorn and aversion of mankind: although, as we have seen, they did not act till it seemed that their position compelled them to do so, and then proceeded with all the gentleness and moderation which were compatible with judicial forms.

Sect. 5.—*The Heliocentric Theory confirmed on Physical considerations.*—(*Prelude to Kepler's Astronomical Discoveries.*)

By physical views, I mean, as I have already said, those which depend on the causes of the motions of matter, as, for instance, the consideration of the nature and laws of the force by which bodies fall downwards. Such considerations were necessarily and immediately brought under notice by the examination of the Copernican theory; but the loose and inaccurate notions which prevailed respecting the nature and laws of force, prevented, for some time, all distinct reasoning on this subject, and gave truth little advantage over error. The formation of a new Science, the Science of Motion and its Causes, was requisite, before the heliocentric system could have justice done it with regard to this part of the subject.

This discussion was at first carried on, as was to be expected, in terms of the received, that is, the Aristotelian doctrines. Thus, Copernicus says that terrestrial things appear to be at rest when they have a motion according to nature, that is, a circular motion; and ascend or descend when they have, in addition to this, a rectilinear motion by which they endeaver to get into their own place. But his disciples soon began to question the Aristotelian dogmas, and to seek for sounder views by the use of their own reason. "The great argument against this system," says Mæstlin, "is that heavy bodies are said to move to the centre of the universe, and light bodies from the centre. But I would ask, where do we get this experience of heavy and light bodies? and how is our knowledge on these subjects extended so far that we can reason with certainty concerning the centre of the whole universe? Is not the only residence and home of all the things which are heavy and light to us, the earth and the air which surrounds it? and what is the earth and the ambient air, with respect to the immensity of the universe? It is a point, a punctule, or something, if there be any thing, still less. As our light and heavy bodies tend to the centre of our earth, it is credible that the sun, the moon, and the other lights, have a similar affection, by which they remain round as we see them; but none of these centres is necessarily the centre of the universe."

The most obvious and important physical difficulty attendant upon the supposition of the motion of the earth was thus stated: If the earth move, how is it that a stone, dropped from the top of a high tower, falls exactly at the foot of the tower? since the tower being carried from west to east by the diurnal revolution of the earth, the stone must be left behind to the west of the place from which it was let fall. The proper answer to this was, that the motion which the falling body received from its tendency downwards was *compounded* with the motion which, before it fell, it had in virtue of the earth's rotation: but this answer could not be clearly made or apprehended, till Galileo and his pupils had established the laws of such Compositions of motion arising from different forces. Rothman, Kepler, and other defenders of the Copernican system, gave their reply somewhat at a venture, when they asserted that the motion of the earth was communicated to bodies at its surface. Still, the facts which indicate and establish this truth are obvious, when the subject is steadily considered; and the Copernicans soon found that they had the superiority of argument on this point as well as others. The attacks upon the Copernican system by Durret, Morin, Riccioli, and the defence of it by Galileo, Lansberg, Gassendi,[13] left on all candid reasoners a clear impression in favor of the system. Morin attempted to stop the motion of the earth, which he called breaking its wings; his *Alæ Terræ Fractæ* was published in 1643, and answered by Gassendi. And Riccioli, as late as 1653, in his *Almagestum Novum*, enumerated fifty-seven Copernican arguments, and pretended to refute them all: but such reasonings now made no converts; and by this time the mechanical objections to the motion of the earth were generally seen to be baseless, as we shall relate when we come to speak of the progress of Mechanics as a distinct science. In the mean time, the beauty and simplicity of the heliocentric theory were perpetually winning the admiration even of those who, from one cause or other, refused their assent to it. Thus Riccioli, the last of its considerable opponents, allows its superiority in these respects; and acknowledges (in 1653) that the Copernican belief appears rather to increase than diminish under the condemnation of the decrees of the Cardinals. He applies to it the lines of Horace:[14]

---

13 Del. *A. M.* vol. i. p. 594.

14 *Almag. Nov.* p. 102.

Per damna per cædes, ab ipso
Sumit opes animumque ferro.

Untamed its pride, unchecked its course,
From foes and wounds it gathers force.

We have spoken of the influence of the motion of the earth on the motions of bodies at its surface; but the notion of a physical connection among the parts of the universe was taken up by Kepler in another point of view, which would probably have been considered as highly fantastical, if the result had not been, that it led to by far the most magnificent and most certain train of truths which the whole expanse of human knowledge can show. I speak of the persuasion of the existence of numerical and geometrical laws connecting the distances, times, and forces of the bodies which revolve about the central sun. That steady and intense conviction of this governing principle, which made its development and verification the leading employment of Kepler's most active and busy life, cannot be considered otherwise than as an example of profound sagacity. That it was connected, though dimly and obscurely, with the notion of a central agency or influence of some sort, emanating from the sun, cannot be doubted. Kepler, in his first essay of this kind, the *Mysterium Cosmographicum*, says, "The motion of the earth, which Copernicus had proved by *mathematical* reasons, I wanted to prove by *physical*, or, if you prefer it, metaphysical." In the twentieth chapter of that work, he endeavors to make out some relation between the distances of the Planets from the Sun and their velocities. The inveterate yet vague notions of forces which preside in this attempt, may be judged of by such passages as the following:—"We must suppose one of two things; either that the moving spirits, in proportion as they are more removed from the sun, are more feeble; or that there is one moving spirit in the centre of all the orbits, namely, in the sun, which urges each body the more vehemently in proportion as it is nearer; but in more distant spaces languishes in consequence of the remoteness and attenuation of its virtue."

We must not forget, in reading such passages, that they were written under a belief that force was requisite to keep up, as well as to change the motion of each planet; and that a body, moving in a circle, would *stop* when the force of the central point ceased, instead of moving off in a tangent to the circle, as we now know it would do. The force which Kepler supposes is a tangential force, in the direction of the body's motion, and nearly perpendicular to the radius; the force which modern philosophy has established, is in the direction of the radius, and nearly perpendicular to the body's path. Kepler was right no further than in his suspicion of a connection between the cause of motion and the distance from the centre; not only was his knowledge imperfect in all particulars, but his most general conception of the mode of action of a cause of motion was erroneous.

With these general convictions and these physical notions in his mind, Kepler endeavored to detect numerical and geometrical relations among the parts of the solar system. After extraordinary labor, perseverance, and ingenuity, he was eminently successful in discovering such relations; but the glory and merit of interpreting them according to their physical meaning, was reserved for his greater successor, Newton.

CHAPTER IV.

Inductive Epoch of Kepler.

Sect. 1.—*Intellectual Character of Kepler.*

SEVERAL persons,[1] especially in recent times, who have taken a view of the discoveries of Kepler, appear to have been surprised and somewhat discontented that conjectures, apparently so fanciful and arbitrary as his, should have led to important discoveries. They seem to have been alarmed at the *Moral* that their readers might draw, from the tale of a Quest of Knowledge, in which the Hero, though fantastical and self-willed, and violating in his conduct, as they conceived, all right rule and sound philosophy, is rewarded with the most signal triumphs. Perhaps one or two reflections may in some measure reconcile us to this result.

---

[1] Laplace, *Précis de l'Hist. d'Ast.* p. 94. "Il est affligeant pour l'esprit humain de voir ce grand homme, même dans ses derniers ouvrages, se complaire avec délices dans ses chimériques spéculations, et les regarder comme l'âme et la vie de l'astronomie."

Hist. of Ast., L. U. K., p. 53. "This success [of Kepler] may well inspire with dismay those who are accustomed to consider experiment and rigorous induction as the only means to interrogate nature with success."

Life of Kepler, L. U. K., p. 14, "Bad philosophy." P. 15, "Kepler's miraculous good fortune in seizing truths across the wildest and most absurd theories." P. 54, "The danger of attempting to follow his method in the pursuit of truth."

In the first place, we may observe that the leading thought which suggested and animated all Kepler's attempts was true, and we may add, sagacious and philosophical; namely, that there must be *some* numerical or geometrical relations among the times, distances, and velocities of the revolving bodies of the solar system. This settled and constant conviction of an important truth regulated all the conjectures, apparently so capricious and fanciful, which he made and examined, respecting particular relations in the system.

In the next place, we may venture to say, that advances in knowledge are not commonly made without the previous exercise of some boldness and license in guessing. The discovery of new truths requires, undoubtedly, minds careful and scrupulous in examining what is suggested; but it requires, no less, such as are quick and fertile in suggesting. What is Invention, except the talent of rapidly calling before us many possibilities, and selecting the appropriate one? It is true, that when we have rejected all the inadmissible suppositions, they are quickly forgotten by most persons; and few think it necessary to dwell on these discarded hypotheses, and on the process by which they were condemned, as Kepler has done. But all who discover truths must have reasoned upon many errors, to obtain each truth; every accepted doctrine must have been one selected out of many candidates. In making many conjectures, which on trial proved erroneous, Kepler was no more fanciful or unphilosophical than other discoverers have been. Discovery is not a "cautious" or "rigorous" process, in the sense of abstaining from such suppositions. But there are great differences in different cases, in the facility with which guesses are proved to be errors, and in the degree of attention with which the error and the proof are afterwards dwelt on. Kepler certainly was remarkable for the labor which he gave to such self-refutations, and for the candor and copiousness with which he narrated them; his works are in this way extremely curious and amusing; and are a very instructive exhibition of the mental process of discovery. But in this respect, I venture to believe, they exhibit to us the usual process (somewhat caricatured) of inventive minds: they rather exemplify the *rule* of genius than (as has generally been hitherto taught) the *exception*. We may add, that if many of Kepler's guesses now appear fanciful and absurd, because time and observation have refuted them, others, which were at the time equally gratuitous, have been confirmed by succeeding discoveries in a manner which makes them appear marvellously sagacious; as, for instance, his assertion of the rotation of the sun on his axis, before the invention of the telescope, and his opinion that the obliquity of the ecliptic was decreasing, but would, after a long-continued diminution, stop, and then increase again.[2] Nothing can be more just, as well as more poetically happy, than Kepler's picture of the philosopher's pursuit of scientific truth, conveyed by means of an allusion to Virgil's shepherd and shepherdess:

> Malo me Galatea petit, lasciva puella
> Et fugit ad salices et se cupit ante videri.
>
> Coy yet inviting, Galatea loves
> To sport in sight, then plunge into the groves;
> The challenge given, she darts along the green,
> Will not be caught, yet would not run unseen.

We may notice as another peculiarity of Kepler's reasonings, the length and laboriousness of the processes by which he discovered the errors of his first guesses. One of the most important talents requisite for a discoverer, is the ingenuity and skill which devises means for rapidly testing false suppositions as they offer themselves. This talent Kepler did not possess: he was not even a good arithmetical calculator, often making mistakes, some of which he detected and laments, while others escaped him to the last. But his defects in this respect were compensated by his courage and perseverance in undertaking and executing such tasks; and, what was still more admirable, he never allowed the labor he had spent upon any conjecture to produce any reluctance in abandoning the hypothesis, as soon as he had evidence of its inaccuracy. The only way in which he rewarded himself for his trouble, was by describing to the world, in his lively manner, his schemes, exertions, and feelings.

The *mystical* parts of Kepler's opinions, as his belief in astrology, his persuasion that the earth was an animal, and many of the loose moral and spiritual as well as sensible analyses by which he represented to himself the powers which he supposed to prevail in the universe, do not appear to have interfered with his discovery, but rather to have stimulated his invention, and animated his exertions. Indeed, where there are clear scientific ideas on one subject in the mind, it does not appear that mysticism on others is at all unfavorable to the successful prosecution of research.

I conceive, then, that we may consider Kepler's character as containing the general features of the character of a scientific discoverer,

---

[2] Bailly, *A. M.* iii. 175.

though some of the features are exaggerated, and some too feebly marked. His spirit of invention was undoubtedly very fertile and ready, and this and his perseverance served to remedy his deficiency in mathematical artifice and method. But the peculiar physiognomy is given to his intellectual aspect by his dwelling in a most prominent manner on those erroneovs trains of thought which other persons conceal from the world, and often themselves forget, because they find means of stopping them at the outset. In the beginning of his book (*Argumenta Capitum*) he says, "if Christopher Columbus, if Magellan, if the Portuguese, when they narrate their wanderings, are not only excused, but if we do not wish these passages omitted, and should lose much pleasure if they were, let no one blame me for doing the same." Kepler's talents were a kindly and fertile soil, which he cultivated with abundant toil and vigor; but with great scantiness of agricultural skill and implements. Weeds and the grain throve and flourished side by side almost undistinguished; and he gave a peculiar appearance to his harvest, by gathering and preserving the one class of plants with as much care and diligence as the other.

Sect. 2.—Kepler's Discovery of his Third Law.

I SHALL now give some account of Kepler's speculations and discoveries. The first discovery which he attempted, the relation among the successive distances of the planets from the sun, was a failure; his doctrine being without any solid foundation, although propounded by him with great triumph, in a work which he called *Mysterium Cosmographicum*, and which was published in 1596. The account which he gives of the train of his thoughts on this subject, namely, the various suppositions assumed, examined, and rejected, is curious and instructive, for the reasons just stated; but we shall not dwell upon these essays, since they led only to an opinion now entirely abandoned. The doctrine which professed to give the true relation of the orbits of the different planets, was thus delivered:[3] "The orbit of the earth is a circle: round the sphere to which this circle belongs, describe a dodecahedron; the sphere including this will give the orbit of Mars. Round Mars describe a tetrahedron; the circle including this will be the orbit of Jupiter. Describe a cube round Jupiter's orbit; the circle including this will be the orbit of Saturn. Now inscribe in the Earth's orbit an icosahedron; the circle inscribed in it will be the orbit of Venus. In-

---

[3] L. U. K. Kepler, 6.

scribe an octahedron in the orbit of Venus; the circle inscribed in it will be Mercury's orbit. This is the reason of the number of the planets." The five kinds of polyhedral bodies here mentioned are the only "Regular Solids."

But though this part of the *Mysterium Cosmographicum* was a failure, the same researches continued to occupy Kepler's mind; and twenty-two years later led him to one of the important rules known to us as "Kepler's Laws;" namely, to the rule connecting the mean distances of the planets from the sun with the times of their revolutions. This rule is expressed in mathematical terms, by saying that the squares of the periodic times are in the same proportion as the cubes of the distances; and was of great importance to Newton in leading him to the law of the sun's attractive force. We may properly consider this discovery as the sequel of the train of thought already noticed. In the beginning of the *Mysterium*, Kepler had said, "In the year 1595, I brooded with the whole energy of my mind on the subject of the Copernican system. There were three things in particular of which I pertinaciously sought the causes why they are not other than they are; the number, the size, and the motion of the orbits." We have seen the nature of his attempt to account for the two first of these points. He had also made some essays to connect the motions of the planets with their distances, but with his success in this respect he was not himself completely satisfied. But in the fifth book of the *Harmonice Mundi*, published in 1619, he says, "What I prophesied two-and-twenty years ago as soon as I had discovered the Five Solids among the Heavenly Bodies; what I firmly believed before I had seen the *Harmonics* of Ptolemy; what I promised my friends in the title of this book (*On the most perfect Harmony of the Celestial Motions*), which I named before I was sure of my discovery; what sixteen years ago I regarded as a thing to be sought; that for which I joined Tycho Brahe, for which I settled in Prague, for which I have devoted the best part of my life to astronomical contemplations; at length I have brought to light, and have recognized its truth beyond my most sanguine expectations."

The rule thus referred to is stated in the third Chapter of this fifth Book. "It is," he says, "a most certain and exact thing that the proportion which exists between the periodic times of any two planets is precisely the sesquiplicate of the proportion of their mean distances; that is, of the radii of the orbits. Thus, the period of the earth is one year, that of Saturn thirty years; if any one trisect the proportion, that is, take the cube root of it, and double the proportion so found, that is, square it, he will find the exact proportion of the distances of the Earth and of Saturn from the sun. For the cube root of 1 is 1, and the square of this is 1; and the cube root of 30 is greater than 3, and therefore the square of it is greater than 9. And Saturn at his mean distance from the sun is at a little more than 9 times the mean distance of the Earth."

When we now look back at the time and exertions which the establishment of this law cost Kepler, we are tempted to imagine that he was strangely blind in not seeing it sooner. His object, we might reason, was to discover a law connecting the distances and the periodic times. What law of connection could be more simple and obvious, we might say, than that one of these quantities should vary as some *power* of the other, or as some *root;* or as some combination of the two, which in a more general view, may still be called a *power?* And if the problem had been viewed in this way, the question must have occurred, to *what* power of the periodic times are the distances proportional? And the answer must have been, the trial being made, that they are proportional to the square of the cube root. This *ex-post-facto* obviousness of discoveries is a delusion to which we are liable with regard to many of the most important principles. In the case of Kepler, we may observe, that the process of connecting two classes of quantities by comparing their *powers*, is obvious only to those who are familiar with general algebraical views; and that in Kepler's time, algebra had not taken the place of geometry, as the most usual vehicle of mathematical reasoning. It may be added, also, that Kepler always sought his *formal* laws by means of *physical* reasonings; and these, though vague or erroneous, determined the nature of the mathematical connection which he assumed. Thus in the *Mysterium* he had been led by his notions of moving virtue of the sun to this conjecture, among others—that, in the planets, the increase of the periods will be double of the difference of the distances; which supposition he found to give him an approach to the actual proportion of the distances, but one not sufficiently close to satisfy him.

The greater part of the fifth Book of the *Harmonics of the Universe* consists in attempts to explain various relations among the distances, times, and eccentricities of the planets, by means of the ratios which belong to certain concords and discords. This portion of the work is so complex and laborious, that probably few modern readers have had courage to go through it. Delambre acknowledged that his patience often failed him during the task;[4] and subscribes to the judgment of Bailly: "After this sublime effort, Kepler replunges himself in the relations of music to the motions, the distance, and the eccentricities of the planets. In all these harmonic ratios there is not one true relation; in a crowd of ideas there is not one truth: he becomes a man after being a spirit of light." Certainly these speculations are of no value, but we may look on them with toleration, when we recollect that Newton has sought for analogies between the spaces occupied by the prismatic colors and the notes of the gamut.[5] The numerical relations of Concords are so peculiar that we can easily suppose them to have other bearings than those which first offer themselves.

It does not belong to my present purpose to speak at length of the speculations concerning the forces producing the celestial motions by which Kepler was led to this celebrated law, or of those which he deduced from it, and which are found in the *Epitome Astronomiæ Copernicanæ*, published in 1622. In that work also (p. 554), he extended this law, though in a loose manner, to the satellites of Jupiter. These *physical* speculations were only a vague and distant prelude to Newton's discoveries; and the law, as a *formal* rule, was complete in itself. We must now attend to the history of the other two laws with which Kepler's name is associated.

Sect. 3.—*Kepler's Discovery of his First and Second Laws.—Elliptical Theory of the Planets.*

THE propositions designated as Kepler's First and Second Laws are these: That the orbits of the planets are elliptical; and, That the areas described, or *swept*, by lines drawn from the sun to the planet, are proportional to the times employed in the motion.

The occasion of the discovery of these laws was the attempt to reconcile the theory of Mars to the theory of eccentrics and epicycles; the event of it was the complete overthrow of that theory, and the establishment, in its stead, of the Elliptical Theory of the planets. Astronomy was now ripe for such a change. As soon as Copernicus had taught men that the orbits of the planets were to be referred to the sun, it obviously became a question, what was the true form of these orbits, and the rule of motion of each planet in its own orbit. Copernicus represented the motions in longitude by means of eccen-

---

[4] *A. M.* a. 358.

[5] *Optics*, b. ii. p. iv. Obs. 5.

trics and epicycles, as we have already said; and the motions in latitude by certain *librations*, or alternate elevations and depressions of epicycles. If a mathematician had obtained a collection of true positions of a planet, the form of the orbit and the motion of the star would have been determined with reference to the sun as well as to the earth; but this was not possible, for though the *geocentric* position, or the direction in which the planet was seen, could be observed, its distance from the earth was not known. Hence, when Kepler attempted to determine the orbit of a planet, he combined the observed geocentric places with successive modifications of the theory of epicycles, till at last he was led, by one step after another, to change the epicyclical into the elliptical theory. We may observe, moreover, that at every step he endeavored to support his new suppositions by what he called, in his fanciful phraseology, "sending into the field a reserve of new physical reasonings on the rout and dispersion of the veterans;"[6] that is, by connecting his astronomical hypotheses with new imaginations, when the old ones became untenable. We find, indeed, that this is the spirit in which the pursuit of knowledge is generally carried on with success; those men arrive at truth who eagerly endeavor to connect remote points of their knowledge, not those who stop cautiously at each point till something compels them to go beyond it.

Kepler joined Tycho Brahe at Prague in 1600, and found him and Longomontanus busily employed in correcting the theory of Mars; and he also then entered upon that train of researches which he published in 1609 in his extraordinary work *On the Motions of Mars.* In this work, as in others, he gives an account, not only of his success, but of his failures, explaining, at length, the various suppositions which he had made, the notions by which he had been led to invent or to entertain them, the processes by which he had proved their

---

[6] I will insert this passage, as a specimen of Kepler's fanciful mode of narrating the defeats which he received in the war which he carried on with Mars. "Dum in hunc modum de Martis motibus triumpho, eique ut planè devicto tabularum carceres et equationum compedes necto, diversis nuntiatur locis, futilem victoriam ut bellum totâ mole recrudescere. Nam domi quidem hostis ut captivus contemptus, rupit omnia equationum vincula, carceresque tabularum effregit. Foris speculatores profligerunt meas causarum physicarum arcessitas copias earumque jugum excusserunt resumtâ libertate. Jamque parum abfuit quia hostis fugitivus sese cum rebellibus suis conjungeret meque in desperationem adigeret: nisi raptim, nova rationum physicarum subsidia, fusis et palantibus veteribus, submisissem, et qua se captivus proripuisset, omni diligentia, edoctus vestigiis ipsius nullâ morâ interpositâ inhæsisserem."

falsehood, and the alternations of hope and sorrow, of vexation and triumph, through which he had gone. It will not be necessary for us to cite many passages of these kinds, curious and amusing as they are.

One of the most important truths contained in the motions of Mars is the discovery that the plane of the orbit of the planet should be considered with reference to the sun itself, instead of referring it to any of the other centres of motion which the eccentric hypothesis introduced: and that, when so considered, it had none of the librations which Ptolemy and Copernicus had attributed to it. The fourteenth chapter of the second part asserts, "Plana eccentricorum esse ἀτάλαντα;" that the planes are *unlibrating;* retaining always the same inclination to the ecliptic, and the same *line of nodes.* With this step Kepler appears to have been justly delighted. "Copernicus," he says, "not knowing the value of what he possessed (his system), undertook to represent Ptolemy, rather than nature, to which, however, he had approached more nearly than any other person. For being rejoiced that the quantity of the latitude of each planet was increased by the approach of the earth to the planet, according to his theory, he did not venture to reject the rest of Ptolemy's increase of latitude, but in order to express it, devised librations of the planes of the eccentric, depending not upon its own eccentric, but (most improbably) upon the orbit of the earth, which has nothing to do with it. I always fought against this impertinent tying together of two orbits, even before I saw the observations of Tycho; and I therefore rejoice much that in this, as in others of my preconceived opinions, the observations were found to be on my side." Kepler established his point by a fair and laborious calculation of the results of observations of Mars made by himself and Tycho Brahe; and had a right to exult when the result of these calculations confirmed his views of the symmetry and simplicity of nature.

We may judge of the difficulty of casting off the theory of eccentrics and epicycles, by recollecting that Copernicus did not do it at all, and that Kepler only did it after repeated struggles; the history of which occupies thirty-nine Chapters of his book. At the end of them he says, "This prolix disputation was necessary, in order to prepare the way to the natural form of the equations, of which I am now to treat.[7] My first error was, that the path of a planet is a perfect circle;—an opinion which was a more mischievous thief of my time,

---

[7] *De Stellâ Martis*, iii. 40.

in proportion as it was supported by the authority of all philosophers, and apparently agreeable to metaphysics." But before he attempts to correct this erroneous part of his hypothesis, he sets about discovering the law according to which the different parts of the orbit are described in the case of the earth, in which case the eccentricity is so small that the effect of the oval form is insensible. The result of this inquiry was[8] the Rule, that the time of describing any arc of the orbit is proportional to the area intercepted between the curve and two lines drawn from the sun to the extremities of the arc. It is to be observed that this rule, at first, though it had the recommendation of being selected after the unavoidable abandonment of many, which were suggested by the notions of those times, was far from being adopted upon any very rigid or cautious grounds. A rule had been proved at the apsides of the orbit, by calculation from observations, and had then been extended by conjecture to other parts of the orbit; and the rule of the areas was only an approximate and inaccurate mode of representing this rule, employed for the purpose of brevity and convenience, in consequence of the difficulty of applying, geometrically, that which Kepler now conceived to be the true rule, and which required him to find the sum of the lines drawn from the sun to *every* point of the orbit. When he proceeded to apply this rule to Mars, in whose orbit the oval form is much more marked, additional difficulties came in his way; and here again the true supposition, that the *oval* is of that special kind called *ellipse*, was adopted at first only in order to simplify calculation,[9] and the deviation from exactness in the result was attributed to the inaccuracy of those approximate processes. The supposition of the oval had already been forced upon Purbach in the case of Mercury, and upon Reinhold in the case of the Moon. The centre of the epicycle was made to describe an egg-shaped figure in the former case, and a lenticular figure in the latter.[10]

It may serve to show the kind of labor by which Kepler was led to his result, if we here enumerate, as he does in his forty-seventh Chapter,[11] six hypotheses, on which he calculated the longitude of Mars, in order to see which best agreed with observation.

1. The simple eccentricity.
2. The bisection of the eccentricity, and the duplication of the superior part of the equation.

---

[8] *De Stellâ Martis*, p. 194.
[9] Ib. iv. c. 47.
[10] L. U. K. Kepler, p. 30.
[11] *De Stellâ Martis*, p. 228.

3. The bisection of the eccentricity, and a stationary point of equations, after the manner of Ptolemy.

4. The vicarious hypothesis by a free section of the eccentricity made to agree as nearly as possible with the truth.

5. The physical hypothesis on the supposition of a perfect circle.

6. The physical hypothesis on the supposition of a perfect ellipse.

By the physical hypothesis, he meant the doctrine that the time of a planet's describing any part of its orbit is proportional to the distance of the planet from the sun, for which supposition, as we have said, he conceived that he had assigned physical reasons.

The two last hypotheses came the nearest to the truth, and differed from it only by about eight minutes, the one in excess and the other in defect. And, after being much perplexed by this remaining error, it at last occurred to him[12] that he might take another ellipsis, exactly intermediate between the former one and the circle, and that this must give the path and the motion of the planet. Making this assumption, and taking the areas to represent the times, he now saw[13] that both the longitude and the distances of Mars would agree with observation to the requisite degree of accuracy. The rectification of the former hypothesis, when thus stated, may, perhaps, appear obvious. And Kepler informs us that he had nearly been anticipated in this step (c. 55). "David Fabricius, to whom I had communicated my hypothesis of cap. 45, was able, by his observations, to show that it erred in making the distances too short at mean longitudes; of which he informed me by letter while I was laboring, by repeated efforts, to discover the true hypothesis. So nearly did he get the start of me in detecting the truth." But this was less easy than it might seem. When Kepler's first hypothesis was enveloped in the complex construction requisite in order to apply it to each point of the orbit, it was far more difficult to see where the error lay, and Kepler hit upon it only by noticing the coincidences of certain numbers, which, as he says, raised him as if from sleep, and gave him a new light. We may observe, also, that he was perplexed to reconcile this new view, according to which the planet described an exact ellipse, with his former opinion, which represented the motion by means of libration in an epicycle. "This," he says, "was my greatest trouble, that, though I considered and reflected till I was almost mad, I could not find why the planet to which, with so much probability, and with such an exact

---

[12] *De Stellâ Martis*, c. 58. [13] Ibid. p. 235.

accordance of the distances, libration in the diameter of the epicycle was attributed, should, according to the indication of the equations, go in an elliptical path. What an absurdity on my part! as if libration in the diameter might not be a way to the ellipse!"

Another scruple respecting this theory arose from the impossibility of solving, by any geometrical construction, the problem to which Kepler was thus led, namely, "To divide the area of a semicircle in a given ratio, by a line drawn from any point of the diameter." This is still termed "Kepler's Problem," and is, in fact, incapable of exact geometrical solution. As, however, the calculation can be performed, and, indeed, was performed by Kepler himself, with a sufficient degree of accuracy to show that the elliptical hypothesis is true, the insolubility of this problem is a mere mathematical difficulty in the deductive process, to which Kepler's induction gave rise.

Of Kepler's physical reasonings we shall speak more at length on another occasion. His numerous and fanciful hypotheses had discharged their office, when they had suggested to him his many lines of laborious calculation, and encouraged him under the exertions and disappointments to which these led. The result of this work was the formal laws of the motion of Mars, established by a clear induction, since they represented, with sufficient accuracy, the best observations. And we may allow that Kepler was entitled to the praise which he claims in the motto on his first leaf. Ramus had said that if any one would construct an astronomy without hypothesis, he would be ready to resign to him his professorship in the University of Paris. Kepler quotes this passage, and adds, "it is well, Ramus, that you have run from this pledge, by quitting life and your professorship;[14] if you held it still, I should, with justice, claim it." This was not saying too much, since he had entirely overturned the hypothesis of eccentrics and epicycles, and had obtained a theory which was a mere representation of the motions and distances as they were observed.

---

[14] Ramus perished in the Massacre of St. Bartholomew.

CHAPTER V.

Sequel to the Epoch of Kepler. Reception, Verification, and Extension of the Elliptical Theory.

Sect. 1.—*Application of the Elliptical Theory to the Planets.*

THE extension of Kepler's discoveries concerning the orbit of Mars to the other planets, obviously offered itself as a strong probability, and was confirmed by trial. This was made in the first place upon the orbit of Mercury; which planet, in consequence of the largeness of its eccentricity, exhibits more clearly than the others the circumstances of the elliptical motion. These and various other supplementary portions of the views to which Kepler's discoveries had led, appeared in the latter part of his *Epitome Astronomiæ Copernicanæ*, published in 1622.

The real verification of the new doctrine concerning the orbits and motions of the heavenly bodies was, of course, to be found in the construction of tables of those motions, and in the continued comparison of such tables with observation. Kepler's discoveries had been founded, as we have seen, principally on Tycho's observations. Longomontanus (so called as being a native of Langberg in Denmark), published in 1621, in his *Astronomia Danica*, tables founded upon the theories as well as the observations of his countryman. Kepler[1] in 1627 published his tables of the planets, which he called *Rudolphine Tables*, the result and application of his own theory. In 1633, Lansberg, a Belgian, published also *Tabulæ Perpetuæ*, a work which was ushered into the world with considerable pomp and pretension, and in which the author cavils very keenly at Kepler and Brahe. We may judge of the impression made upon the astronomical world in general by these rival works, from the account which our countryman Jeremy Horrox has given of their effect on him. He had been seduced by the magnificent promises of Lansberg, and the praises of his admirers, which are prefixed to the work, and was persuaded that the common opinion which preferred Tycho and Kepler to him was a prejudice. In 1636, however, he became acquainted with Crabtree, another young astrono-

[1] Rheticus, *Narratio*, p. 98.

mer, who lived in the same part of Lancashire. By him Horrox was warned that Lansberg was not to be depended on; that his hypotheses were vicious, and his observations falsified or forced into agreement with his theories. He then read the works and adopted the opinions of Kepler; and after some hesitation which he felt at the thought of attacking the object of his former idolatry, he wrote a dissertation on the points of difference between them. It appears that, at one time, he intended to offer himself as the umpire who was to adjudge the prize of excellence among the three rival theories of Longomontanus, Kepler, and Lansberg; and, in allusion to the story of ancient mythology, his work was to have been called *Paris Astronomicus;* we easily see that he would have given the golden apple to the Keplerian goddess. Succeeding observations confirmed his judgment: and the *Rudolphine Tables*, thus published seventy-six years after the Prutenic, which were founded on the doctrines of Copernicus, were for a long time those universally used.

Sect. 2.—*Application of the Elliptical Theory to the Moon.*

THE reduction of the Moon's motions to rule was a harder task than the formation of planetary tables, if accuracy was required; for the Moon's motion is affected by an incredible number of different and complex inequalities, which, till their law is detected, appear to defy all theory. Still, however, progress was made in this work. The most important advances were due to Tycho Brahe. In addition to the first and second inequalities of the moon (the *Equation of the Centre*, known very early, and the *Evection*, which Ptolemy had discovered), Tycho proved that there was another inequality, which he termed the *Variation*,[2] which depended on the moon's position with respect to the sun, and which at its maximum was forty minutes and a half, about a quarter of the evection. He also perceived, though not very distinctly, the necessity of another correction of the moon's place depending on the sun's longitude, which has since been termed the *Annual Equation.*

These steps concerned the Longitude of the Moon; Tycho also made important advances in the knowledge of the Latitude. The Inclination of the Orbit had hitherto been assumed to be the same at all

---

[2] We have seen (chap. iii.), that Aboul-Wefa, in the tenth century, had already noticed this inequality; but his discovery had been entirely forgotten long before the time of Tycho, and has only recently been brought again into notice.

times; and the motion of the Node had been supposed uniform. He found that the inclination increased and diminished by twenty minutes, according to the position of the line of nodes; and that the nodes, though they regress upon the whole, sometimes go forwards and sometimes go backwards.

Tycho's discoveries concerning the moon are given in his *Progymnasmata*, which was published in 1603, two years after the author's death. He represents the Moon's motion in longitude by means of certain combinations of epicycles and eccentrics. But after Kepler had shown that such devices are to be banished from the planetary system, it was impossible not to think of extending the elliptical theory to the moon. Horrox succeeded in doing this; and in 1638 sent this essay to his friend Crabtree. It was published in 1673, with the numerical elements requisite for its application added by Flamsteed. Flamsteed had also (in 1671–2) compared this theory with observation, and found that it agreed far more nearly than the *Philolaic Tables* of Bullialdus, or the *Carolinian Tables* of Street (*Epilogus ad Tabulas*). Moreover Horrox, by making the centre of the ellipse revolve in an epicycle, gave an explanation of the evection, as well as of the equation of the centre.[3]

Modern astronomers, by calculating the effects of the perturbing forces of the solar system, and comparing their calculations with observation, have added many new corrections or equations to those known at the time of Horrox; and since the Motions of the heavenly bodies were even then affected by these variations as yet undetected, it is clear that the Tables of that time must have shown some errors when compared with observation. These errors much perplexed astronomers, and naturally gave rise to the question whether the motions of the heavenly bodies really were exactly regular, or whether they were not affected by accidents as little reducible to rule as wind and weather. Kepler had held the opinion of the *casualty* of such errors; but Horrox, far more philosophically, argues against this opinion, though he

[3] Horrox (*Horrockes* as he himself spelt his name) gave a first sketch of his theory in letters to his friend Crabtree in 1638: in which the variation of the eccentricity is not alluded to. But in Crabtree's letter to Gascoigne in 1642, he gives Horrox's rule concerning it; and Flamsteed in his *Epilogue* to the Tables, published by Wallis along with Horrox's works in 1673, gave an explanation of the theory which made it amount very nearly to a revolution of the centre of the ellipse in an epicycle. Halley afterwards made a slight alteration; but hardly, I think, enough to justify Newton's assertion (*Princip.* Lib. iii. Prop. 35, Schol.), "Halleius centrum ellipseos in epicyclo locavit." See Baily's *Flamsteed*, p. 683.

allows that he is much embarrassed by the deviations. His arguments show a singularly clear and strong apprehension of the features of the case, and their real import. He says,[4] "these errors of the tables are alternately in excess and defect; how could this constant compensation happen if they were casual? Moreover, the alternation from excess to defect is most rapid in the Moon, most slow in Jupiter and Saturn, in which planets the error continues sometimes for years. If the errors were casual, why should they not last as long in the Moon as in Saturn? But if we suppose the tables to be right in the mean motions, but wrong in the equations, these facts are just what must happen; since Saturn's inequalities are of long period, while those of the Moon are numerous, and rapidly changing." It would be impossible, at the present moment, to reason better on this subject; and the doctrine, that all the apparent irregularities of the celestial motions are really regular, was one of great consequence to establish at this period of the science.

Sect. 3.—*Causes of the further Progress of Astronomy.*

We are now arrived at the time when theory and observation sprang forwards with emulous energy. The physical theories of Kepler, and the reasonings of other defenders of the Copernican theory, led inevitably, after some vagueness and perplexity, to a sound science of Mechanics; and this science in time gave a new face to Astronomy. But in the mean time, while mechanical mathematicians were generalizing from the astronomy already established, astronomers were accumulating new facts, which pointed the way to new theories and new generalizations. Copernicus, while he had established the permanent length of the year, had confirmed the motion of the sun's apogee, and had shown that the eccentricity of the earth's orbit, and the obliquity of the ecliptic, were gradually, though slowly, diminishing. Tycho had accumulated a store of excellent observations. These, as well as the laws of the motions of the moon and planets already explained, were materials on which the Mechanics of the Universe was afterwards to employ its most matured powers. In the mean time, the telescope had opened other new subjects of notice and speculation; not only confirming the Copernican doctrine by the phases of Venus, and the analogical examples of Jupiter and Saturn, which with their Satellites

---

[4] *Astron. Kepler.* Proleg. p. 17.

appeared like models of the Solar System; but disclosing unexpected objects, as the Ring of Saturn, and the Spots of the Sun. The art of observing made rapid advances, both by the use of the telescope, and by the sounder notions of the construction of instruments which Tycho introduced. Copernicus had laughed at Rheticus, when he was disturbed about single minutes; and declared that if he could be sure to ten minutes of space, he should be as much delighted as Pythagoras was when he discovered the property of the right-angled triangle. But Kepler founded the revolution which he introduced on a quantity less than this. "Since," he says,[5] "the Divine Goodness has given us in Tycho an observer so exact that this error of eight minutes is impossible, we must be thankful to God for this, and turn it to account. And these eight minutes, which we must not neglect, will, of themselves, enable us to reconstruct the whole of astronomy." In addition to other improvements, the art of numerical calculation made an inestimable advance by means of Napier's invention of Logarithms; and the progress of other parts of pure mathematics was proportional to the calls which astronomy and physics made upon them.

The exactness which observation had attained enabled astronomers both to verify and improve the existing theories, and to study the yet unsystematized facts. The science was, therefore, forced along by a strong impulse on all sides, and its career assumed a new character. Up to this point, the history of European Astronomy was only the sequel of the history of Greek Astronomy; for the heliocentric system, as we have seen, had had a place among the guesses, at least, of the inventive and acute intellects of the Greek philosophers. But the discovery of Kepler's Laws, accompanied, as from the first they were, with a conviction that the relations thus brought to light were the effects and exponents of physical causes, led rapidly and irresistibly to the Mechanical Science of the skies, and collaterally, to the Mechanical Science of the other parts of Nature: Sound, and Light, and Heat; and Magnetism, and Electricity, and Chemistry. The history of these Sciences, thus treated, forms the sequel of the present work, and will be the subject of the succeeding volumes. And since, as I have said, our main object in this work is to deduce, from the history of science, the philosophy of scientific discovery, it may be regarded as fortunate for our purpose that the history, after this point, so far changes its aspect as to offer new materials for such speculations. The details of

---

[5] *De Stellâ Martis*, c. 19.

a history of astronomy, such as the history of astronomy since Newton has been, though interesting to the special lovers of that science, would be too technical, and the features of the narrative too monotonous and unimpressive, to interest the general reader, or to suggest a comprehensive philosophy of science. But when we pass from the Ideas of Space and Time to the Ideas of Force and Matter, of Mediums by which action and sensation are produced, and of the Intimate Constitution of material bodies, we have new fields of inquiry opened to us. And when we find that in these fields, as well as in astronomy, there are large and striking trains of unquestioned discovery to be narrated, we may gird ourselves afresh to the task of writing, and I hope, of reading, the remaining part of the History of the Inductive Sciences, in the trust that it will in some measure help us to answer the important questions, What is Truth? and, How is it to be discovered?

BOOK VI.

THE MECHANICAL SCIENCES.

HISTORY OF MECHANICS,

INCLUDING

FLUID MECHANICS.

ΚΡΑΤΟΣ ΒΙΑ ΤΕ, *σφῷν μὲν ἐντολὴ Διὸς*
Ἔχει Τέλος δὴ, κ' ουδὲν ἐμποδὼν ἔτι.

ÆSCHYLUS. *Prom. Vinct.* 13.

You, FORCE and POWER, have done your destined task;
And naught impedes the work of other hands.

INTRODUCTION.

WE enter now upon a new region of the human mind. In passing from Astronomy to Mechanics we make a transition from the *formal* to the *physical* sciences;—from time and space to force and matter;—from *phenomena* to *causes*. Hitherto we have been concerned only with the paths and orbits, the periods and cycles, the angles and distances, of the objects to which our sciences applied, namely, the heavenly bodies. How these motions are produced;—by what agencies, impulses, powers, they are determined to be what they are;—of what nature are the objects themselves;—are speculations which we have hitherto not dwelt upon. The history of such speculations now comes before us; but, in the first place, we must consider the history of speculations concerning motion in general, terrestrial as well as celestial. We must first attend to Mechanics, and afterwards return to Physical Astronomy.

In the same way in which the development of Pure Mathematics, which began with the Greeks, was a necessary condition of the progress of Formal Astronomy, the creation of the science of Mechanics now became necessary to the formation and progress of Physical Astronomy. Geometry and Mechanics were studied for their own sakes; but they also supplied ideas, language, and reasoning to other sciences. If the Greeks had not cultivated Conic Sections, Kepler could not have superseded Ptolemy; if the Greeks had cultivated Dynamics,[1] Kepler might have anticipated Newton.

---

[1] *Dynamics* is the science which treats of the Motions of Bodies; *Statics* is the science which treats of the Pressure of Bodies which are in equilibrium, and therefore at rest.

CHAPTER I.

Prelude to the Epoch of Galileo.

Sect. 1.—*Prelude to the Science of Statics.*

SOME steps in the science of Motion, or rather in the science of Equilibrium, had been made by the ancients, as we have seen. Archimedes established satisfactorily the doctrine of the Lever, some important properties of the Centre of Gravity, and the fundamental proposition of Hydrostatics. But this beginning led to no permanent progress. Whether the distinction between the principles of the doctrine of Equilibrium and of Motion was clearly seen by Archimedes, we do not know; but it never was caught hold of by any of the other writers of antiquity, or by those of the Stationary Period. What was still worse, the point which Archimedes had won was not steadily maintained.

We have given some examples of the general ignorance of the Greek philosophers on such subjects, in noticing the strange manner in which Aristotle refers to mathematical properties, in order to account for the equilibrium of a lever, and the attitude of a man rising from a chair. And we have seen, in speaking of the indistinct ideas of the Stationary Period, that the attempts which were made to extend the statical doctrine of Archimedes, failed, in such a manner as to show that his followers had not clearly apprehended the idea on which his reasoning altogether depended. The clouds which he had, for a moment, cloven in his advance, closed after him, and the former dimness and confusion settled again on the land.

This dimness and confusion, with respect to all subjects of mechanical reasoning, prevailed still, at the period we now have to consider; namely, the period of the first promulgation of the Copernican opinions. This is so important a point that I must illustrate it further.

Certain general notions of the connection of cause and effect in motion, exist in the human mind at all periods of its development, and are implied in the formation of language and in the most familiar employments of men's thoughts. But these do not constitute a *science* of Mechanics, any more than the notions of *square* and *round* make a Geometry, or the notions of *months* and *years* make an Astronomy. The unfolding these Notions into distinct Ideas, on which can be founded principles and reasonings, is further requisite, in order to produce a science; and, with respect to the doctrines of Motion, this was long in coming to pass; men's thoughts remained long entangled in their primitive and unscientific confusion.

We may mention one or two features of this confusion, such as we find in authors belonging to the period now under review.

We have already, in speaking of the Greek School Philosophy, noticed the attempt to explain some of the differences among Motions, by classifying them into Natural Motions and Violent Motions; and we have spoken of the assertion that heavy bodies fall quicker in proportion to their greater weight. These doctrines were still retained: yet the views which they implied were essentially erroneous and unsound; for they did not refer distinctly to a measurable Force as the cause of all motion or change of motion; and they confounded the causes which *produce*, and those which *preserve*, motion. Hence such principles did not lead immediately to any advance of knowledge, though efforts were made to apply them, in the cases both of terrestrial Mechanics and of the motions of the heavenly bodies.

The effect of the Inclined Plane was one of the first, as it was one of the most important, propositions, on which modern writers employed themselves. It was found that a body, when supported on a sloping surface, might be sustained or raised by a force or exertion which would not have been able to sustain or raise it without such support. And hence, *The Inclined Plane* was placed in the list of Mechanical Powers, or simple machines by which the efficacy of forces is increased: the question was, in what proportion this increase of efficiency takes place. It is easily seen that the force requisite to sustain a body is smaller, as the slope on which it rests is smaller; Cardan (whose work, *De Proportionibus Numerorum, Motuum, Ponderum*, &c., was published in 1545) asserts that the force is double when the angle of inclination is double, and so on for other proportions: this is probably a guess, and is an erroneous one. Guido Ubaldi, of Marchmont, published at Pesaro, in 1577, a work which he called *Mechanicorum Liber*, in which he endeavors to prove that an acute wedge will produce a greater mechanical effect than an obtuse one, without determining in what proportion. There is, he observes, "a certain repugnance" between the direction in which the side of the wedge tends to move the obstacle, and the direction in which it really does move. Thus the Wedge and the Inclined Plane are connected in principle. He also refers the Screw to the Inclined Plane and the Wedge, in a manner which shows a just apprehension of the question. Benedetti (1585) treats the Wedge in a different manner: not exact, but still showing some powers of thought on mechanical subjects. Michael Varro, whose *Tractatus de Motu* was published at Geneva in 1584, deduces the wedge from the composition of hypothetical motions, in a way which may appear to some persons an anticipation of the doctrine of the Composition of Forces.

There is another work on subjects of this kind, of which several editions were published in the sixteenth century, and which treats this matter in nearly the same way as Varro, and in favor of which a claim has been made[1] (I think an unfounded one), as if it contained the true principle of this problem. The work is "Jordanus Nemorarius *De Ponderositate*." The date and history of this author were probably even then unknown; for in 1599, Benedetti, correcting some of the errors of Tartalea, says they are taken "a Jordano quodam antiquo." The book was probably a kind of school-book, and much used; for an edition printed at Frankfort, in 1533, is stated to be *Cum gratia et privilegio Imperiali, Petro Apiano mathematico Ingolstadiano ad xxx annos concesso.* But this edition does not contain the Inclined Plane. Though those who compiled the work assert in words something like the inverse proportion of Weights and their Velocites, they had not learnt at that time how to apply this maxim to the Inclined Plane; nor were they ever able to render a sound reason for it. In the edition of Venice, 1565, however, such an application is attempted. The reasonings are founded on the Aristotelian assumption, "that bodies descend more quickly in proportion as they are heavier." To this principle are added some others; as, that "a body is heavier in proportion as it descends more directly to the centre," and that, in proportion as a body descends more obliquely, the intercepted part of the direct descent is smaller. By means of these principles, the "descending force" of bodies, on inclined planes, was compared, by a process, which, so far as it forms a line of proof at all, is a somewhat curious example of confused and vicious reasoning. When two bodies are supported on two inclined planes, and are connected by a string passing over the junction of the planes, so that when one descends the other ascends,

[1] Mr. Drinkwater's *Life of Galileo*, in the Lib. Usef. Kn. p. 88.

they must move through equal spaces on the planes; but on the plane which is more oblique (that is, more nearly horizontal), the vertical descent will be smaller in the same proportion in which the plane is longer. Hence, by the Aristotelian principle, the weight of the body on the longer plane is less; and, to produce an equality of effect, the body must be greater in the same proportion. We may observe that the Aristotelian principle is not only false, but is here misapplied; for its genuine meaning is, that when bodies *fall freely* by gravity, they move quicker in proportion as they are heavier; but the rule is here applied to the motions which bodies *would* have, if they were moved by a force extraneous to their gravity. The proposition was supposed by the Aristotelians to be true of *actual* velocities; it is applied by Jordanus to *virtual* velocities, without his being aware what he was doing. This confusion being made, the result is got at by taking for granted that bodies *thus* proved to be equally *heavy*, have equal powers of descent on the inclined planes; whereas, in the previous part of the reasoning, the weight was supposed to be proportional to the descent in the vertical direction. It is obvious, in all this, that though the author had adopted the false Aristotelian principle, he had not settled in his own mind whether the motions of which it spoke were actual or virtual motions;—motions in the direction of the inclined plane, or of the intercepted parts of the vertical, corresponding to these; nor whether the "descending force" of a body was something different from its weight. We cannot doubt that, if he had been required to point out, with any exactness, the cases to which his reasoning applied, he would have been unable to do so; not possessing any of those clear fundamental Ideas of Pressure and Force, on which alone any real knowledge on such subjects must depend. The whole of Jordanus's reasoning is an example of the confusion of thought of his period, and of nothing more. It no more supplied the want of some man of genius, who should give the subject a real scientific foundation, than Aristotle's knowledge of the proportion of the weights on the lever superseded the necessity of Archimedes' proof of it.

We are not, therefore, to wonder that, though this pretended theorem was copied by other writers, as by Tartalea, in his *Quesiti et Inventioni Diversi*, published in 1554, no progress was made in the real solution of any one mechanical problem by means of it. Guido Ubaldi, who, in 1577, writes in such a manner as to show that he had taken a good hold of his subject for his time, refers to Pappus's solution of the problem of the Inclined Plane, but makes no mention of that of Jordanus and Tartalea.[2] No progress was likely to occur, till the mathematicians had distinctly recovered the genuine Idea of Pressure, as a Force producing equilibrium, which Archimedes had possessed, and which was soon to reappear in Stevinus.

The properties of the Lever had always continued known to mathematicians, although, in the dark period, the superiority of the proof given by Archimedes had not been recognized. We are not to be surprised, if reasonings like those of Jordanus were applied to demonstrate the theories of the Lever with apparent success. Writers on Mechanics were, as we have seen, so vacillating in their mode of dealing with words and propositions, that their maxims could be made to prove any thing which was already known to be true.

We proceed to speak of the beginning of the real progress of Mechanics in modern times.

Sect. 2.—Revival of the Scientific Idea of Pressure.—Stevinus.—Equilibrium of Oblique Forces.

THE doctrine of the Centre of Gravity was the part of the mechanical speculations of Archimedes which was most diligently prosecuted after his time. Pappus and others, among the ancients, had solved some new problems on this subject, and Commandinus, in 1565, published *De Centro Gravitatis Solidorum.* Such treatises contained, for the most part, only mathematical consequences of the doctrines of Archimedes; but the mathematicians also retained a steady conviction of the mechanical property of the Centre of Gravity, namely, that all the weight of the body might be collected there, without any change in the mechanical results; a conviction which is closely connected with our fundamental conceptions of mechanical action. Such a principle, also, will enable us to determine the result of many simple mechanical arrangements; for instance, if a mathematician of those days had been asked whether a solid ball could be made of such a form, that, when placed on a horizontal plane, it should go on rolling forwards without limit merely by the effect of its own weight, he would probably have answered, that it could not; for that the centre of gravity of the ball would seek the lowest position it could find, and that, when it had found this, the ball could have no tendency to roll any further. And, in making this assertion, the supposed reasoner would not be an-

[2] Ubaldi mentions and blames Jordanus's way of treating the Lever. (See his Preface.)

ticipating any wider proofs of the impossibility of a *perpetual motion*, drawn from principles subsequently discovered, but would be referring the question to certain fundamental convictions, which, whether put into Axioms or not, inevitably accompany our mechanical conceptions.

In the same way, Stevinus of Bruges, in 1586, when he published his *Beghinselen der Waaghconst* (Principles of Equilibrium), had been asked why a loop of chain, hung over a triangular beam, could not, as he asserted it could not, go on moving round and round perpetually, by the action of its own weight, he would probably have answered, that the weight of the chain, if it produced motion at all, must have a tendency to bring it into some certain position, and that when the chain had reached this position, it would have no tendency to go any further; and thus he would have reduced the impossibility of such a perpetual motion, to the conception of gravity, as a force tending to produce equilibrium; a principle perfectly sound and correct.

Upon this principle thus applied, Stevinus did establish the fundamental property of the Inclined Plane. He supposed a loop of string, loaded with fourteen equal balls at equal distances, to hang over a triangular support which was composed of two inclined planes with a horizontal base, and whose sides, being unequal in the proportion of two to one, supported four and two balls respectively. He showed that this loop must hang at rest, because any motion would only bring it into the same condition in which it was at first; and that the festoon of eight balls which hung down below the triangle might be removed without disturbing the equilibrium; so that four balls on the longer plane would balance two balls on the shorter plane; or in other words, the weights would be as the lengths of the planes intercepted by the horizontal line.

Stevinus showed his firm possession of the truth contained in this principle, by deducing from it the properties of forces acting in oblique directions under all kinds of conditions; in short, he showed his entire ability to found upon it a complete doctrine of equilibrium; and upon his foundations, and without any additional support, the mathematical doctrines of Statics might have been carried to the highest pitch of perfection they have yet reached. The formation of the science was finished; the mathematical development and exposition of it were alone open to extension and change.

[2d Ed.] ["Simon Stevin of Bruges," as he usually designates himself in the title-page of his work, has lately become an object of general interest in his own country, and it has been resolved to erect a statue in honor of him in one of the public places of his native city. He was born in 1548, as I learn from M. Quetelet's notice of him, and died in 1620. Montucla says that he died in 1633; misled apparently by the preface to Albert Girard's edition of Stevin's works, which was published in 1634, and which speaks of a death which took place in the preceding year; but on examination it will be seen that this refers to Girard, not to Stevin.

I ought to have mentioned, in consideration of the importance of the proposition, that Stevin distinctly states the *triangle of forces;* namely, that three forces which act upon a point are in equilibrium when they are parallel and proportional to the three sides of any plane triangle. This includes the principle of the *Composition of Statical Forces.* Stevin also applies his principle of equilibrium to cordage, pulleys, funicular polygons, and especially to the bits of bridles; a branch of mechanics which he calls *Chalinothlipsis.*

He has also the merit of having seen very clearly, the distinction of statical and dynamical problems. He remarks that the question, What force will *support* a loaded wagon on an inclined plane? is a statical question, depending on simple conditions; but that the question, What force will *move* the wagon? requires additional considerations to be introduced.

In Chapter iv. of this Book, I have noticed Stevin's share in the rediscovery of the *Laws of the Equilibrium of Fluids.* He distinctly explains the *hydrostatic paradox*, of which the discovery is generally ascribed to Pascal.

Earlier than Stevinus, Leonardo da Vinci must have a place among the discoverers of the Conditions of Equilibrium of Oblique Forces. He published no work on this subject; but extracts from his manuscripts have been published by Venturi, in his *Essai sur les Ouvrages Physico-Mathématiques de Leonard da Vinci, avec des Fragmens tirés de ses Manuscrits apportés d'Italie. Paris*, 1797: and by Libri, in his *Hist. des Sc. Math. en Italie*, 1839. I have also myself examined these manuscripts in the Royal Library at Paris.

It appears that, as early as 1499, Leonardo gave a perfectly correct statement of the proportion of the forces exerted by a cord which acts obliquely and supports a weight on a lever. He distinguishes between the real lever, and the *potential levers*, that is, the perpendiculars drawn from the centre upon the directions of the forces. This is quite sound and satisfactory. These views must in all probability have been sufficiently promulgated in Italy to influence the speculations of Galileo;

whose reasonings respecting the lever much resemble those of Leonardo.—Da Vinci also anticipated Galileo in *asserting* that the time of descent of a body down an inclined plane is to the time of descent down its vertical length in the proportion of the length of the plane to the height. But this cannot, I think, have been more than a guess: there is no vestige of a proof given.]

The contemporaneous progress of the other branch of mechanics, the Doctrine of Motion, interfered with this independent advance of Statics; and to that we must now turn. We may observe, however, that true propositions respecting the composition of forces appear to have rapidly diffused themselves. The *Tractatus de Motu* of Michael Varro of Geneva, already noticed, printed in 1584, had asserted, that the forces which balance each other, acting on the sides of a right-angled triangular wedge, are in the proportion of the sides of the triangle; and although this assertion does not appear to have been derived from a distinct idea of pressure, the author had hence rightly deduced the properties of the wedge and the screw. And shortly after this time, Galileo also established the same results on different principles. In his Treatise *Delle Scienze Mecaniche* (1592), he refers the Inclined Plane to the Lever, in a sound and nearly satisfactory manner; imagining a lever so placed, that the motion of a body at the extremity of one of its arms should be in the same direction as it is upon the plane. A slight modification makes this an unexceptionable proof.

Sect. 3.—*Prelude to the Science of Dynamics.—Attempts at the First Law of Motion.*

We have already seen, that Aristotle divided Motions into Natural and Violent. Cardan endeavored to improve this division by making three classes: *Voluntary* Motion, which is circular and uniform, and which is intended to include the celestial motions; *Natural* Motion, which is stronger towards the end, as the motion of a falling body,—this is in a straight line, because it is motion to an end, and nature seeks her ends by the shortest road; and thirdly, *Violent* Motion, including in this term all kinds different from the former two. Cardan was aware that such Violent Motion might be produced by a very small force; thus he asserts, that a spherical body resting on a horizontal plane may be put in motion by any force which is sufficient to cleave the air; for which, however, he erroneously assigns as a reason, the smallness of the point of contact.[3] But the most common mistake of this period was, that of supposing that as force is requisite to move a body, so a perpetual supply of force is requisite to keep it in motion. The whole of what Kepler called his "physical" reasoning, depended upon this assumption. He endeavored to discover the forces by which the motions of the planets about the sun might be produced; but, in all cases, he considered the velocity of the planet as produced by, and exhibiting the effect of, a force which acted in the direction of the motion. Kepler's essays, which are in this respect so feeble and unmeaning, have sometimes been considered as disclosing some distant anticipation of Newton's discovery of the existence and law of central forces. There is, however, in reality, no other connection between these speculations than that which arises from the use of the term *force* by the two writers in two utterly different meanings. Kepler's Forces were certain imaginary qualities which appeared in the actual motion which the bodies had; Newton's Forces were causes which appeared by the change of motion: Kepler's Forces urged the bodies forwards; Newton's deflected the bodies from such a progress. If Kepler's Forces were destroyed, the body would instantly stop; if Newton's were annihilated, the body would go on uniformly in a straight line. Kepler compares the action of his Forces to the way in which a body might be driven round, by being placed among the sails of a windmill; Newton's Forces would be represented by a rope pulling the body to the centre. Newton's Force is merely mutual attraction; Kepler's is something quite different from this; for though he perpetually illustrates his views by the example of a magnet, he warns us that the sun differs from the magnet in this respect, that its force is not attractive, but directive.[4] Kepler's essays may with considerable reason be asserted to be an anticipation of the Vortices of Descartes; but they can with no propriety whatever be said to anticipate Newton's Dynamical Theory.

The confusion of thought which prevented mathematicians from seeing the difference between producing and preserving motion, was, indeed, fatal to all attempts at progress on this subject. We have already noticed the perplexity in which Aristotle involved himself, by his endeavors to find a reason for the continued motion of a stone

[3] In speaking of the force which would draw a body up an inclined plane he observes, that "per communem animi sententiam," when the plane becomes horizontal, the requisite force is nothing.

[4] *Epitome Astron. Copern.* p. 176.

after the moving power had ceased to act; and that he had ascribed it to the effect of the air or other medium in which the stone moves. Tartalea, whose *Nuova Scienza* is dated 1550, though a good *pure* mathematician, is still quite in the dark on mechanical matters. One of his propositions, in the work just mentioned, is (B. i. Prop. 3), "The more a heavy body recedes from the beginning, or approaches the end of violent motion, the slower and more inertly it goes;" which he applies to the horizontal motion of projectiles. In like manner most other writers about this period conceived that a cannon-ball goes forwards till it loses all its projectile motion, and then falls downwards. Benedetti, who has already been mentioned, must be considered as one of the first enlightened opponents of this and other Aristotelian errors or puzzles. In his *Speculationum Liber* (Venice, 1585), he opposes Aristotle's mechanical opinions, with great expressions of respect, but in a very sweeping manner. His chapter xxiv. is headed, "Whether this eminent man was right in his opinion concerning violent and natural motion." And after stating the Aristotelian opinion just mentioned, that the body is impelled by the air, he says that the air must impede rather than impel the body, and that[5] "the motion of the body, separated from the mover, arises by a certain natural impression from the impetuosity (*ex impetuositate*) received from the mover." He adds, that in natural motions this *impetuosity* continually increases by the continued action of the cause,—namely, the propension of going to the place assigned it by nature; and that thus the velocity increases as the body moves from the beginning of its path. This statement shows a clearness of conception with regard to the cause of accelerated motion, which Galileo himself was long in acquiring.

Though Benedetti was thus on the way to the First Law of Motion, —that all motion is uniform and rectilinear, except so far as it is affected by extraneous forces;—this Law was not likely to be either generally conceived, or satisfactorily proved, till the other Laws of Motion, by which the action of Forces is regulated, had come into view. Hence, though a partial apprehension of this principle had preceded the discovery of the Laws of Motion, we must place the establishment of the principle in the period when those Laws were detected and established, the period of Galileo and his followers.

---

[5] P. 184.

CHAPTER II.

Inductive Epoch of Galileo.—Discovery of the Laws of Motion in Simple Cases.

Sect. 1.—*Establishment of the First Law of Motion.*

AFTER mathematicians had begun to doubt or reject the authority of Aristotle, they were still some time in coming to the conclusion, that the distinction of Natural and Violent Motions was altogether untenable;—that the velocity of a body in motion increased or diminished in consequence of the action of extrinsic causes, not of any property of the motion itself;— and that the apparently universal fact, of bodies growing slower and slower, as if by their own disposition, till they finally stopped, from which Motions had been called Violent, arose from the action of external obstacles not immediately obvious, as the friction and the resistance of the air when a ball runs on the ground, and the action of gravity, when it is thrown upwards. But the truth to which they were at last led, was, that such causes would account for *all* the diminution of velocity which bodies experience when apparently left to themselves; and that without such causes, the motion of all bodies would go on forever, in a straight line and with a uniform velocity.

Who first announced this Law in a general form, it may be difficult to point out; its exact or approximate truth was necessarily taken for granted in all complete investigations on the subject of the laws of motion of falling bodies, and of bodies projected so as to describe curves. In Galileo's first attempt to solve the problem of falling bodies, he did not carry his analysis back to the notion of force, and therefore this law does not appear. In 1604 he had an erroneous opinion on this subject; and we do not know when he was led to the true doctrine which he published in his *Discorso*, in 1638. In his third Dialogue he gives the instance of water in a vessel, for the purpose of showing that circular motion has a tendency to continue. And in his first Dialogue on the Copernican System[1] (published in 1630), he asserts

[1] Dial. 1. p. 40.

Circular Motion alone to be naturally uniform, and retains the distinction between Natural and Violent Motion. In the *Dialogues on Mechanics*, however, published in 1638, but written apparently at an earlier period, in treating of Projectiles,[2] he asserts the true Law. "Mobile super planum horizontale projectum mente concipio omni secluso impedimento; jam constat ex his quæ fusius alibi dicta sunt, illius motum equabilem et perpetuum super ipso plano futurum esse, si planum in infinitum extendatur." "Conceive a movable body upon a horizontal plane, and suppose all obstacles to motion to be removed; it is then manifest, from what has been said more at large in another place, that the body's motion will be uniform and perpetual upon the plane, if the plane be indefinitely extended." His pupil Borelli, in 1667 (in the treatise *De Vi Percussionis*), states the proposition generally, that "Velocity is, by its nature, uniform and perpetual;" and this opinion appears to have been, at that time, generally diffused, as we find evidence in Wallis and others. It is commonly said that Descartes was the first to state this generally. His *Principia* were published in 1644; but his proofs of this First Law of Motion are rather of a theological than of a mechanical kind. His reason for this Law is,[3] "the immutability and simplicity of the operation by which God preserves motion in matter. For he only preserves it precisely as it is in that moment in which he preserves it, taking no account of that which may have been previously." Reasoning of this abstract and *à priori* kind, though it may be urged in favor of true opinions after they have been inductively established, is almost equally capable of being called in on the side of error, as we have seen in the case of Aristotle's philosophy. We ought not, however, to forget that the reference to these abstract and *à priori* principles is an indication of the absolute universality and necessity which we look for in complete Sciences, and a result of those faculties by which such Science is rendered possible, and suitable to man's intellectual nature.

The induction by which the First Law of Motion is established, consists, as induction consists in all cases, in conceiving clearly the Law, and in perceiving the subordination of Facts to it. But the Law speaks of bodies not acted upon by any external force,—a case which never occurs in fact; and the difficulty of the step consisted in bringing all the common cases in which motion is gradually extinguished, under the notion of the action of a retarding force. In order to do this,

---

[2] Dial. i. p. 40. [3] *Princip.* p. 34.

Hooke and others showed that, by diminishing the obvious resistances, the retardation also became less; and men were gradually led to a distinct appreciation of the Resistance, Friction, &c., which, in all terrestrial motions, prevent the Law from being evident; and thus they at last established by experiment a Law which cannot be experimentally exemplified. The natural uniformity of motion was proved by examining all kinds of cases in which motion was not uniform. Men culled the abstract Rule out of the concrete Experiment; although the Rule was, in every case, mixed with other Rules, and each Rule could be collected from the Experiment only by supposing the others known. The perfect simplicity which we necessarily seek for in a law of nature, enables us to disentangle the complexity which this combination appears at first sight to occasion.

The First Law of Motion asserts that the motion of a body, when left to itself, will not only be uniform, but rectilinear also. This latter part of the law is indeed obvious of itself, as soon as we conceive a body detached from all special reference to external points and objects. Yet, as we have seen, Galileo asserted that the naturally uniform motion of bodies was that which takes place in a circle. Benedetti, however, in 1585, had entertained sound notions on this subject. In commenting on Aristotle's question, why we obtain an advantage in throwing by using a sling, he says,[4] that the body, when whirled round, tends to go on in a straight line. In Galileo's second Dialogue, he makes one of his interlocutors (Simplicio), when appealed to on this subject, after thinking intently for a little while, give the same opinion; and the principle is, from this time, taken for granted by the authors who treat of the motion of projectiles. Descartes, as might be supposed, gives the same reason for this as for the other part of the law, namely, the immutability of the Deity.

Sect. 2.—Formation and Application of the Notion of Accelerating Force.—Laws of Falling Bodies.

We have seen how rude and vague were the attempts of Aristotle and his followers to obtain a philosophy of bodies falling downwards or thrown in any direction. If the First Law of Motion had been clearly known, it would then, perhaps, have been seen that the way to understand and analyze the motion of any body, is to consider the

---

[4] "Corpus vellet recta iter peragere." *Speculutionum Liber*, p. 160.

Causes of *change* of motion which at each instant operate upon it; and thus men would have been led to the notion of Accelerating Forces, that is, Forces which act upon bodies already in motion, and accelerate, retard, or deflect their motions. It was, however, only after many attempts that they reached this point. They began by considering the *whole motion* with reference to certain ill-defined abstract Notions, instead of considering, with a clear apprehension of the conditions of Causation, the *successive parts* of which the motion consists. Thus, they spoke of the tendency of bodies to the Centre, or to their Own Place;—of Projecting Force, of Impetus, of Retraction;—with little or no profit to knowledge. The indistinctness of their notions may, perhaps, be judged of from their speculations concerning projectiles. Santbach,[5] in 1561, imagined that a body thrown with great velocity, as, for instance, a ball from a cannon, went in a straight line till all its velocity was exhausted, and then fell directly downwards. He has written a treatise on gunnery, founded on this absurd assumption. To this succeeded another doctrine, which, though not much more philosophical than the former, agreed much better with the phenomena. Nicolo Tartalea (*Nuova Scienza*, Venice, 1550; *Quesiti et Inventioni Diversi*, 1554) and Gualtier Rivius (*Architectura*, &c., Basil, 1582) represented the path of a cannon-ball as consisting, first of a straight line in the direction of the original projection, then of an arc of a circle in which it went on till its motion became vertical downwards, and then of a vertical line in which it continued to fall. The latter of these writers, however, was aware that the path must, from the first, be a curve; and treated it as a straight line, only because the curvature is very slight. Even Santbach's figure represents the path of the ball as partially descending before its final fall, but then it descends by *steps*, not in a curve. Santbach, therefore, did not conceive the *Composition* of the effect of gravity with the existing motion, but supposed them to act alternately; Rivius, however, understood this Composition, and saw that gravity must act as a deflecting force at every point of the path. Galileo, in his second Dialogue,[6] makes Simplicius come to the same conclusion. "Since," he says, "there is nothing to support the body, when it quits that which projects it, it cannot be but that its proper gravity must operate," and it must immediately begin to decline downwards.

---

[5] *Problematum Astronomicorum et Geometricorum Sectiones* vii. &c. &c. Auctore Daniele Santbach, Noviomago. Basileæ, 1561.

[6] P. 147.

The Force of Gravity which thus produces deflection and curvature in the path of a body thrown *obliquely*, constantly increases the velocity of a body when it falls *vertically* downwards. The universality of this increase was obvious, both from reasoning and in fact; the law of it could only be discovered by closer consideration; and the full analysis of the problem required a distinct measure of the quantity of Accelerating Force. Galileo, who first solved this problem, began by viewing it as a question of fact, but conjectured the solution by taking for granted that the rule must be the simplest possible. "Bodies," he says,[7] "will fall in the most simple way, because Natural Motions are always the most simple. When a stone falls, if we consider the matter attentively, we shall find that there is no addition, no increase, of the velocity more simple than that which is always added in the same manner," that is, when equal additions take place in equal times; "which we shall easily understand if we attend to the close connection of motion and time." From this Law, thus assumed, he deduced that the spaces described from the beginning of the motion must be as the squares of the times; and, again, assuming that the laws of descent for balls rolling down inclined planes, must be the same as for bodies falling freely, he verified this conclusion by experiment.

It will, perhaps, occur to the reader that this argument, from the simplicity of the assumed law, is somewhat insecure. It is not always easy for us to discern what that greatest simplicity is, which nature adopts in her laws. Accordingly, Galileo was led wrong by this way of viewing the subject before he was led right. He at first supposed, that the Velocity which the body had acquired at any point must be proportional to the *Space* described from the point where the motion began. This false law is as simple in its enunciation as the true law, that the Velocity is proportional to the *Time:* it had been asserted as the true law by M. Varro (*De Motu Tractatus*, Genevæ, 1584), and by Baliani, a gentleman of Genoa, who published it in 1638. It was, however, soon rejected by Galileo, though it was afterwards taken up and defended by Casræus, one of Galileo's opponents. It so happens, indeed, that the false law is not only at variance with fact, but with itself: it involves a mathematical self-contradiction. This circumstance, however, was accidental: it would be easy to state laws of the increase of velocity which should be simple, and yet false in fact, though quite possible in their own nature.

---

[7] *Dial. Sc.* iv. p. 91.

The Law of Velocity was hitherto, as we have seen, treated as a law of phenomena, without reference to the Causes of the law. "The cause of the acceleration of the motions of falling bodies is not," Galileo observes, "a necessary part of the investigation. Opinions are different. Some refer it to the approach to the centre; others say that there is a certain extension of the centrical medium, which, closing behind the body, pushes it forwards. For the present, it is enough for us to demonstrate certain properties of Accelerated Motion, the acceleration being according to the very simple Law, that the Velocity is proportional to the Time. And if we find that the properties of such motion are verified by the motions of bodies descending freely, we may suppose that the assumption agrees with the laws of bodies falling freely by the action of gravity."[8]

It was, however, an easy step to conceive this acceleration as caused by the continual action of Gravity. This account had already been given by Benedetti, as we have seen. When it was once adopted, Gravity was considered as a *constant* or *uniform* force; on this point, indeed, the adherents of the law of Galileo and of that of Casræus were agreed; but the question was, what *is* a Uniform Force? The answer which Galileo was led to give was obviously this;—*that* is a Uniform Force which generates equal velocities in equal successive times; and this principle leads at once to the doctrine, that Forces are to be compared by comparing the Velocities generated by them in equal times.

Though, however, this was a consequence of the rule by which Gravity is represented as a Uniform Force, the subject presents some difficulty at first sight. It is not immediately obvious that we may thus measure forces by the Velocity *added* in a given time, without taking into account the velocity they have already. If we communicate velocity to a body by the hand or by a spring, the effect we produce in a second of time is lessened, when the body has already a velocity which withdraws it from the pressure of the agent. But it appears that this is not so in the case of gravity; the velocity added in one second is the same, whatever downward motion the body already possesses. A body falling from rest acquires a velocity, in one second, of thirty-two feet; and if a cannon-ball were shot downwards with a velocity of 1000 feet a second, it would equally, at the end of one second, have received an accession of 32 feet to its velocity.

This conception of Gravity as a Uniform Force,—as constantly and

---

[8] Gal. *Op.* iii. 91, 92.

equally *increasing* the velocity of a descending body,—will become clear by a little attention; but it undoubtedly presents difficulty at first. Accordingly, we find that Descartes did not accept it. "It is certain," he says, "that a stone is not equally disposed to receive a new motion or increase of velocity when it is already moving very quickly, and when it is moving slowly."

Descartes showed, by other expressions, that he had not caught hold of the true notion of accelerating force. Thus, he says in a letter to Mersenne, "I am astonished at what you tell me, of having found, by experiment, that bodies thrown up in the air take neither more nor less time to rise than to fall again; and you will excuse me if I say that I look upon the experiment as a very difficult one to make accurately." Yet it is clear from the Notion of a Constant Force that (omitting the resistance of the air) this equality must take place; for the Force which will gradually destroy the whole velocity in a certain time in ascending, will, in the same time, generate again the same velocity by the same gradations inverted; and therefore the same space will be passed over in the same time in the descent and in the ascent.

Another difficulty arose from a necessary consequence of the Laws of Falling Bodies thus established;—the proposition, namely, that in acquiring its motion, a body passes through every intermediate degree of velocity, from the smallest conceivable, up to that which it at last acquires. When a body falls from rest, it begins to fall with *no* velocity; the velocity increases with the time; and in one-thousandth part of a second, the body has only acquired one-thousandth part of the velocity which it has at the end of one second.

This is certain, and manifest on consideration; yet there was at first much difficulty raised on the subject of this assertion; and disputes took place concerning the velocity with which a body *begins* to fall. On this subject also Descartes did not form clear notions. He writes to a correspondent, "I have been revising my notes on Galileo, in which I have not said expressly that falling bodies do not pass through every degree of slowness, but I said that this cannot be known without knowing what Weight is, which comes to the same thing; as to your example, I grant that it proves that every degree of velocity is infinitely divisible, but not that a falling body actually passes through all these divisions."

The Principles of the Motion of Falling Bodies being thus established by Galileo, the Deduction of the principal mathematical consequences was, as is usual, effected with great rapidity, and is to be found in his works, and in those of his scholars and successors. The motion of bodies falling freely was, however, in such treatises, generally combined with the motion of bodies Falling along Inclined Planes; a part of the theory of which we have still to speak.

The Notion of Accelerating Force and of its operation, once formed, was naturally applied in other cases than that of bodies falling freely. The different velocities with which heavy and light bodies fall were explained by the different resistance of the air, which diminishes the accelerating force;[9] and it was boldly asserted, that in a vacuum a lock of wool and a piece of lead would fall equally quickly. It was also maintained[10] that any falling body, however large and heavy, would always have its velocity in some degree diminished by the air in which it falls, and would at last be reduced to a state of uniform motion, as soon as the resistance upwards became equal to the accelerating force downwards. Though the law of progress of a body to this limiting velocity was not made out till the *Principia* of Newton appeared, the views on which Galileo made this assertion are perfectly sound, and show that he had clearly conceived the nature and operation of accelerating and retarding force.

When Uniform Accelerating Forces had once been mastered, there remained only mathematical difficulties in the treatment of Variable Forces. A Variable Force was measured by the *Limit* of the increment of the Velocity, compared with the increment of the Time; just as a Variable Velocity was measured by the Limit of the increment of the Space compared with that of the Time.

With this introduction of the Notion of Limits, we are, of course, led to the Higher Geometry, either in its geometrical or its analytical form. The general laws of bodies falling by the action of any Variable Forces were given by Newton in the Seventh Section of the *Principia*. The subject is there, according to Newton's preference of geometrical methods, treated by means of the Quadrature of Curves; the Doctrine of Limits being exhibited in a peculiar manner in the First Section of the work, in order to prepare the way for such applications of it. Leibnitz, the Bernouillis, Euler, and since their time, many other mathematicians, have treated such questions by means of the analytical method of limits, the Differential Calculus. The Rectilinear Motion of bodies acted upon by variable forces is, of course, a simpler problem than their Curvilinear Motion, to which we have now to proceed. But it

---

9 Galileo, iii. 43.

10 iii. 54.

may be remarked that Newton, having established the laws of Curvilinear Motion independently, has, in a great part of his Seventh Section, deduced the simpler case of the Rectilinear Motion from the more complex problem, by reasonings of great ingenuity and beauty.

Sect. 3.—Establishment of the Second Law of Motion.—Curvilinear Motions.

A SLIGHT degree of distinctness in men's mechanical notions enabled them to perceive, as we have already explained, that a body which traces a curved line must be urged by some force, by which it is constantly made to deviate from that rectilinear path, which it would pursue if acted upon by no force. Thus, when a body is made to describe a circle, as when a stone is whirled round in a sling, we find that the string does exert such a force on the stone; for the string is stretched by the effort, and if it be too slender, it may thus be broken. This *centrifugal force* of bodies moving in circles was noticed even by the ancients. The effect of force to produce curvilinear motion also appears in the paths described by projectiles. We have already seen that though Tartalea did not perceive this correctly, Rivius, about the same time, did.

To see that a transverse force would produce a curve, was one step; to determine what the curve is, was another step, which involved the discovery of the Second Law of Motion. This step was made by Galileo. In his *Dialogues on Motion*, he asserts that a body projected horizontally will retain a uniform motion in the horizontal direction, and will have, compounded with this, a uniformly accelerated motion downwards, that is, the motion of a body falling vertically from rest; and will thus describe the curve called a parabola.

The Second Law of Motion consists of this assertion in a general form;—namely, that in all cases the motion which the force will produce is compounded with the motion which the body previously has. This was not obvious; for Cardan had maintained,[11] that "if a body is moved by two motions at once, it will come to the place resulting from their composition slower than by either of them." The proof of the truth of the law to Galileo's mind was, so far as we collect from the Dialogue itself, the simplicity of the supposition, and his clear perception of the causes which, in some cases, produced an obvious deviation in practice

---

[11] *Op.* vol. iv. p. 490.

from this theoretical result. For it may be observed, that the curvilinear paths ascribed to military projectiles by Rivius and Tartalea, and by other writers who followed them, as Digges and Norton in our own country, though utterly different from the theoretical form, the parabola, do, in fact, approach nearer the true paths of a cannon or musket ball than a parabola would do; and this approximation more especially exists in that which at first sight appears most absurd in the old theory; namely, the assertion that the ball, which ascends in a sloping direction, finally descends vertically. In consequence of the resistance of the air, this is really the path of a projectile; and when the velocity is very great, as in military projectiles, the deviation from the parabolic form is very manifest. This cause of discrepancy between the theory, which does not take resistance into the account, and the fact, Galileo perceived; and accordingly he says,[12] that the velocities of the projectiles, in such cases, may be considered as excessive and supernatural. With the due allowance to such causes, he maintained that his theory was verified, and might be applied in practice. Such practical applications of the doctrine of projectiles no doubt had a share in establishing the truth of Galileo's views. We must not forget, however, that the full establishment of this second law of motion was the result of the theoretical and experimental discussions concerning the motion of the earth: its fortunes were involved in those of the Copernican system; and it shared the triumph of that doctrine. This triumph was already decisive, indeed, in the time of Galileo, but not complete till the time of Newton.

Sect. 4.—Generalization of the Laws of Equilibrium.—Principle of Virtual Velocities.

It was known, even as early as Aristotle, that the two weights which balance each other on the lever, if they move at all, move with velocities which are in the inverse proportions of the weights. The peculiar resources of the Greek language, which could state this relation of inverse proportionality in a single word (*ἀντιπέπονθεν*), fixed it in men's minds, and prompted them to generalize from this property. Such attempts were at first made with indistinct ideas, and on conjecture only, and had, therefore, no scientific value. This is the judgment which we must pass on the book of Jordanus Nemorarius, which

---

[12] *Op.* vol. iii. p. 147.

we have already mentioned. Its reasonings are professedly on Aristotelian principles, and exhibit the common Aristotelian absence of all distinct mechanical ideas. But in Varro, whose *Tractatus de Motu* appeared in 1584, we find the principle, in a general form, not satisfactorily proved, indeed, but much more distinctly conceived. This is his first theorem: "Duarum virium connexarum quarum (si moveantur) motus erunt ipsis ἀντιπέπονθῶς proportionales, neutra alteram movebit, sed equilibrium facient." The proof offered of this is, that the resistance to a force is as the motion produced; and, as we have seen, the theorem is rightly applied in the example of the wedge. From this time it appears to have been usual to prove the properties of machines by means of this principle. This is done, for instance, in *Les Raisons des Forces Mouvantes*, the production of Solomon de Caus, engineer to the Elector Palatine, published at Antwerp in 1616; in which the effect of Toothed-Wheels and of the Screw is determined in this manner, but the Inclined Plane is not treated of. The same is the case in Bishop Wilkins's *Mathematical Magic*, in 1648.

When the true doctrine of the Inclined Plane had been established, the laws of equilibrium for all the simple machines or Mechanical Powers, as they had usually been enumerated in books on Mechanics, were brought into view; for it was easy to see that the *Wedge* and the *Screw* involved the same principle as the *Inclined Plane*, and the *Pulley* could obviously be reduced to the *Lever*. It was, also, not difficult for a person with clear mechanical ideas to perceive how any other combination of bodies, on which pressure and traction are exerted, may be reduced to these simple machines, so as to disclose the relation of the forces. Hence by the discovery of Stevinus, all problems of equilibrium were essentially solved.

The conjectural generalization of the property of the lever, which we have just mentioned, enabled mathematicians to express the solution of all these problems by means of one proposition. This was done by saying, that in raising a weight by any machine, we *lose* in Time what we *gain* in Force; the weight raised moves as much *slower* than the power, as it is *larger* than the power. This was explained with great clearness by Galileo, in the preface to his *Treatise on Mechanical Science*, published in 1592.

The motions, however, which we here suppose the parts of the machine to have, are not motions which the forces produce; for at present we are dealing with the case in which the forces balance each other, and therefore produce no motion. But we ascribe to the

Weights and Powers hypothetical motions, arising from some other cause; and then, by the construction of the machine, the velocities of the Weights and Powers must have certain definite ratios. These velocities, being thus hypothetically supposed and not actually produced, are called *Virtual* Velocities. And the general law of equilibrium is, that in any machine, the Weights which balance each other, are reciprocally to each other as their Virtual Velocities. This is called the *Principle of Virtual Velocities.*

This Principle (which was afterwards still further generalized) is, by some of the admirers of Galileo, dwelt upon as one of his great services to Mechanics. But if we examine it more nearly, we shall see that it has not much importance in our history. It is a generalization, but a generalization established rather by enumeration of cases, than by any induction proceeding upon one distinct Idea, like those generalizations of Facts by which Laws are primarily established. It rather serves verbally to conjoin Laws previously known, than to exhibit a connection in them: it is rather a help for the memory than a proof for the reason.

The Principle of Virtual Velocities is so far from implying any clear possession of mechanical ideas, that any one who knows the property of the Lever, whether he is capable of seeing the reason for it or not, can see that the greater weight moves slower in the exact proportion of its greater magnitude. Accordingly, Aristotle, whose entire want of sound mechanical views we have shown, has yet noticed this truth. When Galileo treats of it, instead of offering any reasons which could independently establish this principle, he gives his readers a number of analogies and illustrations, many of them very loose ones. Thus the raising a great weight by a small force, he illustrates by supposing the weight broken into many small parts, and conceiving those parts raised one by one. By other persons, the analogy, already intimated, of gain and loss is referred to as an argument for the principle in question. Such images may please the fancy, but they cannot be accepted as mechanical reasons.

Since Galileo neither first enunciated this rule, nor ever proved it as an independent principle of Mechanics, we cannot consider the discovery of it as one of his mechanical achievements. Still less can we compare his reference to this principle with Stevinus's proof of the Inclined Plane; which, as we have seen, was rigorously inferred from the sound axiom, that a body cannot put itself in motion. If we were to assent to the really self-evident axioms of Stevinus, only in virtue of the unproved verbal generalization of Galileo, we should be in great danger of allowing ourselves to be referred successively from one truth to another, without any reasonable hope of ever arriving at any thing ultimate and fundamental.

But though this Principle of Virtual Velocity cannot be looked upon as a great discovery of Galileo, it is a highly useful rule; and the various forms under which he and his successors urged it, tended much to dissipate the vague wonder with which the effects of machines had been looked upon; and thus to diffuse sounder and clearer notions on such subjects.

The Principle of Virtual Velocities also affected the progress of mechanical science in another way: it suggested some of the analogies by the aid of which the Third Law of Motion was made out; leading to the adoption of the notion of *Momentum* as the arithmetical product of weight and velocity. Since on a machine on which a weight of two pounds at one part balances three pounds at another part, the former weight would move through three inches while the latter would move through two inches; we see (since three multiplied into two is equal to two multiplied into three) that the *Product* of the weight and the velocity is the same for the two balancing weights; and if we call this Product *Momentum*, the Law of Equilibrium is, that when two weights balance on a machine, the Momentum of the two would be the same, if they were put in motion.

The Notion of Momentum was here employed in connection with Virtual Velocities; but it also came under consideration in treating of Actual Velocities, as we shall soon see.

Sect. 5.—Attempts at the Third Law of Motion.—Notion of Momentum.

In the questions we have hitherto had to consider respecting Motion, no regard is had to the Size of the body moved, but only to the Velocity and Direction of the motion. We must now trace the progress of knowledge respecting the mode in which the Mass of the body influences the effect of Force. This is a more difficult and complex branch of the subject; but it is one which requires to be noticed, as obviously as the former. Questions belonging to this department of Mechanics, as well as to the others, occur in Aristotle's Mechanical Problems. "Why," says he, "is it, that neither very small nor very large bodies go far when we throw them; but, in order that this may happen, the thing thrown must have a certain proportion to the agent which throws it? Is it that what is thrown or pushed must react[13] against that which pushes it; and that a body so large as not to yield at all, or so small as to yield entirely, and not to react, produces no throw or push?" The same confusion of ideas prevailed after his time; and mechanical questions were in vain discussed by means of general and abstract terms, employed with no distinct and steady meaning; such as *impetus*, *power*, *momentum*, *virtue*, *energy*, and the like. From some of these speculations we may judge how thorough the confusion in men's heads had become. Cardan perplexes himself with the difficulty, already mentioned, of the comparison of the forces of bodies at rest and in motion. If the Force of a body depends on its velocity, as it appears to do, how is it that a body at rest has any Force at all, and how can it resist the slightest effort, or exert any pressure? He flatters himself that he solves the question, by asserting that bodies at rest have an occult motion. "Corpus movetur occulto motu quiescendo."—Another puzzle, with which he appears to distress himself rather more wantonly, is this: "If one man can draw half of a certain weight, and another man also one half; when the two act together, these proportions should be compounded; so that they ought to be able to draw one half of one half, or one quarter only." The talent which ingenious men had for getting into such perplexities, was certainly at one time very great. Arriaga,[14] who wrote in 1639, is troubled to discover how several flat weights, lying one upon another on a board, should produce a greater pressure than the lowest one alone produces, since that alone touches the board. Among other solutions, he suggests that the board affects the upper weight, which it does not touch, by determining its *ubication*, or *whereness*.

Aristotle's doctrine, that a body ten times as heavy as another, will fall ten times as fast, is another instance of the confusion of Statical and Dynamical Forces: the Force of the greater body, while *at rest*, is ten times as great as that of the other; but the Force as measured by the *velocity* produced, is equal in the two cases. The two bodies would fall downwards with the same rapidity, except so far as they are affected by accidental causes. The merit of proving this by experiment, and thus refuting the Aristotelian dogma, is usually ascribed to Galileo, who made his experiment from the famous leaning tower of Pisa, about 1590. But others about the same time had not over-

---

[13] ἀντερείδειν. [14] Rod. de Arriaga, *Cursus Philosophicus*. Paris, 1639.

looked so obvious a fact.—F. Piccolomini, in his *Liber Scientiæ de Natura*, published at Padua, in 1597, says, "On the subject of the motion of heavy and light bodies, Aristotle has put forth various opinions, which are contrary to sense and experience, and has delivered rules concerning the proportion of quickness and slowness, which are palpably false. For a stone twice as great does *not* move twice as fast." And Stevinus, in the Appendix to his Statics, describes his having made the experiment, and speaks with great correctness of the apparent deviations from the rule, arising from the resistance of the air. Indeed, the result followed by very obvious reasoning; for ten bricks, in contact with each other, side by side, would obviously fall in the same time as one; and these might be conceived to form a body ten times as large as one of them. Accordingly, Benedetti, in 1585, reasons in this manner with regard to bodies of different size, though he retains Aristotle's error as to the different velocity of bodies of different density.

The next step in this subject is more clearly due to Galileo; he discovered the true proportion which the Accelerating Force of a body falling down an inclined plane bears to the Accelerating Force of the same body falling freely. This was at first a happy conjecture; it was then confirmed by experiments, and, finally, after some hesitation, it was referred to its true principle, the Third Law of Motion, with proper elementary simplicity. The Principle here spoken of is this:—that for the same body, the Dynamical effect of force is as the Statical effect; that is, the Velocity which any force generates in a given time when it puts the body in motion, is proportional to the Pressure which the same force produces in a body at rest. The Principle, so stated, appears very simple and obvious; yet this was not the form in which it suggested itself either to Galileo or to other persons who sought to prove it. Galileo, in his *Dialogues on Motion*, assumes, as his fundamental proposition on this subject, one much less evident than that we have quoted, but one in which that is involved. His Postulate is,[15] that when the same body falls down different planes of the same height, the velocities acquired are equal. He confirms and illustrates this by a very ingenious experiment on a pendulum, showing that the weight swings to the same height whatever path it be compelled to follow. Torricelli, in his treatise published 1644, says that he had heard that Galileo had, towards the end of his life, proved his assump-

[15] *Opere*, iii. 96.

tion, but that, not having seen the proof, he will give his own. In this he refers us to the right principle, but appears not distinctly to conceive the proof, since he estimates *momentum* indiscriminately by the statical Pressure of a body, and by its Velocity when in motion; as if these two quantities were self-evidently equal. Huyghens, in 1673, expresses himself dissatisfied with the proof by which Galileo's assumption was supported in the later editions of his works. His own proof rests on this principle;—that if a body fall down one inclined plane, and proceed up another with the velocity thus acquired, it cannot, under any circumstances, ascend to a higher position than that from which it fell. This principle coincides very nearly with Galileo's experimental illustration. In truth, however, Galileo's principle, which Huyghens thus slights, may be looked upon as a satisfactory statement of the true law; namely, that, in the same body, the velocity produced is as the pressure which produces it. "We are agreed," he says,[16] "that, in a movable body, the *impetus*, *energy*, *momentum*, or *propension to motion*, is as great as is the *force* or *least resistance* which suffices to *support* it." The various terms here used, both for dynamical and statical Force, show that Galileo's ideas were not confused by the ambiguity of any one term, as appears to have happened to some mathematicians. The principle thus announced, is, as we shall see, one of great extent and value; and we read with interest the circumstances of its discovery, which are thus narrated.[17] When Viviani was studying with Galileo, he expressed his dissatisfaction at the want of any clear reason for Galileo's postulate respecting the equality of velocities acquired down inclined planes of the same heights; the consequence of which was, that Galileo, as he lay, the same night, sleepless through indisposition, discovered the proof which he had long sought in vain, and introduced it in the subsequent editions. It is easy to see, by looking at the proof, that the discoverer had had to struggle, not for intermediate steps of reasoning between remote notions, as in a problem of geometry, but for a clear possession of ideas which were near each other, and which he had not yet been able to bring into contact, because he had not yet a sufficiently firm grasp of them. Such terms as Momentum and Force had been sources of confusion from the time of Aristotle; and it required considerable steadiness of thought to compare the forces of bodies at rest and in motion under the obscurity and vacillation thus produced.

---

[16] Galileo, *Op.* iii. 104.

[17] Drinkwater, *Life of Galileo*, p. 59.

The term *Momentum* had been introduced to express the force of bodies in motion, before it was known what that effect was. Galileo, in his *Discorso intorno alle Cose che stanno in su l'Acqua*, says, that "Momentum is the force, efficacy, or virtue, with which the motion moves and the body moved resists, depending not upon weight only, but upon the velocity, inclination, and any other cause of such virtue." When he arrived at more precision in his views, he determined, as we have seen, that, in the same body, the Momentum is *proportional* to the Velocity; and, hence it was easily seen that in different bodies it was proportional to the Velocity and Mass jointly. The principle thus enunciated is capable of very extensive application, and, among other consequences, leads to a determination of the results of the mutual Percussion of Bodies. But though Galileo, like others of his predecessors and contemporaries, had speculated concerning the problem of Percussion, he did not arrive at any satisfactory conclusion; and the problem remained for the mathematicians of the next generation to solve.

We may here notice Descartes and his Laws of Motion, the publication of which is sometimes spoken of as an important event in the history of Mechanics. This is saying far too much. The *Principia* of Descartes did little for physical science. His assertion of the Laws of Motion, in their most general shape, was perhaps an improvement in form; but his Third Law is false in substance. Descartes claimed several of the discoveries of Galileo and others of his contemporaries; but we cannot assent to such claims, when we find that, as we shall see, he did not understand, or would not apply, the Laws of Motion when he had them before him. If we were to compare Descartes with Galileo, we might say, that of the mechanical truths which were easily attainable in the beginning of the seventeenth century, Galileo took hold of as many, and Descartes of as few, as was well possible for a man of genius.

[2d Ed.] [The following remarks of M. Libri appear to be just. After giving an account of the doctrines put forth on the subject of Astronomy, Mechanics, and other branches of science, by Leonardo da Vinci, Fracastoro, Maurolycus, Commandinus, Benedetti, he adds (*Hist. des Sciences Mathématiques en Italie*, t. iii. p. 131): "This short analysis is sufficient to show that, at the period at which we are arrived, Aristotle no longer reigned unquestioned in the Italian Schools. If we had to write the history of philosophy, we should prove by a multitude of facts that it was the Italians who overthrew the ancient idol of philosophers. Men go on incessantly repeating that the struggle was begun by Descartes, and they proclaim him the legislator of modern philosophers. But when we examine the philosophical writings of Fracastoro, of Benedetti, of Cardan, and above all, those of Galileo; when we see on all sides energetic protests raised against the peripatetic doctrines; we ask, what there remained for the inventor of vortices to do, in overturning the natural philosophy of Aristotle? In addition to this, the memorable labors of the School of Cosenza, of Telesius, of Giordano Bruno, of Campanella; the writings of Patricius, who was, besides, a good geometer; of Nizolius, whom Leibnitz esteemed so highly, and of the other metaphysicians of the same epoch, —prove that the ancient philosophy had already lost its empire on that side the Alps, when Descartes threw himself upon the enemy now put to the rout. The yoke was cast off in Italy, and all Europe had only to follow the example, without its being necessary to give a new impulse to real science."

In England, we are accustomed to hear Francis Bacon, rather than Descartes, spoken of as the first great antagonist of the Aristotelian schools, and the legislator of modern philosophy. But it is true, both of one and the other, that the overthrow of the ancient system had been effectively begun before their time by the practical discoverers here mentioned, and others who, by experiment and reasoning, established truths inconsistent with the received Aristotelian doctrines. Gilbert in England, Kepler in Germany, as well as Benedetti and Galileo in Italy, gave a powerful impulse to the cause of real knowledge, before the influence of Bacon and Descartes had produced any general effect. What Bacon really did was this;—that by the august image which he presented of a future Philosophy, the rival of the Aristotelian, and far more powerful and extensive, he drew to it the affections and hopes of all men of comprehensive and vigorous minds, as well as of those who attended to special trains of discovery. He announced a New Method, not merely a correction of special current errors; he thus converted the Insurrection into a Revolution, and established a new philosophical Dynasty. Descartes had, in some degree, the same purpose; and, in addition to this, he not only proclaimed himself the author of a New Method, but professed to give a complete system of the results of the Method. His physical philosophy was put forth as complete and demonstrative, and thus involved the vices of the ancient dogmatism. Telesius and Campanella had also grand notions of an entire reform in the method of philosophizing, as I have noticed in the *Philosophy of the Inductive Sciences*, Book xii.]

CHAPTER III.

SEQUEL TO THE EPOCH OF GALILEO.—PERIOD OF VERIFICATION AND DEDUCTION.

THE evidence on which Galileo rested the truth of the Laws of Motion which he asserted, was, as we have seen, the simplicity of the laws themselves, and the agreement of their consequences with facts; proper allowances being made for disturbing causes. His successors took up and continued the task of making repeated comparisons of the theory with practice, till no doubt remained of the exactness of the fundamental doctrines: they also employed themselves in simplifying, as much as possible, the mode of stating these doctrines, and in tracing their consequences in various problems by the aid of mathematical reasoning. These employments led to the publication of various Treatises on Falling Bodies, Inclined Planes, Pendulums, Projectiles, Spouting Fluids, which occupied a great part of the seventeenth century.

The authors of these treatises may be considered as the School of Galileo. Several of them were, indeed, his pupils or personal friends. Castelli was his disciple and astronomical assistant at Florence, and afterwards his correspondent. Torricelli was at first a pupil of Castelli, but became the inmate and amanuensis of Galileo in 1641, and succeeded him in his situation at the court of Florence on his death, which took place a few months afterwards. Viviani formed one of his family during the three last years of his life; and surviving him and his contemporaries (for Viviani lived even into the eighteenth century), has a manifest pleasure and pride in calling himself the last of the disciples of Galileo. Gassendi, an eminent French mathematician and professor, visited him in 1628; and it shows us the extent of his reputation when we find Milton referring thus to his travels in Italy:[1] "There it was that I found and visited the famous Galileo, grown old, a prisoner in the Inquisition, for thinking in astronomy otherwise than the Franciscan and Dominican licensers thought."

Besides the above writers, we may mention, as persons who pursued and illustrated Galileo's doctrines, Borelli, who was professor at Florence and Pisa; Mersenne, the correspondent of Descartes, who was

---

[1] *Speech for the Liberty of Unlicensed Printing.*

professor at Paris; Wallis, who was appointed Savilian professor at Oxford in 1649, his predecessor being ejected by the parliamentary commissioners. It is not necessary for us to trace the progress of purely mathematical inventions, which constitute a great part of the works of these authors; but a few circumstances may be mentioned.

The question of the proof of the Second Law of Motion was, from the first, identified with the controversy respecting the truth of the Copernican System; for this law supplied the true answer to the most formidable of the objections against the motion of the earth; namely, that if the earth were moving, bodies which were dropt from an elevated object would be left behind by the place from which they fell. This argument was reproduced in various forms by the opponents of the new doctrine; and the answers to the argument, though they belong to the history of Astronomy, and form part of the Sequel to the Epoch of Copernicus, belong more peculiarly to the history of Mechanics, and are events in the sequel to the Discoveries of Galileo. So far, indeed, as the mechanical controversy was concerned, the advocates of the Second Law of Motion appealed, very triumphantly, to experiment. Gassendi made many experiments on this subject publicly, of which an account is given in his *Epistolæ tres de Motu Impresso a Motore Translato.*[2] It appeared in these experiments, that bodies let fall downwards, or cast upwards, forwards, or backwards, from a ship, or chariot, or man, whether at rest, or in any degree of motion, had always the same motion relatively to the *motor*. In the application of this principle to the system of the world, indeed, Gassendi and other philosophers of his time were greatly hampered; for the deference which religious scruples required, did not allow them to say that the earth really moved, but only that the physical reasons against its motion were invalid. This restriction enabled Riccioli and other writers on the geocentric side to involve the subject in metaphysical difficulties; but the conviction of men was not permanently shaken by these, and the Second Law of Motion was soon assumed as unquestioned.

The Laws of the Motion of Falling Bodies, as assigned by Galileo, were confirmed by the reasonings of Gassendi and Fermat, and the experiments of Riccioli and Grimaldi; and the effect of resistance was pointed out by Marsenne and Dechales. The parabolic motion of Projectiles was more especially illustrated by experiments on the jet which spouts from an orifice in a vessel full of fluid. This mode of experimenting

---

[2] Mont. ii. 199.

is well adapted to attract notice, since the curve described, which is transient and invisible in the case of a single projectile, becomes permanent and visible when we have a continuous stream. The doctrine of the motions of fluids has always been zealously cultivated by the Italians. Castelli's treatise, *Della Misura dell' Acque Corrente* (1638), is the first work on this subject, and Montucla with justice calls him "the creator of a new branch of hydraulics;"[3] although he mistakenly supposed the velocity of efflux to be as the depth of the orifice from the surface. Marsenne and Torricelli also pursued this subject, and after them, many others.

Galileo's belief in the near approximation of the curve described by a cannon-ball or musket-ball to the theoretical parabola, was somewhat too obsequiously adopted by succeeding practical writers on artillery. They underrated, as he had done, the effect of the resistance of the air, which is in fact so great as entirely to change the form and properties of the curve. Notwithstanding this, the parabolic theory was employed, as in Anderson's *Art of Gunnery* (1674); and Blondel, in his *Art de jeter les Bombes* (1683), not only calculated Tables on this supposition, but attempted to answer the objections which had been made respecting the form of the curve described. It was not till a later period (1740), when Robins made a series of careful and sagacious experiments on artillery, and when some of the most eminent mathematicians calculated the curve, taking into account the resistance, that the Theory of Projectiles could be said to be verified in fact.

The Third Law of Motion was still in some confusion when Galileo died, as we have seen. The next great step made in the school of Galileo was the determination of the Laws of the motions of bodies in their Direct Impact, so far as this impact affects the motion of translation. The difficulties of the problem of Percussion arose, in part, from the heterogeneous nature of Pressure (of a body at rest), and Momentum (of a body in motion); and, in part, from mixing together the effects of percussion on the parts of a body, as, for instance, cutting, bruising, and breaking, with its effect in moving the whole.

The former difficulty had been seen with some clearness by Galileo himself. In a posthumous addition to his *Mechanical Dialogues*, he says, "There are two kinds of resistance in a movable body, one internal, as when we say it is more difficult to lift a weight of a thousand pounds than a weight of a hundred; another respecting space, as

---

[3] Mont. ii. 201.

when we say that it requires more force to throw a stone one hundred paces than fifty."[4] Reasoning upon this difference, he comes to the conclusion that "the Momentum of percussion is infinite, since there is no resistance, however great, which is not overcome by a force of percussion, however small."[5] He further explains this by observing that the resistance to percussion must occupy some portion of time, although this portion may be insensible. This correct mode of removing the apparent incongruity of continuous and instantaneous force, was a material step in the solution of the problem.

The Laws of the mutual Impact of bodies were erroneously given by Descartes in his *Principia;* and appear to have been first correctly stated by Wren, Wallis, and Huyghens, who about the same time (1669) sent papers to the Royal Society of London on the subject. In these solutions, we perceive that men were gradually coming to apprehend the Third Law of Motion in its most general sense; namely, that the Momentum (which is proportional to the Mass of the body and its Velocity jointly) may be taken for the measure of the effect; so that this Momentum is as much diminished in the striking body by the resistance it experiences, as it is increased in the body struck by the Impact. This was sometimes expressed by saying that "the Quantity of Motion remains unaltered," *Quantity of Motion* being used as synonymous with *Momentum.* Newton expressed it by saying that "Action and Reaction are equal and opposite," which is still one of the most familiar modes of expressing the Third Law of Motion.

In this mode of stating the Law, we see an example of a propensity which has prevailed very generally among mathematicians; namely, a disposition to present the fundamental laws of rest and of motion as if they were equally manifest, and, indeed, identical. The close analogy and connection which exists between the principles of equilibrium and of motion, often led men to confound the evidence of the two; and this confusion introduced an ambiguity in the use of words, as we have seen in the case of Momentum, Force, and others. The same may be said of *Action* and *Reaction*, which have both a statical and a dynamical signification. And by this means, the most general statements of the laws of motion are made to coincide with the most general statical propositions. For instance, Newton deduced from his principles the conclusion, that by the mutual action of bodies, the motion of their centre of gravity cannot be affected. Marriotte, in his *Traité de la*

---

[4] *Op.* iii. 210. [5] iii. 211.

Percussion (1684), had asserted this proposition for the case of direct impact. But by the reasoners of Newton's time, the dynamical proposition, that the motion of the centre of gravity is not altered by the *actual* free motion and impact of bodies, was associated with the statical proposition, that when bodies are in equilibrium, the centre of gravity cannot be made to ascend or descend by the *virtual* motions of the bodies. This latter is a proposition which was assumed as self-evident by Torricelli; but which may more philosophically be proved from elementary statical principles.

This disposition to identify the elementary laws of equilibrium and of motion, led men to think too slightingly of the ancient solid and sufficient foundation of Statics, the doctrine of the lever. When the progress of thought had opened men's minds to a more general view of the subject, it was considered as a blemish in the science to found it on the properties of one particular machine. Descartes says in his Letters, that "it is ridiculous to prove the pulley by means of the lever." And Varignon was led by similar reflections to the project of his *Nouvelle Mécanique*, in which the whole of statics should be founded on the composition of forces. This project was published in 1687; but the work did not appear till 1725, after the death of the author. Though the attempt to reduce the equilibrium of all machines to the composition of forces, is philosophical and meritorious, the attempt to reduce the composition of Pressures to the composition of *Motions*, with which Varignon's work is occupied, was a retrograde step in the subject, so far as the progress of distinct mechanical ideas was concerned.

Thus, at the period at which we have now arrived, the Principles of Elementary Mechanics were generally known and accepted; and there was in the minds of mathematicians a prevalent tendency to reduce them to the most simple and comprehensive form of which they admitted. The execution of this simplification and extension, which we term the generalization of the laws, is so important an event, that though it forms part of the natural sequel of Galileo, we shall treat of it in a separate chapter. But we must first bring up the history of the mechanics of fluids to the corresponding point.

CHAPTER IV.

Discovery of the Mechanical Principles of Fluids.

Sect. 1.—*Rediscovery of the Laws of Equilibrium of Fluids.*

WE have already said, that the true laws of the equilibrium of fluids were discovered by Archimedes, and rediscovered by Galileo and Stevinus; the intermediate time having been occupied by a vagueness and confusion of thought on physical subjects, which made it impossible for men to retain such clear views as Archimedes had disclosed. Stevinus must be considered as the earliest of the authors of this rediscovery; for his work (*Principles of Statik and Hydrostatik*) was published in Dutch about 1585; and in this, his views are perfectly distinct and correct. He restates the doctrines of Archimedes, and shows that, as a consequence of them, it follows that the pressure of a fluid on the bottom of a vessel may be much greater than the weight of the fluid itself: this he proves, by imagining some of the upper portions of the vessel to be filled with fixed solid bodies, which take the place of the fluid, and yet do not alter the pressure on the base. He also shows what will be the pressure on any portion of a base in an oblique position; and hence, by certain mathematical artifices which make an approach to the Infinitesimal Calculus, he finds the whole pressure on the base in such cases. This mode of treating the subject would take in a large portion of our elementary Hydrostatics as the science now stands. Galileo saw the properties of fluids no less clearly, and explained them very distinctly, in 1612, in his *Discourse on Floating Bodies.* It had been maintained by the Aristotelians, that *form* was the cause of bodies floating; and collaterally, that ice was *condensed* water; apparently from a confusion of thought between *rigidity* and *density.* Galileo asserted, on the contrary, that ice is *rarefied* water, as appears by its floating: and in support of this, he proved, by various experiments, that the floating of bodies does not depend on their form. The happy genius of Galileo is the more remarkable in this case, as the controversy was a good deal perplexed by the mixture of phenomena of another kind, due to what is usually called *capillary* or *molecular attraction.* Thus it is a fact, that a *ball* of ebony sinks in water, while a *flat slip* of the same material lies on the surface; and it required considerable sagacity to separate such cases from the general rule. Galileo's opinions were attacked by various writers, as Nozzolini, Vincenzio di Grazia, Ludovico delle Colombe; and defended by his pupil Castelli, who published a reply in 1615. These opinions were generally adopted and diffused; but somewhat later, Pascal pursued the subject more systematically, and wrote his *Treatise of the Equilibrium of Fluids*, in 1653; in which he shows that a fluid, inclosed in a vessel, necessarily presses equally in all directions, by imagining two *pistons*, or sliding plugs, applied at different parts, the surface of one being centuple that of the other: it is clear, as he observes, that the force of one man acting at the first piston, will balance the force of one hundred men acting at the other. "And thus," says he, "it appears that a vessel full of water is a new Principle of Mechanics, and a new Machine which will multiply force to any degree we choose." Pascal also referred the equilibrium of fluids to the "principle of virtual velocities," which regulates the equilibrium of other machines. This, indeed, Galileo had done before him. It followed from this doctrine, that the pressure which is exercised by the lower parts of a fluid arises from the weight of the upper parts.

In all this there was nothing which was not easily assented to; but the extension of these doctrines to the air required an additional effort of mechanical conception. The pressure of the air on all sides of us, and its weight above us, were two truths which had never yet been apprehended with any kind of clearness. Seneca, indeed,[1] talks of the "gravity of the air," and of its power of diffusing itself when condensed, as the causes of wind; but we can hardly consider such propriety of phraseology in him as more than a chance; for we see the value of his philosophy by what he immediately adds: "Do you think that we have forces by which we move ourselves, and that the air is left without any power of moving? when even water has a motion of its own, as we see in the growth of plants." We can hardly attach much value to such a recognition of the gravity and elasticity of the air.

Yet the effects of these causes were so numerous and obvious, that the Aristotelians had been obliged to invent a principle to account for them; namely, "Nature's Horror of a Vacuum." To this principle were referred many familiar phenomena, as suction, breathing, the

[1] *Quæst. Nat.* v. 5.

action of a pair of bellows, its drawing water if immersed in water, its refusing to open when the vent is stopped up. The action of a cupping instrument, in which the air is rarefied by fire; the fact that water is supported when a full inverted bottle is placed in a basin; or when a full tube, open below and closed above, is similarly placed; the running out of the water, in this instance, when the top is opened; the action of a siphon, of a syringe, of a pump; the adhesion of two polished plates, and other facts, were all explained by the *fuga vacui.* Indeed, we must contend that the principle was a very good one, inasmuch as it brought together all these facts which are really of the same kind, and referred them to a common cause. But when urged as an ultimate principle, it was not only *unphilosophical*, but *imperfect* and *wrong.* It was *unphilosophical*, because it introduced the notion of an emotion, Horror, as an account of physical facts; it was *imperfect*, because it was at best only a law of phenomena, not pointing out any physical cause; and it was *wrong*, because it gave an unlimited extent to the effect. Accordingly, it led to mistakes. Thus Mersenne, in 1644, speaks of a siphon which shall go over a mountain, being ignorant then that the effect of such an instrument was limited to a height of thirty-four feet. A few years later, however, he had detected this mistake; and in his third volume, published in 1647, he puts his siphon in his *emendanda*, and speaks correctly of the weight of air as supporting the mercury in the tube of Torricelli. It was, indeed, by finding this horror of a vacuum to have a limit at the height of thirty-four feet, that the true principle was suggested. It was discovered that when attempts were made to raise water higher than this, Nature tolerated a vacuum above the water which rose. In 1643, Torricelli tried to produce this vacuum at a smaller height, by using, instead of water, the heavier fluid, quicksilver; an attempt which shows that the true explanation, the balance of the weight of the water by another pressure, had already suggested itself. Indeed, this appears from other evidence. Galileo had already taught that the air has weight; and Baliani, writing to him in 1630, says,[2] "If we were in a vacuum, the weight of the air above our heads would be felt." Descartes also appears to have some share in this discovery; for, in a letter of the date of 1631, he explains the suspension of mercury in a tube, closed at top, by the pressure of the column of air reaching to the clouds.

---

[2] Drinkwater's *Galileo*, p. 90.

Still men's minds wanted confirmation in this view; and they found such confirmation, when, in 1647, Pascal showed practically, that if we alter the length of the superincumbent column of air by going to a high place, we alter the weight which it will support. This celebrated experiment was made by Pascal himself on a church-steeple in Paris, the column of mercury in the Torricellian tube being used to compare the weights of the air; but he wrote to his brother-in-law, who lived near the high mountain of Puy de Dôme in Auvergne, to request him to make the experiment there, where the result would be more decisive. "You see," he says, "that if it happens that the height of the mercury at the top of the hill be less than at the bottom (which I have many reasons to believe, though all those who have thought about it are of a different opinion), it will follow that the weight and pressure of the air are the sole cause of this suspension, and not the horror of a vacuum: since it is very certain that there is more air to weigh on it at the bottom than at the top; while we cannot say that nature abhors a vacuum at the foot of a mountain more than on its summit."—M. Perrier, Pascal's correspondent, made the observation as he had desired, and found a difference of three inches of mercury, "which," he says, "ravished us with admiration and astonishment."

When the least obvious case of the operation of the pressure and weight of fluids had thus been made out, there were no further difficulties in the progress of the theory of Hydrostatics. When mathematicians began to consider more general cases than those of the action of gravity, there arose differences in the way of stating the appropriate principles: but none of these differences imply any different conception of the fundamental nature of fluid equilibrium.

Sect. 2.—Discovery of the Laws of Motion of Fluids.

The art of conducting water in pipes, and of directing its motion for various purposes, is very old. When treated systematically, it has been termed *Hydraulics:* but *Hydrodynamics* is the general name of the science of the laws of the motions of fluids, under those or other circumstances. The Art is as old as the commencement of civilization: the Science does not ascend higher than the time of Newton, though attempts on such subjects were made by Galileo and his scholars.

When a fluid spouts from an orifice in a vessel, Castelli saw that the velocity of efflux depends on the depth of the orifice below the surface: but he erroneously judged the velocity to be exactly proportional to the depth. Torricelli found that the fluid, under the inevitable causes of defect which occur in the experiment, would spout nearly to the height of the surface: he therefore inferred, that the full velocity is that which a body would acquire in falling through the depth; and that it is consequently proportional to the square root of the depth.—This, however, he stated only as a result of experience, or law of phenomena, at the end of his treatise, *De Motu Naturaliter Accelerato*, printed in 1643.

Newton treated the subject theoretically in the *Principia* (1687); but we must allow, as Lagrange says, that this is the least satisfactory passage of that great work. Newton, having made his experiments in another manner than Torricelli, namely, by measuring the *quantity* of the efflux instead of its velocity, found a result inconsistent with that of Torricelli. The velocity inferred from the quantity discharged, was only that due to *half* the depth of the fluid.

In the first edition of the *Principia*,[3] Newton gave a train of reasoning by which he theoretically demonstrated his own result, going upon the principle, that the momentum of the issuing fluid is equal to the momentum which the column vertically over the orifice would generate by its gravity. But Torricelli's experiments, which had given the velocity due to the whole depth, were confirmed on repetition: how was this discrepancy to be explained?

Newton explained the discrepancy by observing the contraction which the jet, or vein of water, undergoes, just after it leaves the orifice, and which he called the *vena contracta*. At the orifice, the velocity is that due to half the height; at the vena contracta it is that due to the whole height. The former velocity regulates the quantity of the discharge; the latter, the path of the jet.

This explanation was an important step in the subject; but it made Newton's original proof appear very defective, to say the least. In the second edition of the *Principia* (1714), Newton attacked the problem in a manner altogether different from his former investigation. He there assumed, that when a round vessel, containing fluid, has a hole in its bottom, the descending fluid may be conceived to be a conoidal mass, which has its base at the surface of the fluid, and its narrow end at the orifice. This portion of the fluid he calls the *cataract;* and supposes that while this part descends, the surrounding

[3] B. ii. Prop. xxxvii.

parts remain immovable, as if they were frozen; in this way he finds a result agreeing with Torricelli's experiments on the velocity of the efflux.

We must allow that the assumptions by which this result is obtained are somewhat arbitrary; and those which Newton introduces in attempting to connect the problem of issuing fluids with that of the resistance to a body moving in a fluid, are no less so. But even up to the present time, mathematicians have not been able to reduce problems concerning the motions of fluids to mathematical principles and calculations, without introducing some steps of this arbitrary kind. And one of the uses of experiments on this subject is, to suggest those hypotheses which may enable us, in the manner most consonant with the true state of things, to reduce the motions of fluids to those general laws of mechanics, to which we know they must be subject.

Hence the science of the Motion of Fluids, unlike all the other primary departments of Mechanics, is a subject on which we still need experiments, to point out the fundamental principles. Many such experiments have been made, with a view either to compare the results of deduction and observation, or, when this comparison failed, to obtain purely empirical rules. In this way the resistance of fluids, and the motion of water in pipes, canals, and rivers, has been treated. Italy has possessed, from early times, a large body of such writers. The earlier works of this kind have been collected in sixteen quarto volumes. Lecchi and Michelotti about 1765, Bidone more recently, have pursued these inquiries. Bossut, Buat, Hachette, in France, have labored at the same task, as have Coulomb and Prony, Girard and Poncelet. Eytelwein's German treatise (*Hydraulik*), contains an account of what others and himself have done. Many of these trains of experiments, both in France and Italy, were made at the expense of governments, and on a very magnificent scale. In England less was done in this way during the last century, than in most other countries. The *Philosophical Transactions*, for instance, scarcely contain a single paper on this subject founded on experimental investigations.[4] Dr. Thomas Young, who was at the head of his countrymen in so many branches of science, was one of the first to call back attention to this: and Mr. Rennie and others have recently made valuable experiments. In many of the questions now spoken of, the accordance which engineers are able to obtain, between their calculated and observed results,

---

[4] Rennie, *Report to Brit. Assoc.*

is very great: but these calculations are performed by means of empirical formulæ, which do not connect the facts with their causes, and still leave a wide space to be traversed, in order to complete the science.

In the mean time, all the other portions of Mechanics were reduced to general laws, and analytical processes; and means were found of including Hydrodynamics, notwithstanding the difficulties which attend its special problems, in this common improvement of form. This progress we must relate.

[2d Ed.] [The hydrodynamical problems referred to above are, the laws of a fluid issuing from a vessel, the laws of the motion of water in pipes, canals, and rivers, and the laws of the resistance of fluids. To these may be added, as an hydrodynamical problem important in theory, in experiment, and in the comparison of the two, the laws of waves. Newton gave, in the *Principia*, an explanation of the waves of water (Lib. ii. Prop. 44), which appears to proceed upon an erroneous view of the nature of the motion of the fluid: but in his solution of the problem of sound, appeared, for the first time, a correct view of the propagation of an undulation in a fluid. The history of this subject, as bearing upon the theory of sound, is given in Book viii.: but I may here remark, that the laws of the motion of waves have been pursued experimentally by various persons, as Bremontier (*Recherches sur le Mouvement des Ondes*, 1809), Emy (*Du Mouvement des Ondes*, 1831), the Webers (*Wellenlehre*, 1825); and by Mr. Scott Russell (*Reports of the British Association*, 1844). The analytical theory has been carried on by Poisson, Cauchy, and, among ourselves, by Prof. Kelland (*Edin. Trans.*), and Mr. Airy (in the article *Tides*, in the *Encyclopædia Metropolitana*). And though theory and experiment have not yet been brought into complete accordance, great progress has been made in that work, and the remaining chasm between the two is manifestly due only to the incompleteness of both.]

Perhaps the most remarkable case of fluid motion recently discussed, is one which Mr. Scott Russell has presented experimentally; and which, though novel, is easily seen to follow from known principles; namely, the *Great Solitary Wave.* A wave may be produced, which shall move along a canal unaccompanied by any other wave: and the simplicity of this case makes the mathematical conditions and consequences more simple than they are in most other problems of Hydrodynamics.

CHAPTER V.

Generalization of the Principles of Mechanics.

Sect. 1.—*Generalization of the Second Law of Motion.—Central Forces.*

THE Second Law of Motion being proved for constant Forces which act in parallel lines, and the Third Law for the Direct Action of bodies, it still required great mathematical talent, and some inductive power, to see clearly the laws which govern the motion of any number of bodies, acted upon by each other, and by any forces, anyhow varying in magnitude and direction. This was the task of the generalization of the laws of motion.

Galileo had convinced himself that the velocity of projection, and that which gravity alone would produce, are "both maintained, without being altered, perturbed, or impeded in their mixture." It is to be observed, however, that the truth of this result depends upon a particular circumstance, namely, that gravity, at all points, acts in lines, which, as to sense, are parallel. When we have to consider cases in which this is not true, as when the force tends to the centre of a circle, the law of composition cannot be applied in the same way; and, in this case, mathematicians were met by some peculiar difficulties.

One of these difficulties arises from the apparent inconsistency of the statical and dynamical measures of force. When a body moves in a circle, the force which urges the body to the centre is only a *tendency* to motion; for the body does not, in fact, approach to the centre; and this mere tendency to motion is combined with an actual motion, which takes place in the circumference. We appear to have to compare two things which are heterogeneous. Descartes had noticed this difficulty, but without giving any satisfactory solution of it.[1] If we combine the actual motion to or from the centre with the traverse motion about the centre, we obtain a result which is false on mechanical principles. Galileo endeavored in this way to find the curve described by a body which falls towards the earth's centre, and is, at the same time, carried

[1] *Princip.* P. iii. 59.

round by the motion of the earth; and obtained an erroneous result. Kepler and Fermat attempted the same problem, and obtained solutions different from that of Galileo, but not more correct.

Even Newton, at an early period of his speculations, had an erroneous opinion respecting this curve, which he imagined to be a kind of spiral. Hooke animadverted upon this opinion when it was laid before the Royal Society of London in 1679, and stated, more truly, that, supposing no resistance, it would be "an eccentric ellipsoid," that is, a figure resembling an ellipse. But though he had made out the approximate form of the curve, in some unexplained way, we have no reason to believe that he possessed any means of determining the mathematical properties of the curve described in such a case. The perpetual composition of a central force with the previous motion of the body, could not be successfully treated without the consideration of the Doctrine of Limits, or something equivalent to that doctrine. The first example which we have of the right solution of such a problem occurs, so far as I know, in the Theorems of Huyghens concerning Circular Motion, which were published, without demonstration, at the end of his *Horologium Oscillatorium*, in 1673. It was there asserted that when equal bodies describe circles, if the times are equal, the centrifugal forces will be as the diameters of the circles; if the velocities are equal, the forces will be reciprocally as the diameters, and so on. In order to arrive at these propositions, Huyghens must, virtually at least, have applied the Second Law of Motion to the limiting elements of the curve, according to the way in which Newton, a few years later, gave the demonstration of the theorems of Huyghens in the *Principia*.

The growing persuasion that the motions of the heavenly bodies about the sun might be explained by the action of central forces, gave a peculiar interest to these mechanical speculations, at the period now under review. Indeed, it is not easy to state separately, as our present object requires us to do, the progress of Mechanics, and the progress of Astronomy. Yet the distinction which we have to make is, in its nature, sufficiently marked. It is, in fact, no less marked than the distinction between speaking logically and speaking truly. The framers of the science of motion were employed in establishing those notions, names, and rules, in conformity to which *all* mechanical *truth must* be expressed; but *what was the truth* with regard to the mechanism of the universe remained to be determined by other means. Physical Astronomy, at the period of which we speak, eclipsed and overlaid theoretical Mechanics, as, a little previously, Dynamics had eclipsed and superseded Statics.

The laws of variable force and of curvilinear motion were not much pursued, till the invention of Fluxions and of the Differential Calculus again turned men's minds to these subjects, as easy and interesting exercises of the powers of these new methods. Newton's *Principia*, of which the first two Books are purely dynamical, is the great exception to this assertion; inasmuch as it contains correct solutions of a great variety of the most general problems of the science; and indeed is, even yet, one of the most complete treatises which we possess upon the subject.

We have seen that Kepler, in his attempts to explain the curvilinear motions of the planets by means of a central force, failed, in consequence of his belief that a continued transverse action of the central body was requisite to keep up a continual motion. Galileo had founded his theory of projectiles on the principle that such an action was not necessary; yet Borelli, a pupil of Galileo, when, in 1666, he published his theory of the Medicean Stars (the satellites of Jupiter), did not keep quite clear of the same errors which had vitiated Kepler's reasonings. In the same way, though Descartes is sometimes spoken of as the first promulgator of the First Law of Motion, yet his theory of Vortices must have been mainly suggested by a want of an entire confidence in that law. When he represented the planets and satellites as owing their motions to oceans of fluid diffused through the celestial spaces, and constantly whirling round the central bodies, he must have felt afraid of trusting the planets to the operation of the laws of motion in free space. Sounder physical philosophers, however, began to perceive the real nature of the question. As early as 1666, we read, in the Journals of the Royal Society, that "there was read a paper of Mr. Hooke's explicating the inflexion of a direct motion into a curve by a supervening attractive principle;" and before the publication of the *Principia* in 1687, Huyghens, as we have seen, in Holland, and, in our own country, Wren, Halley, and Hooke, had made some progress in the true mechanics of circular motion,[2] and had distinctly contemplated the problem of the motion of a body in an ellipse by a central force, though they could not solve it. Halley went to Cambridge in 1684,[3] for the express purpose of consulting Newton upon the subject of the production of the elliptical motion of the planets by means of a central

[2] Newt. *Princip.* Schol. to Prop. iv. [3] Sir D. Brewster's *Life of Newton*, p. 154.

force, and, on the 10th of December,[4] announced to the Royal Society that he had seen Mr. Newton's book, *De Motu Corporum*. The feeling that mathematicians were on the brink of discoveries such as are contained in this work was so strong, that Dr. Halley was requested to remind Mr. Newton of his promise of entering them in the Register of the Society, "for securing the invention to himself till such time as he can be at leisure to publish it." The manuscript, with the title *Philosophiæ Naturalis Principia Mathematica*, was presented to the society (to which it was dedicated) on the 28th of April, 1686. Dr. Vincent, who presented it, spoke of the novelty and dignity of the subject; and the president (Sir J. Hoskins) added, with great truth, "that the method was so much the more to be prized as it was both invented and perfected at the same time."

The reader will recollect that we are here speaking of the *Principia* as a Mechanical Treatise only; we shall afterwards have to consider it as containing the greatest discoveries of Physical Astronomy. As a work on Dynamics, its merit is, that it exhibits a wonderful store of refined and beautiful mathematical artifices, applied to solve all the most general problems which the subject offered. The *Principia* can hardly be said to contain any new inductive discovery respecting the principles of mechanics; for though Newton's *Axioms or Laws of Motion*, which stand at the beginning of the book, are a much clearer and more general statement of the grounds of Mechanics than had yet appeared, they do not involve any doctrines which had not been previously stated or taken for granted by other mathematicians.

The work, however, besides its unrivalled mathematical skill, employed in tracing out, deductively, the consequences of the laws of motion, and its great cosmical discoveries, which we shall hereafter treat of, had great philosophical value in the history of Dynamics, as exhibiting a clear conception of the new character and functions of that science. In his Preface, Newton says, "Rational Mechanics must be the science of the Motions which result from any Forces, and of the Forces which are required for any Motions, accurately propounded and demonstrated. For many things induce me to suspect, that all natural phenomena may depend upon some Forces by which the particles of bodies are either drawn towards each other, and cohere, or repel and recede from each other: and these Forces being hitherto unknown, philosophers have pursued their researches in vain. And I hope

---

[4] Id. p. 184.

that the principles expounded in this work will afford some light, either to this mode of philosophizing, or to some mode which is more true."

Before we pursue this subject further, we must trace the remainder of the history of the Third Law.

Sect. 2.—Generalization of the Third Law of Motion.—Centre of Oscillation.—Huyghens.

The Third Law of Motion, whether expressed according to Newton's formula (by the equality of Action and Reaction), or in any other of the ways employed about the same time, easily gave the solution of mechanical problems in all cases of *direct* action; that is, when each body acted directly on others. But there still remained the problems in which the action is *indirect*;—when bodies, in motion, act on each other by the intervention of levers, or in any other way. If a rigid rod, passing through two weights, be made to swing about its upper point, so as to form a pendulum, each weight will act and react on the other by means of the rod, considered as a lever turning about the point of suspension. What, in this case, will be the effect of this action and reaction? In what time will the pendulum oscillate by the force of gravity? Where is the point at which a single weight must be placed to oscillate in the same time? in other words, where is the *Centre of Oscillation?*

Such was the problem—an example only of the general problem of indirect action—which mathematicians had to solve. That it was by no means easy to see in what manner the law of the communication of motion was to be extended from simpler cases to those where rotatory motion was produced, is shown by this;—that Newton, in attempting to solve the mechanical problem of the Precession of the Equinoxes, fell into a serious error on this very subject. He assumed that, when a part has to communicate rotatory movement to the whole (as the protuberant portion of the terrestrial spheroid, attracted by the sun and moon, communicates a small movement to the whole mass of the earth), the quantity of the *motion*, "motus," will not be altered by being communicated. This principle is true, if, by *motion*, we understand what is called *moment of inertia*, a quantity in which both the velocity of each particle and its distance from the axis of rotation are taken into account: but Newton, in his calculations of its amount, considered the velocity only; thus making *motion*, in this case, identical with the *momentum* which he introduces in treating of the simpler case of the third law of motion, when the action is direct. This error was retained even in the later editions of the *Principia*.[5]

The question of the centre of oscillation had been proposed by Mersenne somewhat earlier,[6] in 1646. And though the problem was out of the reach of any principles at that time known and understood, some of the mathematicians of the day had rightly solved some cases of it, by proceeding as if the question had been to find the *Centre of Percussion.* The Centre of Percussion is the point about which the momenta of all the parts of a body balance each other, when it is in motion about any axis, and is stopped by striking against an obstacle placed at that centre. Roberval found this point in some easy cases; Descartes also attempted the problem; their rival labors led to an angry controversy: and Descartes was, as in his physical speculations he often was, very presumptuous, though not more than half right.

Huyghens was hardly advanced beyond boyhood when Mersenne first proposed this problem; and, as he says,[7] could see no principle which even offered an opening to the solution, and had thus been repelled at the threshold. When, however, he published his *Horologium Oscillatorium* in 1673, the fourth part of that work was on the Centre of Oscillation or Agitation; and the principle which he then assumed, though not so simple and self-evident as those to which such problems were afterwards referred, was perfectly correct and general, and led to exact solutions in all cases. The reader has already seen repeatedly in the course of this history, complex and derivative principles presenting themselves to men's minds, before simple and elementary ones. The "hypothesis" assumed by Huyghens was this; "that if any weights are put in motion by the force of gravity, they *cannot* move so that the centre of gravity of them all shall rise *higher* than the place from which it descended." This being assumed, it is easy to show that the centre of gravity will, under all circumstances, rise *as high* as its original position; and this consideration leads to a determination of the oscillation of a compound pendulum. We may observe, in the principle thus selected, a conviction that, in all mechanical action, the centre of gravity may be taken as the representative of the whole system. This conviction, as we have seen, may be traced in the axioms of Archimedes and Stevinus; and Huyghens, when he proceeds upon it, undertakes to show,[8] that he assumes only this, that a heavy body cannot, of itself, move upwards.

---

[5] B. iii. Lemma iii. to Prop. xxxix.

[6] Mont. ii. 423.

[7] *Hor. Osc.* Pref.

[8] *Hor. Osc.* p. 121.

Clear as Huyghen's principle appeared to himself, it was, after some time, attacked by the Abbé Catelan, a zealous Cartesian. Catelan also put forth principles which he conceived were evident, and deduced from them conclusions contradictory to those of Huyghens. His principles, now that we know them to be false, appear to us very gratuitous. They are these; "that in a compound pendulum, the sum of the velocities of the component weights is equal to the sum of the velocities which they would have acquired if they had been detached pendulums;" and "that the time of the vibration of a compound pendulum is an arithmetic mean between the times of the vibrations of the weights, moving as detached pendulums." Huyghens easily showed that these suppositions would make the centre of gravity ascend to a greater height than that from which it fell; and after some time, James Bernoulli stept into the arena, and ranged himself on the side of Huyghens. As the discussion thus proceeded, it began to be seen that the question really was, in what manner the Third Law of Motion was to be extended to cases of indirect action; whether by distributing the action and reaction according to statical principles, or in some other way. "I propose it to the consideration of mathematicians," says Bernoulli in 1686, "what law of the communication of velocity is observed by bodies in motion, which are sustained at one extremity by a fixed fulcrum, and at the other by a body also moving, but more slowly. Is the excess of velocity which must be communicated from the one body to the other to be distributed in the same proportion in which a load supported on the lever would be distributed?" He adds, that if this question be answered in the affirmative, Huyghens will be found to be in error; but this is a mistake. The principle, that the action and reaction of bodies thus moving are to be distributed according to the rules of the lever, is true; but Bernoulli mistook, in estimating this action and reaction by the *velocity* acquired at any moment; instead of taking, as he should have done, the *increment* of velocity which gravity tended to impress in the next instant. This was shown by the Marquis de l'Hôpital; who adds, with justice, "I conceive that I have thus fully answered the call of Bernoulli, when he says, I propose it to the consideration of mathematicians, &c."

We may, from this time, consider as known, but not as fully established, the principle that "When bodies in motion affect each other, the action and reaction are distributed according to the laws of Statics;" although there were still found occasional difficulties in the generalization and application of the rule. James Bernoulli, in 1703, gave "a General Demonstration of the Centre of Oscillation, drawn from the nature of the Lever." In this demonstration[9] he takes as a fundamental principle, that bodies in motion, connected by levers, balance, when the products of their momenta and the lengths of the levers are equal in opposite directions. For the proof of this proposition, he refers to Marriotte, who had asserted it of weights acting by percussion,[10] and in order to prove it, had balanced the effect of a weight on a lever by the effect of a jet of water, and had confirmed it by other experiments.[11] Moreover, says Bernoulli, there is no one who denies it. Still, this kind of proof was hardly satisfactory or elementary enough. John Bernoulli took up the subject after the death of his brother James, which happened in 1705. The former published in 1714 his *Meditatio de Naturâ Centri Oscillationis*. In this memoir, he assumes, as his brother had done, that the effects of forces on a lever in motion are distributed according to the common rules of the lever.[12] The principal generalization which he introduced was, that he considered gravity as a force soliciting to motion, which might have different intensities in different bodies. At the same time, Brook Taylor in England solved the problem, upon the same principles as Bernoulli; and the question of priority on this subject was one point in the angry intercourse which, about this time, became common between the English mathematicians and those of the Continent. Hermann also, in his *Phoronomia*, published in 1716, gave a proof which, as he informs us, he had devised before he saw John Bernoulli's. This proof is founded on the statical equivalence of the "*solicitations of gravity*," and the "*vicarious solicitations*" which correspond to the actual motion of each part; or, as it has been expressed by more modern writers, the equilibrium of the *impressed* and *effective forces.*

It was shown by John Bernoulli and Hermann, and was indeed easily proved, that the proposition assumed by Huyghens as the foundation of his solution, was, in fact, a consequence of the elementary principles which belong to this branch of mechanics. But this assumption of Huyghens was an example of a more general proposition, which by some mathematicians at this time had been put forward as an original and elementary law; and as a principle which ought to supersede the usual measure of the forces of bodies in motion; this principle they called "*the Conservation of Vis Viva.*" The attempt to

[9] *Op.* ii. 930.

[10] *Choq. des Corps*, p. 296.

[11] Ib. Prop. xi.

[12] P. 172.

make this change was the commencement of one of the most obstinate and curious of the controversies which form part of the history of mechanical science. The celebrated Leibnitz was the author of the new opinion. In 1686, he published, in the Leipsic Acts, "A short Demonstration of a memorable Error of Descartes and others, concerning the natural law by which they think that God always preserves the same quantity of motion; in which they pervert mechanics." The principle that the same quantity of motion, and therefore of moving force, is always preserved in the world, follows from the equality of action and reaction; though Descartes had, after his fashion, given a theological reason for it; Leibnitz allowed that the quantity of moving force remains always the same, but denied that this force is measured by the quantity of motion or momentum. He maintained that the same force is requisite to raise a weight of one pound through four feet, and a weight of four pounds through one foot, though the momenta in this case are as one to two. This was answered by the Abbé de Conti; who truly observed, that allowing the effects in the two cases to be equal, this did not prove the forces to be equal; since the effect, in the first case, was produced in a double time, and therefore it was quite consistent to suppose the force only half as great. Leibnitz, however, persisted in his innovation; and in 1695 laid down the distinction between *vires mortuæ*, or pressures, and *vires vivæ*, the name he gave to his own measure of force. He kept up a correspondence with John Bernoulli, whom he converted to his peculiar opinions on this subject; or rather, as Bernoulli says,[13] made him think for himself, which ended in his proving directly that which Leibnitz had defended by indirect reasons. Among other arguments, he had pretended to show (what is certainly not true), that if the common measure of forces be adhered to, a perpetual motion would be possible. It is easy to collect many cases which admit of being very simply and conveniently reasoned upon by means of the *vis viva*, that is, by taking the force to be proportional to the *square* of the velocity, and not to the velocity itself. Thus, in order to give the arrow *twice* the velocity, the bow must be *four* times as strong; and in all cases in which no account is taken of the time of producing the effect, we may conveniently use similar methods.

But it was not till a later period that the question excited any general notice. The Academy of Sciences of Paris in 1724 proposed

---

[13] *Op.* iii. 40.

as a subject for their prize dissertation the laws of the impact of bodies. Bernoulli, as a competitor, wrote a treatise, upon Leibnitzian principles, which, though not honored with the prize, was printed by the Academy with commendation.[14] The opinions which he here defended and illustrated were adopted by several mathematicians; the controversy extended from the mathematical to the literary world, at that time more attentive than usual to mathematical disputes, in consequence of the great struggle then going on between the Cartesian and the Newtonian system. It was, however, obvious that by this time the interest of the question, so far as the progress of Dynamics was concerned, was at an end; for the combatants all agreed as to the results in each particular case. The Laws of Motion were now established; and the question was, by means of what definitions and abstractions could they be best expressed;—a metaphysical, not a physical discussion, and therefore one in which "the paper philosophers," as Galileo called them, could bear a part. In the first volume of the *Transactions of the Academy of St. Petersburg*, published in 1728, there are three Leibnitzian memoirs by Hermann, Bullfinger, and Wolff. In England, Clarke was an angry assailant of the German opinion, which S'Gravesande maintained. In France, Mairan attacked the *vis viva* in 1728; "with strong and victorious reasons," as the Marquise du Chatelet declared, in the first edition of her *Treatise on Fire*.[15] But shortly after this praise was published, the Chateau de Cirey, where the Marquise usually lived, became a school of Leibnitzian opinions, and the resort of the principal partisans of the *vis viva*. "Soon," observes Mairan, "their language was changed; the *vis viva* was enthroned by the side of the *monads*." The Marquise tried to retract or explain away her praises; she urged arguments on the other side. Still the question was not decided; even her friend Voltaire was not converted. In 1741 he read a memoir *On the Measure and Nature of Moving Forces*, in which he maintained the old opinion. Finally, D'Alembert in 1743 declared it to be, as it truly was, a mere question of words; and by the turn which Dynamics then took, it ceased to be of any possible interest or importance to mathematicians.

The representation of the laws of motion and of the reasonings depending on them, in the most general form, by means of analytical language, cannot be said to have been fully achieved till the time of D'Alembert; but as we have already seen, the discovery of these laws

---

[14] *Discours sur les Loix de la Communication du Mouvement.* [15] Mont. iii. 640.

had taken place somewhat earlier; and that law which is more particularly expressed in D'Alembert's Principle (*the equality of the action gained and lost*) was, it has been seen, rather led to by the general current of the reasoning of mathematicians about the end of the seventeenth century than discovered by any one. Huyghens, Marriotte, the two Bernoullis, L'Hôpital, Taylor, and Hermann, have each of them their name in the history of this advance; but we cannot ascribe to any of them any great real inductive sagacity shown in what they thus contributed, except to Huyghens, who first seized the principle in such a form as to find the centre of oscillation by means of it. Indeed, in the steps taken by the others, language itself had almost made the generalization for them at the time when they wrote; and it required no small degree of acuteness and care to distinguish the old cases, in which the law had already been applied, from the new cases, in which they had to apply it.

CHAPTER VI.

SEQUEL TO THE GENERALIZATION OF THE PRINCIPLES OF MECHANICS.—PERIOD OF MATHEMATICAL DEDUCTION.—ANALYTICAL MECHANICS.

WE have now finished the history of the discovery of Mechanical Principles, strictly so called. The three Laws of Motion, generalized in the manner we have described, contain the materials of the whole structure of Mechanics; and in the remaining progress of the science, we are led to no new truth which was not implicitly involved in those previously known. It may be thought, therefore, that the narrative of this progress is of comparatively small interest. Nor do we maintain that the application and development of principles is a matter of so much importance to the philosophy of science, as the advance towards and to them. Still, there are many circumstances in the latter stages of the progress of the science of Mechanics, which well deserve notice, and make a rapid survey of that part of its history indispensable to our purpose.

The Laws of Motion are expressed in terms of Space and Number; the development of the consequences of these laws must, therefore, be performed by means of the reasonings of mathematics; and the science of Mechanics may assume the various aspects which belong to the different modes of dealing with mathematical quantities. Mechanics, like pure mathematics, may be geometrical or may be analytical; that is, it may treat space either by a direct consideration of its properties, or by a symbolical representation of them: Mechanics, like pure mathematics, may proceed from special cases, to problems and methods of extreme generality;—may summon to its aid the curious and refined relations of symmetry, by which general and complex conditions are simplified;—may become more powerful by the discovery of more powerful analytical artifices;—may even have the generality of its principles further expanded, inasmuch as symbols are a more general language than words. We shall very briefly notice a series of modifications of this kind.

1. *Geometrical Mechanics, Newton, &c.*—The first great systematical Treatise on Mechanics, in the most general sense, is the two first Books of the *Principia* of Newton. In this work, the method employed is predominantly geometrical: not only space is not represented symbolically, or by reference to number; but numbers, as, for instance, those which measure time and force, are represented by spaces; and the laws of their changes are indicated by the properties of curve lines. It is well known that Newton employed, by preference, methods of this kind in the exposition of his theorems, even where he had made the discovery of them by analytical calculations. The intuitions of space appeared to him, as they have appeared to many of his followers, to be a more clear and satisfactory road to knowledge, than the operations of symbolical language. Hermann, whose *Phoronomia* was the next great work on this subject, pursued a like course; employing curves, which he calls "the scale of velocities," "of forces," &c. Methods nearly similar were employed by the two first Bernoullis, and other mathematicians of that period; and were, indeed, so long familiar, that the influence of them may still be traced in some of the terms which are used on such subjects; as, for instance, when we talk of "reducing a problem to quadratures," that is, to the finding the area of the curves employed in these methods.

2. *Analytical Mechanics. Euler.*—As analysis was more cultivated, it gained a predominancy over geometry; being found to be a far more powerful instrument for obtaining results; and possessing a beauty and an evidence, which, though different from those of geometry, had great attractions for minds to which they became familiar. The person who did most to give to analysis the generality and sym-

metry which are now its pride, was also the person who made Mechanics analytical; I mean Euler. He began his execution of this task in various memoirs which appeared in the *Transactions of the Academy of Sciences at St. Petersburg*, commencing with its earliest volumes; and in 1736, he published there his *Mechanics, or the Science of Motion analytically expounded; in the way of a Supplement to the Transactions of the Imperial Academy of Sciences.* In the preface to this work, he says, that though the solutions of problems by Newton and Hermann were quite satisfactory, yet he found that he had a difficulty in applying them to new problems, differing little from theirs; and that, therefore, he thought it would be useful to extract an analysis out of their synthesis.

3. *Mechanical Problems.*—In reality, however, Euler has done much more than merely give analytical methods, which may be applied to mechanical problems: he has himself applied such methods to an immense number of cases. His transcendent mathematical powers, his long and studious life, and the interest with which he pursued the subject, led him to solve an almost inconceivable number and variety of mechanical problems. Such problems suggested themselves to him on all occasions. One of his memoirs begins, by stating that, happening to think of the line of Virgil,

> Anchora de prorâ jacitur stant litore puppes;
> The anchor drops, the rushing keel is staid;

he could not help inquiring what would be the nature of the ship's motion under the circumstances here described. And in the last few days of his life, after his mortal illness had begun, having seen in the newspapers some statements respecting balloons, he proceeded to calculate their motions; and performed a difficult integration, in which this undertaking engaged him. His Memoirs occupy a very large portion of the *Petropolitan Transactions* during his life, from 1728 to 1783; and he declared that he should leave papers which might enrich the publications of the Academy of Petersburg for twenty years after his death;—a promise which has been more than fulfilled; for, up to 1818, the volumes usually contain several Memoirs of his. He and his contemporaries may be said to have exhausted the subject; for there are few mechanical problems which have been since treated, which they have not in some manner touched upon.

I do not dwell upon the details of such problems; for the next great step in Analytical Mechanics, the publication of D'Alembert's Principle in 1743, in a great degree superseded their interest. The Transactions of the Academies of Paris and Berlin, as well as St. Petersburg, are filled, up to this time, with various questions of this kind. They require, for the most part, the determination of the motions of several bodies, with or without weight, which pull or push each other by means of threads, or levers, to which they are fastened, or along which they can slide; and which, having a certain impulse given them at first, are then left to themselves, or are compelled to move in given lines and surfaces. The postulate of Huyghens, respecting the motion of the centre of gravity, was generally one of the principles of the solution; but other principles were always needed in addition to this; and it required the exercise of ingenuity and skill to detect the most suitable in each case. Such problems were, for some time, a sort of trial of strength among mathematicians: the principle of D'Alembert put an end to this kind of challenges, by supplying a direct and general method of resolving, or at least of throwing into equations, any imaginable problem. The mechanical difficulties were in this way reduced to difficulties of pure mathematics.

4. *D'Alembert's Principle.*—D'Alembert's Principle is only the expression, in the most general form, of the principle upon which John Bernoulli, Hermann, and others, had solved the problem of the centre of oscillation. It was thus stated, "The motion *impressed* on each particle of any system by the forces which act upon it, may be resolved into two, the *effective* motion, and the motion gained or *lost*: the effective motions will be the real motions of the parts, and the motions gained and lost will be such as would keep the system at rest." The distinction of *statics*, the doctrine of equilibrium, and *dynamics*, the doctrine of motion, was, as we have seen, fundamental; and the difference of difficulty and complexity in the two subjects was well understood, and generally recognized by mathematicians. D'Alembert's principle reduces every dynamical question to a statical one; and hence, by means of the conditions which connect the possible motions of the system, we can determine what the actual motions must be. The difficulty of determining the laws of equilibrium, in the application of this principle in complex cases is, however, often as great as if we apply more simple and direct considerations.

5. *Motion in Resisting Media. Ballistics.*—We shall notice more particularly the history of some of the problems of mechanics. Though John Bernoulli always spoke with admiration of Newton's *Principia*, and of its author, he appears to have been well disposed to point out

real or imagined blemishes in the work. Against the validity of Newton's determination of the path described by a body projected in any part of the solar system, Bernoulli urges a cavil which it is difficult to conceive that a mathematician, such as he was, could seriously believe to be well founded. On Newton's determination of the path of a body in a resisting medium, his criticism is more just. He pointed out a material error in this solution: this correction came to Newton's knowledge in London, in October, 1712, when the impression of the second edition of the *Principia* was just drawing to a close, under the care of Cotes at Cambridge; and Newton immediately cancelled the leaf and corrected the error.[1]

This problem of the motion of a body in a resisting medium, led to another collision between the English and the German mathematicians. The proposition to which we have referred, gave only an indirect view of the nature of the curve described by a projectile in the air; and it is probable that Newton, when he wrote the *Principia*, did not see his way to any direct and complete solution of this problem. At a later period, in 1718, when the quarrel had waxed hot between the admirers of Newton and Leibnitz, Keill, who had come forward as a champion on the English side, proposed this problem to the foreigners as a challenge. Keill probably imagined that what Newton had not discovered, no one of his time would be able to discover. But the sedulous cultivation of analysis by the Germans had given them mathematical powers beyond the expectations of the English; who, whatever might be their talents, had made little advance in the effective use of general methods; and for a long period seemed to be fascinated to the spot, in their admiration of Newton's excellence. Bernoulli speedily solved the problem; and reasonably enough, according to the law of honor of such challenges, called upon the challenger to produce his solution. Keill was unable to do this; and after some attempts at procrastination, was driven to very paltry evasions. Bernoulli then published his solution, with very just expressions of scorn towards his antagonist. And this may, perhaps, be considered as the first material addition which was made to the *Principia* by subsequent writers.

6. *Constellation of Mathematicians.*— We pass with admiration along the great series of mathematicians, by whom the science of theoretical mechanics has been cultivated, from the time of Newton to our own. There is no group of men of science whose fame is

---

[1] MS. Correspondence in Trin. Coll. Library.

higher or brighter. The great discoveries of Copernicus, Galileo, Newton, had fixed all eyes on those portions of human knowledge on which their successors employed their labors. The certainty belonging to this line of speculation seemed to elevate mathematicians above the students of other subjects; and the beauty of mathematical relations, and the subtlety of intellect which may be shown in dealing with them, were fitted to win unbounded applause. The successors of Newton and the Bernoullis, as Euler, Clairaut, D'Alembert, Lagrange, Laplace, not to introduce living names, have been some of the most remarkable men of talent which the world has seen. That their talent is, for the most part, of a different kind from that by which the laws of nature were discovered, I shall have occasion to explain elsewhere; for the present, I must endeavor to arrange the principal achievements of those whom I have mentioned.

The series of persons is connected by social relations. Euler was the pupil of the first generation of Bernoullis, and the intimate friend of the second generation; and all these extraordinary men, as well as Hermann, were of the city of Basil, in that age a spot fertile of great mathematicians to an unparalleled degree. In 1740, Clairaut and Maupertuis visited John Bernoulli, at that time the Nestor of mathematicians, who died, full of age and honors, in 1748. Euler, several of the Bernoullis, Maupertuis, Lagrange, among other mathematicians of smaller note, were called into the north by Catharine of Russia and Frederic of Prussia, to inspire and instruct academies which the brilliant fame then attached to science, had induced those monarchs to establish. The prizes proposed by these societies, and by the French Academy of Sciences, gave occasion to many of the most valuable mathematical works of the century.

7. *The Problem of Three Bodies.*—In 1747, Clairaut and D'Alembert sent, on the same day, to this body, their solutions of the celebrated "Problem of Three Bodies," which, from that time, became the great object of attention of mathematicians;—the bow in which each tried his strength, and endeavored to shoot further than his predecessors.

This problem was, in fact, the astronomical question of the effect produced by the attraction of the sun, in disturbing the motions of the moon about the earth; or by the attraction of one planet, disturbing the motion of another planet about the sun; but being expressed generally, as referring to one body which disturbs any two others, it became a mechanical problem, and the history of it belongs to the present subject.

One consequence of the synthetical form adopted by Newton in the *Principia*, was, that his successors had the problem of the solar system to begin entirely anew. Those who would not do this, made no progress, as was long the case with the English. Clairaut says, that he tried for a long time to make some use of Newton's labors; but that, at last, he resolved to take up the subject in an independent manner. This, accordingly, he did, using analysis throughout, and following methods not much different from those still employed. We do not now speak of the comparison of this theory with observation, except to remark, that both by the agreements and by the discrepancies of this comparison, Clairaut and other writers were perpetually driven on to carry forwards the calculation to a greater and greater degree of accuracy.

One of the most important of the cases in which this happened, was that of the movement of the Apogee of the Moon; and in this case, a mode of approximating to the truth, which had been depended on as nearly exact, was, after having caused great perplexity, found by Clairaut and Euler to give only half the truth. This same Problem of Three Bodies was the occasion of a memoir of Clairaut, which gained the prize of the Academy of St. Petersburg in 1751; and, finally, of his *Théorie de la Lune*, published in 1765. D'Alembert labored at the same time on the same problem; and the value of their methods, and the merit of the inventors, unhappily became a subject of controversy between those two great mathematicians. Euler also, in 1753, published a *Theory of the Moon*, which was, perhaps, more useful than either of the others, since it was afterwards the basis of Mayer's method, and of his Tables. It is difficult to give the general reader any distinct notion of these solutions. We may observe, that the quantities which determine the moon's position, are to be determined by means of certain algebraical equations, which express the mechanical conditions of the motion. The operation, by which the result is to be obtained, involves the process of integration; which, in this instance, cannot be performed in an immediate and definite manner; since the quantities thus to be operated on depend upon the moon's position, and thus require us to know the very thing which we have to determine by the operation. The result must be got at, therefore, by successive approximations: we must first find a quantity near the truth; and then, by the help of this, one nearer still; and so on; and, in this manner, the moon's place will be given by a converging series of terms. The form of these terms depends upon the relations of position between the sun and moon, their apogees, the moon's nodes, and other quantities; and by the variety of combinations of which these admit, the terms become very numerous and complex. The magnitude of the terms depends also upon various circumstances; as the relative force of the sun and earth, the relative times of the solar and lunar revolutions, the eccentricities and inclinations of the two orbits. These are combined so as to give terms of different orders of magnitudes; and it depends upon the skill and perseverance of the mathematician how far he will continue this series of terms. For there is no limit to their number: and though the methods of which we have spoken do theoretically enable us to calculate as many terms as we please, the labor and the complexity of the operations are so serious that common calculators are stopped by them. None but very great mathematicians have been able to walk safely any considerable distance into this avenue,—so rapidly does it darken as we proceed. And even the possibility of doing what has been done, depends upon what we may call accidental circumstances; the smallness of the inclinations and eccentricities of the system, and the like. "If nature had not favored us in this way," Lagrange used to say, "there would have been an end of the geometers in this problem." The expected return of the comet of 1682 in 1759, gave a new interest to the problem, and Clairaut proceeded to calculate the case which was thus suggested. When this was treated by the methods which had succeeded for the moon, it offered no prospect of success, in consequence of the absence of the favorable circumstances just referred to, and, accordingly, Clairaut, after obtaining the six equations to which he reduces the solution,[2] adds, "Integrate them who can" (Intègre maintenant qui pourra). New methods of approximation were devised for this case.

The problem of three bodies was not prosecuted in consequence of its analytical beauty, or its intrinsic attraction; but its great difficulties were thus resolutely combated from necessity; because in no other way could the theory of universal gravitation be known to be true or made to be useful. The construction of *Tables of the Moon*, an object which offered a large pecuniary reward, as well as mathematical glory, to the successful adventurer, was the main purpose of these labors.

The *Theory of the Planets* presented the Problem of Three Bodies in a new form, and involved in peculiar difficulties; for the approxima-

---

[2] *Journal des Sçavans*, Aug. 1759.

tions which succeed in the Lunar Theory fail here. Artifices somewhat modified are required to overcome the difficulties of this case.

Euler had investigated, in particular, the motions of Jupiter and Saturn, in which there was a secular acceleration and retardation, known by observation, but not easily explicable by theory. Euler's memoirs, which gained the prize of the French Academy, in 1748 and 1752, contained much beautiful analysis; and Lagrange published also a theory of Jupiter and Saturn, in which he obtained results different from those of Euler. Laplace, in 1787, showed that this inequality arose from the circumstance that two of Saturn's years are very nearly equal to five of Jupiter's.

The problems relating to Jupiter's *Satellites*, were found to be even more complex than those which refer to the planets: for it was necessary to consider each satellite as disturbed by the other three at once; and thus there occurred the Problem of *Five* Bodies. This problem was resolved by Lagrange.[3]

Again, the newly-discovered *small Planets*, Juno, Ceres, Vesta, Pallas, whose orbits almost coincide with each other, and are more inclined and more eccentric than those of the ancient planets, give rise, by their perturbations, to new forms of the problem, and require new artifices.

In the course of these researches respecting Jupiter, Lagrange and Laplace were led to consider particularly the *secular Inequalities* of the solar system; that is, those inequalities in which the duration of the cycle of change embraces very many revolutions of the bodies themselves. Euler in 1749 and 1755, and Lagrange[4] in 1766, had introduced the method of the *Variation of the Elements* of the orbit; which consists in tracing the effect of the perturbing forces, not as directly altering the place of the planet, but as producing a change from one instant to another, in the dimensions and position of the Elliptical orbit which the planet describes.[5] Taking this view, he deter-

---

[3] Bailly, *Ast. Mod.* iii. 178.

[4] Gautier, *Prob. de Trois Corps*, p. 155.

[5] In the first edition of this History, I had ascribed to Lagrange the invention of the Method of Variation of Elements in the theory of Perturbations. But justice to Euler requires that we should assign this distinction to him; at least, next to Newton, whose mode of representing the paths of bodies by means of a *Revolving Orbit*, in the Ninth Section of the *Principia*, may be considered as an anticipation of the method of variation of elements. In the fifth volume of the *Mécanique Céleste*, livre xv. p. 305, is an abstract of Euler's paper of 1749; where Laplace adds, "C'est le premier essai de la méthode de la variation des constantes arbitraires." And in page 310 is an abstract of the paper of 1756: and speaking of the method, Laplace

mines the secular changes of each of the *elements* or determining quantities of the orbit. In 1773, Laplace also attacked this subject of secular changes, and obtained expressions for them. On this occasion, he proved the celebrated proposition that, "the mean motions of the planets are invariable:" that is, that there is, in the revolutions of the system, no progressive change which is not finally stopped and reversed; no increase, which is not, after some period, changed into decrease; no retardation which is not at last succeeded by acceleration; although, in some cases, millions of years may elapse before the system reaches the turning-point. Thomas Simpson noticed the same consequence of the laws of universal attraction. In 1774 and 1776, Lagrange[6] still labored at the secular equations; extending his researches to the nodes and inclinations; and showed that the invariability of the mean motions of the planets, which Laplace had proved, neglecting the fourth powers of the eccentricities and inclinations of the orbits,[7] was true, however far the approximation was carried, so long as the squares of the disturbing masses were neglected. He afterwards improved his methods;[8] and, in 1783, he endeavored to extend the calculation of the changes of the elements to the periodical equations, as well as the secular.

8. *Mécanique Céleste, &c.*—Laplace also resumed the consideration of the secular changes; and, finally, undertook his vast work, the *Mécanique Céleste*, which he intended to contain a complete view of the existing state of this splendid department of science. We may see, in the exultation which the author obviously feels at the thought of erecting this monument of his age, the enthusiasm which had been excited by the splendid course of mathematical successes of which I have given a sketch. The two first volumes of this great work appeared in 1799. The third and fourth volumes were published in 1802 and 1805 respectively. Since its publication, little has been added to the solution of the great problems of which it treats. In 1808, Laplace presented to the French Bureau des Longitudes, a Supplement to the *Mécanique Céleste;* the object of which was to improve still further

---

says, "It consists in regarding the elements of the elliptical motion as variable in virtue of the perturbing forces. Those elements are, 1, the axis major; 2, the epoch of the body being at the apse; 3, the eccentricity; 4, the movement of the apse; 5, the inclination; 6, the longitude of the node;" and he then proceeds to show how Euler did this. It is possible that Lagrange knew nothing of Euler's paper. See *Méc. Cél.* vol. v. p. 312. But Euler's conception and treatment of the method are complete, so that he must be looked upon as the author of it.

[6] Gautier, p. 104. [7] Ib. p. 184. [8] Ib p. 196.

the mode of obtaining the secular variations of the elements. Poisson and Lagrange proved the invariability of the major axes of the orbits, as far as the second order of the perturbing forces. Various other authors have since labored at this subject. Burckhardt, in 1808, extended the perturbing function as far as the sixth order of the eccentricities. Gauss, Hansen, and Bessel, Ivory, MM. Lubbock, Plana, Pontécoulant, and Airy, have, at different periods up to the present time, either extended or illustrated some particular part of the theory, or applied it to special cases; as in the instance of Professor Airy's calculation of an inequality of Venus and the earth, of which the period is 240 years. The approximation of the Moon's motions has been pushed to an almost incredible extent by M. Damoiseau, and, finally, Plana has once more attempted to present, in a single work (three thick quarto volumes), all that has hitherto been executed with regard to the theory of the Moon.

I give only the leading points of the progress of analytical dynamics. Hence I have not spoken in detail of the theory of the Satellites of Jupiter, a subject on which Lagrange gained a prize for a Memoir, in 1766, and in which Laplace discovered some most curious properties in 1784. Still less have I referred to the purely speculative question of *Tautochronous Curves* in a resisting medium, though it was a subject of the labors of Bernoulli, Euler, Fontaine, D'Alembert, Lagrange, and Laplace. The reader will rightly suppose that many other curious investigations are passed over in utter silence.

[2d Ed.] [Although the analytical calculations of the great mathematicians of the last century had determined, in a demonstrative manner, a vast series of inequalities to which the motions of the sun, moon, and planets were subject in virtue of their mutual attraction, there were still unsatisfactory points in the solutions thus given of the great mechanical problems suggested by the System of the Universe. One of these points was the want of any evident mechanical significance in the successive members of these series. Lindenau relates that Lagrange, near the end of his life, expressed his sorrow that the methods of approximation employed in Physical Astronomy rested on arbitrary processes, and not on any insight into the results of mechanical action. But something was subsequently done to remove the ground of this complaint. In 1818, Gauss pointed out that secular equations may be conceived to result from the disturbing body being distributed along its orbit so as to form a ring, and thus made the result conceivable more distinctly than as a mere result of calculation. And it appears to me that Professor Airy's treatise entitled *Gravitation*, published at Cambridge in 1834, is of great value in supplying similar modes of conception with regard to the mechanical origin of many of the principal inequalities of the solar system.

Bessel in 1824, and Hansen in 1828, published works which are considered as belonging, along with those of Gauss, to a new era in physical astronomy.[9] Gauss's *Theoria Motuum Corporum Celestium*, which had Lalande's medal assigned to it by the French Institute, had already (1810) resolved all problems concerning the determination of the place of a planet or comet in its orbit in function of the elements. The value of Hansen's labors respecting the Perturbations of the Planets was recognized by the Astronomical Society of London, which awarded to them its gold medal.

The investigations of M. Damoiseau, and of MM. Plana and Carlini, on the Problem of the Lunar Theory, followed nearly the same course as those of their predecessors. In these, as in the *Mécanique Céleste*, and in preceding works on the same subject, the Moon's co-ordinates (time, radius vector, and latitude) were expressed in function of her true longitude. The integrations were effected in series, and then by reversion of the series, the longitude was expressed in function of the time; and then in the same manner the other two co-ordinates. But Sir John Lubbock and M. Pontécoulant have made the *mean* longitude of the moon, that is, the time, the independent variable, and have expressed the moon's co-ordinates in terms of sines and cosines of angles increasing proportionally to the time. And this method has been adopted by M. Poisson (*Mem. Inst.* xiii. 1835, p. 212). M. Damoiseau, like Laplace and Clairaut, had deduced the successive coefficients of the lunar inequalities by numerical equations. But M. Plana expresses explicitly each coefficient in general terms of the letters expressing the constants of the problem, arranging them according to the order of the quantities, and substituting numbers at the end of the operation only. By attending to this arrangement, MM. Lubbock and Pontécoulant have verified or corrected a large portion of the terms contained in the investigations of MM. Damoiseau and Plana. Sir John Lubbock has calculated the polar co-ordinates of the Moon directly; M. Poisson, on the other hand, has obtained the variable elliptical elements; M. Pontécoulant conceives that the method of variation or arbitrary con-

---

9 *Abhand. der Akad. d. Wissensch. zu Berlin.* 1824; and *Disquisitiones circa Theoriam Perturbationum.* See Jahn. *Gesch. der Astron.* p. 84.

stants may most conveniently be reserved for secular inequalities and inequalities of long periods.

MM. Lubbock and Pontécoulant have made the mode of treating the Lunar Theory and the Planetary Theory agree with each other, instead of following two different paths in the calculation of the two problems, which had previously been done.

Prof. Hansen, also, in his *Fundamenta Nova Investigationis Orbitæ veræ quam Luna perlustrat* (*Gothæ*, 1838), gives a general method, including the Lunar Theory and the Planetary Theory as two special cases. To this is annexed a solution of the *Problem of Four Bodies.*

I am here speaking of the Lunar and Planetary Theories as Mechanical Problems only. Connected with this subject, I will not omit to notice a very general and beautiful method of solving problems respecting the motion of systems mutually attracting bodies, given by Sir W. R. Hamilton, in the *Philosophical Transactions* for 1834–5 ("On a General Method in Dynamics"). His method consists in investigating the *Principal Function* of the co-ordinates of the bodies: this function being one, by the differentiation of which, the co-ordinates of the bodies of the system may be found. Moreover, an approximate value of this function being obtained, the same formulæ supply a means of successive approximation without limit.]

9. *Precession. Motion of Rigid Bodies.*—The series of investigations of which I have spoken, extensive and complex as it is, treats the moving bodies as points only, and takes no account of any peculiarity of their form or motion of their parts. The investigation of the motion of a body of any magnitude and form, is another branch of analytical mechanics, which well deserves notice. Like the former branch, it mainly owed its cultivation to the problems suggested by the solar system. Newton, as we have seen, endeavored to calculate the effect of the attraction of the sun and moon in producing the *precession of the equinoxes;* but in doing this he made some mistakes. In 1747, D'Alembert solved this problem by the aid of his "Principle;" and it was not difficult for him to show, as he did in his *Opuscules,* in 1761, that the same method enabled him to determine the motion of a body of any figure acted upon by any forces. But, as the reader will have observed in the course of this narrative, the great mathematicians of this period were always nearly abreast of each other in their advances. —Euler,[10] in the mean time, had published, in 1751, a solution of the

[10] *Ac. Berl.* 1745, 1750.

problem of the precession; and in 1752, a memoir which he entitled, *Discovery of a New Principle of Mechanics*, and which contains a solution of the general problem of the alteration of rotary motion by forces. D'Alembert noticed with disapprobation the assumption of priority which this title implied, though allowing the merit of the memoir. Various improvements were made in these solutions; but the final form was given them by Euler; and they were applied to a great variety of problems in his *Theory of the Motion of Solid and Rigid Bodies*, which was written[11] about 1760, and published in 1765. The formulæ in this work were much simplified by the use of a discovery of Segner, that every body has three axes which were called Principal Axes, about which alone (in general) it would permanently revolve. The equations which Euler and other writers had obtained, were attacked as erroneous by Landen in the Philosophical Transactions for 1785; but I think it is impossible to consider this criticism otherwise than as an example of the inability of the English mathematicians of that period to take a steady hold of the analytical generalizations to which the great Continental authors had been led. Perhaps one of the most remarkable calculations of the motion of a rigid body is that which Lagrange performed with regard to the *Moon's Libration;* and by which he showed that the Nodes of the Moon's Equator and those of her Orbit must always coincide.

10. *Vibrating Strings.*—Other mechanical questions, unconnected with astronomy, were also pursued with great zeal and success. Among these was the problem of a vibrating string, stretched between two fixed points. There is not much complexity in the mechanical conceptions which belong to this case, but considerable difficulty in reducing them to analysis. Taylor, in his *Method of Increments*, published in 1716, had annexed to his work a solution of this problem; obtained on suppositions, limited indeed, but apparently conformable to the most common circumstances of practice. John Bernoulli, in 1728, had also treated the same problem. But it assumed an interest altogether new, when, in 1747, D'Alembert published his views on the subject; in which he maintained that, instead of one kind of curve only, there were an infinite number of different curves, which answered the conditions of the question. The problem, thus put forward by one great mathematician, was, as usual, taken up by the others, whose names the reader is now so familiar with in such an association. In

---

[11] See the preface to the book.

1748, Euler not only assented to the generalization of D'Alembert, but held that it was not necessary that the curves so introduced should be defined by any algebraical condition whatever. From this extreme indeterminateness D'Alembert dissented; while Daniel Bernoulli, trusting more to physical and less to analytical reasonings, maintained that both these generalizations were inapplicable in fact, and that the solution was really restricted, as had at first been supposed, to the form of the trochoid, and to other forms derivable from that. He introduced, in such problems, the "Law of Coexistent Vibrations," which is of eminent use in enabling us to conceive the results of complex mechanical conditions, and the real import of many analytical expressions. In the mean time, the wonderful analytical genius of Lagrange had applied itself to this problem. He had formed the Academy of Turin, in conjunction with his friends Saluces and Cigna; and the first memoir in their Transactions was one by him on this subject: in this and in subsequent writings he has established, to the satisfaction of the mathematical world, that the functions introduced in such cases are not necessarily continuous, but are arbitrary to the same degree that the motion is so practically; though capable of expression by a series of circular functions. This controversy, concerning the degree of lawlessness with which the conditions of the solution may be assumed, is of consequence, not only with respect to vibrating strings, but also with respect to many problems, belonging to a branch of Mechanics which we now have to mention, the Doctrine of Fluids.

11. *Equilibrium of Fluids. Figure of the Earth. Tides.*—The application of the general doctrines of Mechanics to fluids was a natural and inevitable step, when the principles of the science had been generalized. It was easily seen that a fluid is, for this purpose, nothing more than a body of which the parts are movable amongst each other with entire facility; and that the mathematician must trace the consequences of this condition upon his equations. This accordingly was done, by the founders of mechanics, both for the cases of the equilibrium and of motion. Newton's attempt to solve the problem of the *figure of the earth*, supposing it fluid, is the first example of such an investigation: and this solution rested upon principles which we have already explained, applied with the skill and sagacity which distinguished all that Newton did.

We have already seen how the generality of the principle, that fluids press equally in all directions, was established. In applying it to calculation, Newton took for his fundamental principle, the equal weight of columns of the fluid reaching to the centre; Huyghens took, as his basis, the prependicularity of the resulting force at each point to the surface of the fluid; Bouguer conceived that both principles were necessary; and Clairaut showed that the equilibrium of *all* canals is requisite. He also was the first mathematician who deduced from this principle the Equations of Partial Differentials by which these laws are expressed; a step which, as Lagrange says,[12] changed the face of Hydrostatics, and made it a new science. Euler simplified the mode of obtaining the Equations of Equilibrium for any forces whatever; and put them in the form which is now generally adopted in our treatises.

The explanation of the *Tides*, in the way in which Newton attempted it in the third book of the *Principia*, is another example of a hydrostatical investigation: for he considered only the form that the ocean would have if it were at rest. The memoirs of Maclaurin, Daniel Bernoulli, and Euler, on the question of the Tides, which shared among them the prize of the Academy of Sciences in 1740, went upon the same views.

The *Treatise of the Figure of the Earth*, by Clairaut, in 1743, extended Newton's solution of the same problem, by supposing a solid nucleus covered with a fluid of different density. No peculiar novelty has been introduced into this subject, except a method employed by Laplace for determining the attractions of spheroids of small eccentricity, which is, as Professor Airy has said,[13] "a calculus the most singular in its nature, and the most powerful in its effects, of any which has yet appeared."

12. *Capillary Action.*—There is only one other problem of the statics of fluids on which it is necessary to say a word,—the doctrine of Capillary Attraction. Daniel Bernoulli,[14] in 1738, states that he passes over the subject, because he could not reduce the facts to general laws: but Clairaut was more successful, and Laplace and Poisson have since given great analytical completeness to his theory. At present our business is, not so much with the sufficiency of the theory to explain phenomena, as with the mechanical problem of which this is an example, which is one of a very remarkable and important character; namely, to determine the effect of attractions which are exercised by all the particles of bodies, on the hypothesis that the attrac-

---

[12] *Méc. Analyt.* ii. p. 180.

[13] *Enc. Met.* Fig. of Earth, p. 192.

[14] *Hydrodyn.* Pref. p. 5.

tion of each particle, though sensible when it acts upon another particle at an extremely small distance from it, becomes insensible and vanishes the moment this distance assumes a perceptible magnitude. It may easily be imagined that the analysis by which results are obtained under conditions so general and so peculiar, is curious and abstract; the problem has been resolved in some very extensive cases.

13. *Motion of Fluids.*—The only branch of mathematical mechanics which remains to be considered, is that which is, we may venture to say, hitherto incomparably the most incomplete of all,—Hydrodynamics. It may easily be imagined that the mere hypothesis of absolute relative mobility in the parts, combined with the laws of motion and nothing more, are conditions too vague and general to lead to definite conclusions. Yet such are the conditions of the problems which relate to the motion of fluids. Accordingly, the mode of solving them has been, to introduce certain other hypotheses, often acknowledged to be false, and almost always in some measure arbitrary, which may assist in determining and obtaining the solution. The Velocity of a fluid issuing from an orifice in a vessel, and the Resistance which a solid body suffers in moving in a fluid, have been the two main problems on which mathematicians have employed themselves. We have already spoken of the manner in which Newton attacked both these, and endeavored to connect them. The subject became a branch of Analytical Mechanics by the labors of D. Bernoulli, whose *Hydrodynamica* was published in 1738. This work rests upon the Huyghenian principle of which we have already spoken in the history of the centre of oscillation; namely, the equality of the *actual descent* of the particles and the *potential ascent;* or, in other words, the conservation of *vis viva.* This was the first analytical treatise; and the analysis is declared by Lagrange to be as elegant in its steps as it is simple in its results. Maclaurin also treated the subject; but is accused of reasoning in such a way as to show that he had determined upon his result beforehand; and the method of John Bernoulli, who likewise wrote upon it, has been strongly objected to by D'Alembert. D'Alembert himself applied the principle which bears his name to this subject; publishing a *Treatise on the Equilibrium and Motion of Fluids* in 1744, and on the *Resistance of Fluids* in 1753. His *Réflexions sur la Cause Générale des Vents,* printed in 1747, are also a celebrated work, belonging to this part of mathematics. Euler, in this as in other cases, was one of those who most contributed to give analytical elegance to the subject. In addition to the questions which

have been mentioned, he and Lagrange treated the problems of the small vibrations of fluids, both inelastic and elastic;—a subject which leads, like the question of vibrating strings, to some subtle and abstruse considerations concerning the significations of the integrals of partial differential equations. Laplace also took up the subject of waves propagated along the surface of water; and deduced a very celebrated theory of the tides, in which he considered the ocean to be, not in equilibrium, as preceding writers had supposed, but agitated by a constant series of undulations, produced by the solar and lunar forces. The difficulty of such an investigation may be judged of from this, that Laplace, in order to carry it on, is obliged to assume a mechanical proposition, unproved, and only conjectured to be true; namely,[15] that, "in a system of bodies acted upon by forces which are periodical, the state of the system is periodical like the forces." Even with this assumption, various other arbitrary processes are requisite; and it appears still very doubtful whether Laplace's theory is either a better mechanical solution of the problem, or a nearer approximation to the laws of the phenomena, than that obtained by D. Bernoulli, following the views of Newton.

In most cases, the solutions of problems of hydrodynamics are not satisfactorily confirmed by the results of observation. Poisson and Cauchy have prosecuted the subject of waves, and have deduced very curious conclusions by a very recondite and profound analysis. The assumptions of the mathematician here do not represent the conditions of nature; the rules of theory, therefore, are not a good standard to which we may refer the aberrations of particular cases; and the laws which we obtain from experiment are very imperfectly illustrated by *à priori* calculation. The case of this department of knowledge, Hydrodynamics, is very peculiar; we have reached the highest point of the science,—the laws of extreme simplicity and generality from which the phenomena flow; we cannot doubt that the ultimate principles which we have obtained are the true ones, and those which really apply to the facts; and yet we are far from being able to apply the principles to explain or find out the facts. In order to do this, we want, in addition to what we have, true and useful principles, intermediate between the highest and the lowest;—between the extreme and almost barren generality of the laws of motion, and the endless varieties and inextricable complexity of fluid motions in special cases.

---

15 *Méc. Cél.* t. ii. p. 218.

The reason of this peculiarity in the science of Hydrodynamics appears to be, that its general principles were not discovered with reference to *the science itself, but by extension* from the sister science of the Mechanics of Solids; they were not obtained by ascending gradually from particulars to truths more and more general, respecting the motions of fluids; but were caught at once, by a perception that the parts of fluids are included in that range of generality which we are entitled to give to the supreme laws of motions of solids. Thus, Solid Dynamics and Fluid Dynamics resemble two edifices which have their highest apartment in common, and though we can explore every part of the former building, we have not yet succeeded in traversing the staircase of the latter, either from the top or from the bottom. If we had lived in a world in which there were no solid bodies, we should probably not have yet discovered the laws of motion; if we had lived in a world in which there were no fluids, we should have no idea how insufficient a complete possession of the general laws of motion may be, to give us a true knowledge of particular results.

14. *Various General Mechanical Principles.*—The generalized laws of motion, the points to which I have endeavored to conduct my history, include in them all other laws by which the motions of bodies can be regulated; and among such, several laws which had been discovered before the highest point of generalization was reached, and which thus served as stepping-stones to the ultimate principles. Such were, as we have seen, the Principles of the Conservation of *vis viva*, the Principle of the Conservation of the Motion of the Centre of Gravity, and the like. These principles may, of course, be deduced from our elementary laws, and were finally established by mathematicians on that footing. There are other principles which may be similarly demonstrated; among the rest, I may mention the Principle of *the Conservation of areas*, which extends to any number of bodies a law analogous to that which Kepler had observed, and Newton demonstrated, respecting the areas described by each planet round the sun. I may mention also, the Principle of the *Immobility of the plane of maximum areas*, a plane which is not disturbed by any mutual action of the parts of any system. The former of these principles was published about the same time by Euler, D. Bernoulli, and Darcy, under different forms, in 1746 and 1747; the latter by Laplace.

To these may be added a law, very celebrated in its time, and the occasion of an angry controversy, *the Principle of least action.* Maupertuis conceived that he could establish *à priori*, by theological arguments, that all mechanical changes must take place in the world so as to occasion the least possible quantity of *action*. In asserting this, it was proposed to measure the Action by the product of Velocity and Space; and this measure being adopted, the mathematicians, though they did not generally assent to Maupertuis' reasonings, found that his principle expressed a remarkable and useful truth, which might be established on known mechanical grounds.

15. *Analytical Generality. Connection of Statics and Dynamics.*—Before I quit this subject, it is important to remark the peculiar character which the science of Mechanics has now assumed, in consequence of the extreme analytical generality which has been given it. Symbols, and operations upon symbols, include the whole of the reasoner's task; and though the relations of space are the leading subjects in the science, the great analytical treatises upon it do not contain a single diagram. The *Mécanique Analytique* of Lagrange, of which the first edition appeared in 1788, is by far the most consummate example of this analytical generality. "The plan of this work," says the author, "is entirely new. I have proposed to myself to reduce the whole theory of this science, and the art of resolving the problems which it includes, to general formulæ, of which the simple development gives all the equations necessary for the solution of the problem."—"The reader will find no figures in the work. The methods which I deliver do not require either constructions, or geometrical or mechanical reasonings; but only algebraical operations, subject to a regular and uniform rule of proceeding." Thus this writer makes Mechanics a branch of Analysis; instead of making, as had previously been done, Analysis an implement of Mechanics.[16] The transcendent generalizing genius of Lagrange, and his matchless analytical skill and elegance, have made this undertaking as successful as it is striking.

The mathematical reader is aware that the language of mathematical symbols is, in its nature, more general than the language of words: and that in this way truths, translated into symbols, often suggest their own generalizations. Something of this kind has happened in Mechanics. The same Formula expresses the general condition of Statics and that of Dynamics. The tendency to generalization which is thus introduced by analysis, makes mathematicians unwilling to acknowl-

---

[16] Lagrange himself terms Mechanics, "An Analytical Geometry of four dimensions." Besides the *three co-ordinates* which determine the place of a body in *space*, the *time* enters as a *fourth co-ordinate*. [Note by Littrow.]

edge a plurality of Mechanical principles; and in the most recent analytical treatises on the subject, all the doctrines are deduced from the single Law of Inertia. Indeed, if we identify Forces with the Velocities which produce them, and allow the Composition of Forces to be applicable to force *so understood*, it is easy to see that we can reduce the Laws of Motion to the Principles of Statics; and this conjunction, though it may not be considered as philosophically just, is verbally correct. If we thus multiply or extend the meanings of the term Force, we make our elementary principles simpler and fewer than before; and those persons, therefore, who are willing to assent to such a use of words, can thus obtain an additional generalization of dynamical principles; and this, as I have stated, has been adopted in several recent treatises. I shall not further discuss here how far this is a real advance in science.

Having thus rapidly gone through the history of Force and Attraction in the abstract, we return to the attempt to interpret the phenomena of the universe by the aid of these abstractions thus established.

But before we do so, we may make one remark on the history of this part of science. In consequence of the vast career into which the Doctrine of Motion has been drawn by the splendid problems proposed to it by Astronomy, the origin and starting-point of Mechanics, namely Machines, had almost been lost out of sight. *Machines* had become the smallest part of *Mechanics*, as *Land-measuring* had become the smallest part of *Geometry*. Yet the application of Mathematics to the doctrine of Machines has led, at all periods of the Science, and especially in our own time, to curious and valuable results. Some of these will be noticed in the *Additions* to this volume.

BOOK VII.

THE MECHANICAL SCIENCES.

(CONTINUED.)

HISTORY

OF

PHYSICAL ASTRONOMY.

DESCEND from heaven, Urania, by that name
If rightly thou art called, whose voice divine
Following, above the Olympian hill I soar,
Above the flight of Pegasean wing.
The meaning, not the name, I call, for thou
Nor of the muses nine, nor on the top
Of old Olympus dwell'st: but heavenly-born,
Before the hills appeared, or fountain flowed,
Thou with Eternal Wisdom didst converse,
Wisdom, thy sister.

Paradise Lost, B. vii.

CHAPTER I.

Prelude to the Inductive Epoch of Newton.

WE have now to contemplate the last and most splendid period of the progress of Astronomy;—the grand completion of the history of the most ancient and prosperous province of human knowledge; —the steps which elevated this science to an unrivalled eminence above other sciences;—the first great example of a wide and complex assemblage of phenomena indubitably traced to their single simple cause;— in short, the first example of the formation of a perfect Inductive Science.

In this, as in other considerable advances in real science, the complete disclosure of the new truths by the principal discoverer, was preceded by movements and glimpses, by trials, seekings, and guesses on the part of others; by indications, in short, that men's minds were already carried by their intellectual impulses in the direction in which the truth lay, and were beginning to detect its nature. In a case so important and interesting as this, it is more peculiarly proper to give some view of this Prelude to the Epoch of the full discovery.

(*Francis Bacon.*) That Astronomy should become Physical Astronomy,—that the motions of the heavenly bodies should be traced to their causes, as well as reduced to rule,—was felt by all persons of active and philosophical minds as a pressing and irresistible need, at the time of which we speak. We have already seen how much this feeling had to do in impelling Kepler to the train of laborious research by which he made his discoveries. Perhaps it may be interesting to point out how strongly this persuasion of the necessity of giving a physical character to astronomy, had taken possession of the mind of Bacon, who, looking at the progress of knowledge with a more comprehensive spirit, and from a higher point of view than Kepler, could have none of his astronomical prejudices, since on that subject he was of a different school, and of far inferior knowledge. In his "Description of the Intellectual Globe," Bacon says that while Astronomy had, up to that time, had it for her business to inquire into the rules of the heavenly motions, and Philosophy into their causes, they had both so far worked without due appreciation of their respective tasks; Philosophy neglecting facts, and Astronomy claiming assent to her mathematical hypotheses, which ought to be considered as mere steps of calculation. "Since, therefore," he continues,[1] "each science has hitherto been a slight and ill-constructed thing, we must assuredly take a firmer stand; our ground being, that these two subjects, which on account of the narrowness of men's views and the traditions of professors have been so long dissevered, are, in fact, one and the same thing, and compose one body of science." It must be allowed that, however erroneous might be the points of Bacon's positive astronomical creed, these general views of the nature and position of the science are most sound and philosophical.

(*Kepler.*) In his attempts to suggest a right physical view of the starry heavens and their relation to the earth, Bacon failed, along with all the writers of his time. It has already been stated that the main cause of this failure was the want of a knowledge of the true theory of motion;—the non-existence of the science of Dynamics. At the time of Bacon and Kepler, it was only just beginning to be possible to reduce the heavenly motions to the laws of earthly motion, because the latter were only just then divulged. Accordingly, we have seen that the whole of Kepler's physical speculations proceed upon an ignorance of the first law of motion, and assume it to be the main problem of the physical astronomer to assign the cause which *keeps up* the motions of the planets. Kepler's doctrine is, that a certain Force or Virtue resides in the sun, by which all bodies within his influence are carried round him. He illustrates[2] the nature of this Virtue in various ways, comparing it to Light, and to the Magnetic Power, which it resembles in the circumstances of operating at a distance, and also in exercising a feebler influence as the distance becomes greater. But it was obvious that these comparisons were very imperfect; for they do not explain how the sun produces in a body at a distance a motion *athwart* the line of emanation; and though Kepler introduced an assumed rotation of the sun on his axis as the cause of this effect, that such a cause could produce the result could not be established by any analogy of terrestrial motions. But another image to which he referred, suggested a much more substantial and conceivable kind of mechanical action by which the celestial motions might be produced, namely, a current of fluid matter circulating round the sun, and carrying the planet with it, like a boat in a stream. In the Table of Contents of the work on the planet Mars, the purport of the chapter to which I have alluded is

---

[1] Vol. ix. 221.

[2] *De Stellâ Martis*, P. 3. c. xxxiv.

stated as follows: "A physical speculation, in which it is demonstrated that the vehicle of that Virtue which urges the planets, circulates through the spaces of the universe after the manner of a river or whirlpool (*vortex*), moving quicker than the planets." I think it will be found, by any one who reads Kepler's phrases concerning the *moving force,—the magnetic nature,—the immaterial virtue* of the sun, that they convey no distinct conception, except so far as they are interpreted by the expressions just quoted. A vortex of fluid constantly whirling round the sun, kept in this whirling motion by the rotation of the sun himself, and carrying the planets round the sun by its revolution, as a whirlpool carries straws, could be readily understood; and though it appears to have been held by Kepler that this current and vortex was immaterial, he ascribes to it the power of overcoming the inertia of bodies, and of putting them and keeping them in motion, the only material properties with which he had any thing to do. Kepler's physical reasonings, therefore, amount, in fact, to the doctrine of Vortices round the central bodies, and are occasionally so stated by himself; though by asserting these vortices to be "an immaterial species," and by the fickleness and variety of his phraseology on the subject, he leaves this theory in some confusion;—a proceeding, indeed, which both his want of sound mechanical conceptions, and his busy and inventive fancy, might have led us to expect. Nor, we may venture to say, was it easy for any one at Kepler's time to devise a more plausible theory than the theory of vortices might have been made. It was only with the formation and progress of the science of Mechanics that this theory became untenable.

(*Descartes.*) But if Kepler might be excused, or indeed admired, for propounding the theory of Vortices at his time, the case was different when the laws of motion had been fully developed, and when those who knew the state of mechanical science ought to have learned to consider the motions of the stars as a mechanical problem, subject to the same conditions as other mechanical problems, and capable of the same exactness of solution. And there was an especial inconsistency in the circumstance of the Theory of Vortices being put forwards by Descartes, who pretended, or was asserted by his admirers, to have been one of the discoverers of the true Laws of Motion. It certainly shows both great conceit and great shallowness, that he should have proclaimed with much pomp this crude invention of the ante-mechanical period, at the time when the best mathematicians of Europe, as Borelli in Italy, Hooke and Wallis in England, Huyghens in Holland, were patiently laboring to bring the mechanical problem of the universe into its most distinct form, in order that it might be solved at last and forever.

I do not mean to assert that Descartes borrowed his doctrines from Kepler, or from any of his predecessors, for the theory was sufficiently obvious; and especially if we suppose the inventor to seek his suggestions rather in the casual examples offered to the sense than in the exact laws of motion. Nor would it be reasonable to rob this philosopher of that credit, of the plausible deduction of a vast system from apparently simple principles, which, at the time, was so much admired; and which undoubtedly was the great cause of the many converts to his views. At the same time we may venture to say that a system of doctrine thus deduced from assumed principles by a long chain of reasoning, and not verified and confirmed at every step by detailed and exact facts, has hardly a chance of containing any truth. Descartes said that he should think it little to show how the world *is* constructed, if he could not also show that it *must* of necessity have been so constructed. The more modest philosophy which has survived the boastings of his school is content to receive all its knowledge of facts from experience, and never dreams of interposing its peremptory *must be* when nature is ready to tell us what *is*. The *à priori* philosopher has, however, always a strong feeling in his favor among men. The deductive form of his speculations gives them something of the charm and the apparent certainty of pure mathematics; and while he avoids that laborious recurrence to experiments, and measures, and multiplied observations, which is irksome and distasteful to those who are impatient to grow wise at once, every fact of which the theory appears to give an explanation, seems to be an unasked and almost an infallible witness in its favor.

My business with Descartes here is only with his physical Theory of Vortices; which, great as was its glory at one time, is now utterly extinguished. It was propounded in his *Principia Philosophiæ*, in 1644. In order to arrive at this theory, he begins, as might be expected of him, from reasonings sufficiently general. He lays it down as a maxim, in the first sentence of his book, that a person who seeks for truth must, once in his life, doubt of all that he most believes. Conceiving himself thus to have stripped himself of all his belief on all subjects, in order to resume that part of it which merits to be retained, he begins with his celebrated assertion, "I think, therefore I am;" which appears to him a certain and immovable principle, by means of which he may proceed to something more. Accordingly, to this he soon adds the idea, and hence the certain existence, of God and his perfections. He then asserts it to be also manifest, that a vacuum in any part of the universe is impossible; the whole must be filled with matter, and the matter must be divided into equal angular parts, this being the most simple, and therefore the most natural supposition.[3] This matter being in motion, the parts are necessarily ground into a spherical form; and the corners thus rubbed off (like filings or saw-dust) form a second and more subtle matter.[4] There is, besides, a third kind of matter, of parts more coarse and less fitted for motion. The first matter makes luminous bodies, as the sun, and the fixed stars; the second is the transparent substance of the skies; the third is the material of opake bodies, as the earth, planets, and comets. We may suppose, also,[5] that the motions of these parts take the form of revolving circular currents,[6] or *vortices*. By this means, the first matter will be collected to the centre of each vortex, while the second, or subtle matter, surrounds it, and, by its centrifugal effort, constitutes light. The planets are carried round the sun by the motion of his vortex,[7] each planet being at such a distance from the sun as to be in a part of the vortex suitable to its solidity and mobility. The motions are prevented from being exactly circular and regular by various causes; for instance, a vortex may be pressed into an oval shape by contiguous vortices. The satellites are, in like manner, carried round their primary planets by subordinate vortices; while the comets have sometimes the liberty of gliding out of one vortex into the one next contiguous, and thus travelling in a sinuous course, from system to system, through the universe.

It is not necessary for us to speak here of the entire deficiency of this system in mechanical consistency, and in a correspondency to observation in details and measures. Its general reception and temporary sway, in some instances even among intelligent men and good mathematicians, are the most remarkable facts connected with it. These may be ascribed, in part, to the circumstance that philosophers were now ready and eager for a physical astronomy commensurate with the existing state of knowledge; they may have been owing also, in some measure, to the character and position of Descartes. He was a man of high claims in every department of speculation, and, in pure mathematics, a genuine inventor of great eminence;—a man of family and a soldier;—an inoffensive philosopher, attacked and persecuted

[3] *Prin.* p. 58. [4] Ib. p. 59. [5] Ib. p. 56.
[6] Ib. p. 61. [7] Ib. c. 140, p. 114.

for his opinions with great bigotry and fury by a Dutch divine, Voet; —the favorite and teacher of two distinguished princesses, and, it is said, the lover of one of them. This was Elizabeth, the daughter of the Elector Frederick, and consequently grand-daughter of our James the First. His other royal disciple, the celebrated Christiana of Sweden, showed her zeal for his instructions by appointing the hour of five in the morning for their interviews. This, in the climate of Sweden, and in the winter, was too severe a trial for the constitution of the philosopher, born in the sunny valley of the Loire; and, after a short residence at Stockholm, he died of an inflammation of the chest in 1650. He always kept up an active correspondence with his friend Mersenne, who was called, by some of the Parisians, "the Resident of Descartes at Paris;" and who informed him of all that was done in the world of science. It is said that he at first sent to Mersenne an account of a system of the universe which he had devised, which went on the assumption of a vacuum; Mersenne informed him that the *vacuum* was no longer the fashion at Paris; upon which he proceeded to remodel his system, and to re-establish it on the principle of a *plenum*. Undoubtedly he tried to avoid promulgating opinions which might bring him into trouble. He, on all occasions, endeavored to explain away the doctrine of the motion of the earth, so as to evade the scruples to which the decrees of the pope had given rise; and, in stating the theory of vortices, he says,[8] "There is no doubt that the world was created at first with all its perfection; nevertheless, it is well to consider how it might have arisen from certain principles, although we know that it did not." Indeed, in the whole of his philosophy, he appears to deserve the character of being both rash and cowardly, "*pusillanimus simul et audax*," far more than Aristotle, to whose physical speculations Bacon applies this description.[9]

Whatever the causes might be, his system was well received and rapidly adopted. Gassendi, indeed, says that he found nobody who had the courage to read the *Principia* through;[10] but the system was soon embraced by the younger professors, who were eager to dispute in its favor. It is said[11] that the University of Paris was on the point of publishing an edict against these new doctrines, and was only prevented from doing so by a pasquinade which is worth mentioning. It was composed by the poet Boileau (about 1684), and professed to be a Request in favor of Aristotle, and an Edict issued from Mount Parnas-

[8] *Prin.* p. 56.

[9] Bacon, *Descriptio Globi Intellectualis.*

[10] Del. *A. M.* ii. 193.

[11] *Enc. Brit.* art. *Cartesianism.*

sus in consequence. It is obvious that, at this time, the cause of Cartesianism was looked upon as the cause of free inquiry and modern discovery, in opposition to that of bigotry, prejudice, and ignorance. Probably the poet was far from being a very severe or profound critic of the truth of such claims. "This petition of the Masters of Arts, Professors and Regents of the University of Paris, humbly showeth, that it is of public notoriety that the sublime and incomparable Aristotle was, without contest, the first founder of the four elements, fire, air, earth, and water; that he did, by special grace, accord unto them a simplicity which belongeth not to them of natural right;" and so on. "Nevertheless, since, a certain time past, two individuals, named Reason and Experience, have leagued themselves together to dispute his claim to the rank which of justice pertains to him, and have tried to erect themselves a throne on the ruins of his authority; and, in order the better to gain their ends, have excited certain factious spirits, who, under the names of Cartesians and Gassendists, have begun to shake off the yoke of their master, Aristotle; and, contemning his authority, with unexampled temerity, would dispute the right which he had acquired of making true pass for false and false for true;"—In fact, this production does not exhibit any of the peculiar tenets of Descartes, although, probably, the positive points of his doctrines obtained a footing in the University of Paris, under the cover of this assault on his adversaries. The Physics of Rohault, a zealous disciple of Descartes, was published at Paris about 1670,[12] and was, for a time, the standard book for students of this subject, both in France and in England. I do not here speak of the later defenders of the Cartesian system, for, in their hands, it was much modified by the struggle which it had to maintain against the Newtonian system.

We are concerned with Descartes and his school only as they form part of the picture of the intellectual condition of Europe just before the publication of Newton's discoveries. Beyond this, the Cartesian speculations are without value. When, indeed, Descartes' countrymen could no longer refuse their assent and admiration to the Newtonian theory, it came to be the fashion among them to say that Descartes had been the necessary precursor of Newton; and to adopt a favorite saying of Leibnitz, that the Cartesian philosophy was the antechamber of Truth. Yet this comparison is far from being happy: it appeared rather as if these suitors had mistaken the door; for those

---

[12] And a second edition in 1672.

who first came into the presence of Truth herself, were those who never entered this imagined antechamber, and those who were in the antechamber first, were the last in penetrating further. In partly the same spirit, Playfair has noted it as a service which Newton perhaps owed to Descartes, that "he had exhausted one of the most tempting forms of error." We shall see soon that this temptation had no attraction for those who looked at the problem in its true light, as the Italian and English philosophers already did. Voltaire has observed, far more truly, that Newton's edifice rested on no stone of Descartes' foundations. He illustrates this by relating that Newton only once read the work of Descartes, and, in doing so, wrote the word "*error*," repeatedly, on the first seven or eight pages; after which he read no more. This volume, Voltaire adds, was for some time in the possession of Newton's nephew.[13]

(*Gassendi.*) Even in his own country, the system of Descartes was by no means universally adopted. We have seen that though Gassendi was coupled with Descartes as one of the leaders of the new philosophy, he was far from admiring his work. Gassendi's own views of the causes of the motions of the heavenly bodies are not very clear, nor even very clearly referrible to the laws of mechanics; although he was one of those who had most share in showing that those laws apply to astronomical motions. In a chapter, headed[14] "Quæ sit motrix siderum causa," he reviews several opinions; but the one which he seems to adopt, is that which ascribes the motion of the celestial globes to certain fibres, of which the action is similar to that of the muscles of animals. It does not appear, therefore, that he had distinctly apprehended, either the continuation of the movements of the planets by the First Law of Motion, or their deflection by the Second Law;—the two main steps on the road to the discovery of the true forces by which they are made to describe their orbits.

(*Leibnitz, &c.*) Nor does it appear that in Germany mathematicians had attained this point of view. Leibnitz, as we have seen, did not assent to the opinions of Descartes, as containing the complete truth; and yet his own views of the physics of the universe do not seem to have any great advantage over these. In 1671 he published *A new physical hypothesis, by which the causes of most phenomena are deduced from a certain single universal motion supposed in our globe;—not to be despised either by the Tychonians or the Copernicans.* He supposes

[13] *Cartesianism*, Enc. Phil.

[14] Gassendi, *Opera*, vol. i. p. 639.

the particles of the earth to have separate motions, which produce collisions, and thus propagate[15] an "agitation of the ether," radiating in all directions; and,[16] "by the rotation of the sun on its axis, concurring with its rectilinear action on the earth, arises the motion of the earth about the sun." The other motions of the solar system are, as we might expect, accounted for in a similar manner; but it appears difficult to invest such an hypothesis with any mechanical consistency.

John Bernoulli maintained to the last the Cartesian hypothesis, though with several modifications of his own, and even pretended to apply mathematical calculation to his principles. This, however, belongs to a later period of our history; to the reception, not to the prelude, of the Newtonian theory.

(*Borelli.*) In Italy, Holland, and England, mathematicians appear to have looked much more steadily at the problem of the celestial motions, by the light which the discovery of the real laws of motion threw upon it. In Borelli's *Theories of the Medicean Planets*, printed at Florence in 1666, we have already a conception of the nature of central action, in which true notions begin to appear. The attraction of a body upon another which revolves about it is spoken of and likened to magnetic action; not converting the attracting force into a transverse force, according to the erroneous views of Kepler, but taking it as a tendency of the bodies to meet. "It is manifest," says he,[17] "that every planet and satellite revolves round some principal globe of the universe as a fountain of virtue, which so draws and holds them that they cannot by any means be separated from it, but are compelled to follow it wherever it goes, in constant and continuous revolutions." And, further on, he describes[18] the nature of the action, as a matter of conjecture indeed, but with remarkable correctness.[19] "We shall account for these motions by supposing, that which can hardly be denied, that the planets have a certain natural appetite for uniting themselves with the globe round which they revolve, and that they really tend, with all their efforts, to approach to such globe; the planets, for instance, to the sun, the Medicean Stars to Jupiter. It is certain, also, that circular motion gives a body a tendency to recede from the centre of such revolution, as we find in a wheel, or a stone whirled in a sling. Let us suppose, then, the planet to endeavor to approach the sun; since, in the mean time, it requires, by the circular motion, a force to recede from the same central body, it comes to pass, that when

---

15 Art. 5. 16 Ib. 8. 17 Cap. 2. 18 Ib. 11. 19 P. 47.

those two opposite forces are equal, each compensates the other, and the planet cannot go nearer to the sun nor further from him than a certain determinate space, and thus appears balanced and floating about him."

This is a very remarkable passage; but it will be observed, at the same time, that the author has no distinct conception of the manner in which the change of direction of the planet's motion is regulated from one instant to another; still less do his views lead to any mode of calculating the distance from the central body at which the planet would be thus balanced, or the space through which it might approach to the centre and recede from it. There is a great interval from Borelli's guesses, even to Huyghens' theorems; and a much greater to the beginning of Newton's discoveries.

(*England.*) It is peculiarly interesting to us to trace the gradual approach towards these discoveries which took place in the minds of English mathematicians; and this we can do with tolerable distinctness. Gilbert, in his work, *De Magnete*, printed in 1600, has only some vague notions that the magnetic virtue of the earth in some way determines the direction of the earth's axis, the rate of its diurnal rotation, and that of the revolution of the moon about it.[20] He died in 1603, and, in his posthumous work, already mentioned (*De Mundo nostro Sublunari Philosophia nova*, 1651), we have already a more distinct statement of the attraction of one body by another.[21] "The force which emanates from the moon reaches to the earth, and, in like manner, the magnetic virtue of the earth pervades the region of the moon: both correspond and conspire by the joint action of both, according to a proportion and conformity of motions; but the earth has more effect, in consequence of its superior mass; the earth attracts and repels the moon, and the moon, within certain limits, the earth; not so as to make the bodies come together, as magnetic bodies do, but so that they may go on in a continuous course." Though this phraseology is capable of representing a good deal of the truth, it does not appear to have been connected, in the author's mind, with any very definite notions of mechanical action in detail. We may probably say the same of Milton's language:

What if the sun
Be centre to the world; and other stars,
By his attractive virtue and their own
Incited, dance about him various rounds?
Par. Lost, B. viii.

[20] Lib. vi. cap. 6, 7.

[21] Ib. ii. c. 19.

Boyle, about the same period, seems to have inclined to the Cartesian hypothesis. Thus, in order to show the advantage of the natural theology which contemplates organic contrivances, over that which refers to astronomy, he remarks: "It may be said, that in bodies inanimate,[22] the contrivance is very rarely so exquisite but that the various motions and occurrences of their parts may, without much improbability, be suspected capable, after many essays, to cast one another into several of those circumvolutions called by Epicurus συστροφὰς, and by Descartes, *vortices;* which being once made, may continue a long time after the manner explained by the latter." Neither Milton nor Boyle, however, can be supposed to have had an exact knowledge of the laws of mechanics; and therefore they do not fully represent the views of their mathematical contemporaries. But there arose about this time a group of philosophers, who began to knock at the door where Truth was to be found, although it was left for Newton to force it open. These were the founders of the Royal Society, Wilkins, Wallis, Seth Ward, Wren, Hooke, and others. The time of the beginning of the speculations and association of these men corresponds to the time of the civil wars between the king and parliament in England; and it does not appear a fanciful account of their scientific zeal and activity, to say, that while they shared the common mental ferment of the times, they sought in the calm and peaceful pursuit of knowledge a contrast to the vexatious and angry struggles which at that time disturbed the repose of society. It was well if these dissensions produced any good to science to balance the obvious evils which flowed from them. Gascoigne, the inventor of the micrometer, a friend of Horrox, was killed in the battle of Marston Moor. Milburne, another friend of Horrox, who like him detected the errors of Lansberg's astronomical tables, left papers on this subject, which were lost by the coming of the Scotch army into England in 1639; in the civil war which ensued, the anatomical collections of Harvey were plundered and destroyed. Most of these persons of whom I have lately had to speak, were involved in the changes of fortune of the Commonwealth, some on one side, and some on the other. Wilkins was made Warden of Wadham by the committee of parliament appointed for reforming the University of Oxford; and was, in 1659, made Master of Trinity College, Cambridge, by Richard Cromwell, but ejected thence the year following, upon the restoration of the

---

[22] Shaw's Boyle's *Works*, ii. 160.

royal sway. Seth Ward, who was a Fellow of Sidney College, Cambridge, was deprived of his Fellowship by the parliamentary committee; but at a later period (1649) he took the engagement to be faithful to the Commonwealth, and became Savilian Professor of Astronomy at Oxford. Wallis held a Fellowship of Queen's College, Cambridge, but vacated it by marriage. He was afterwards much employed by the royal party in deciphering secret writings, in which art he had peculiar skill. Yet he was appointed by the parliamentary commissioners Savilian Professor of Geometry at Oxford, in which situation he was continued by Charles II. after his restoration. Christopher Wren was somewhat later, and escaped these changes. He was chosen Fellow of All-Souls in 1652, and succeeded Ward as Savilian Professor of Astronomy. These men, along with Boyle and several others, formed themselves into a club, which they called the Philosophical, or the Invisible College; and met, from about the year 1645, sometimes in London, and sometimes in Oxford, according to the changes of fortune and residence of the members. Hooke went to Christ Church, Oxford, in 1653, where he was patronized by Boyle, Ward, and Wallis; and when the Philosophical College resumed its meetings in London, after the Restoration, as the Royal Society, Hooke was made "curator of experiments." Halley was of the next generation, and comes after Newton; he studied at Queen's College, Oxford, in 1673; but was at first a man of some fortune, and not engaged in any official situation. His talents and zeal, however, made him an active and effective ally in the promotion of science.

The connection of the persons of whom we have been speaking has a bearing on our subject, for it led, historically speaking, to the publication of Newton's discoveries in physical astronomy. Rightly to propose a problem is no inconsiderable step to its solution; and it was undoubtedly a great advance towards the true theory of the universe to consider the motion of the planets round the sun as a mechanical question, to be solved by a reference to the laws of motion, and by the use of mathematics. So far the English philosophers appear to have gone, before the time of Newton. Hooke, indeed, when the doctrine of gravitation was published, asserted that he had discovered it previously to Newton; and though this pretension could not be maintained, he certainly had perceived that the thing to be done was, to determine the effect of a central force in producing curvilinear motion; which effect, as we have already seen, he illustrated by experiment as early as 1666. Hooke had also spoken more clearly on this subject in *An Attempt to prove the Motion of the Earth from Observations*, published in 1674. In this, he distinctly states that the planets would move in straight lines, if they were not deflected by central forces; and that the central attractive power increases in approaching the centre in certain degrees, dependent on the distance. "Now what these degrees are," he adds, "I have not yet experimentally verified;" but he ventures to promise to any one who succeeds in this undertaking, a discovery of the cause of the heavenly motions. He asserted, in conversation, to Halley and Wren, that he had solved this problem, but his solution was never produced. The proposition that the attractive force of the sun varies inversely as the square of the distance from the centre, had already been divined, if not fully established. If the orbits of the planets were circles, this proportion of the forces might be deduced in the same manner as the propositions concerning circular motion, which Huyghens published in 1673; yet it does not appear that Huyghens made this application of his principles. Newton, however, had already made this step some years before this time. Accordingly, he says in a letter to Halley, on Hooke's claim to this discovery,[23] "When Huygenius put out his *Horologium Oscillatorium*, a copy being presented to me, in my letter of thanks I gave those rules in the end thereof a particular commendation for their usefulness in computing the forces of the moon from the earth, and the earth from the sun." He says, moreover, "I am almost confident by circumstances, that Sir Christopher Wren knew the duplicate proportion when I gave him a visit; and then Mr. Hooke, by his book *Cometa*, will prove the last of us three that knew it." Hooke's *Cometa* was published in 1678. These inferences were all connected with Kepler's law, that the times are in the sesquiplicate ratio of the major axes of the orbits. But Halley had also been led to the duplicate proportion by another train of reasoning, namely, by considering the force of the sun as an emanation, which must become more feeble in proportion to the increased spherical surface over which it is diffused, and therefore in the inverse proportion of the square of the distances.[24] In this view of the matter, however, the difficulty was to determine what would be the motion of a body acted on by such a force, when the orbit is not circular but oblong. The investigation of this case was a problem which, we can

[23] *Biog. Brit.*, art. *Hooke.*

[24] Bullialdus, in 1645, had asserted that the force by which the sun "prehendit et harpagat," takes hold of and grapples the planets, must be as the inverse square of the distance.

easily conceive, must have appeared of very formidable complexity while it was unsolved, and the first of its kind. Accordingly Halley, as his biographer says, "finding himself unable to make it out in any geometrical way, first applied to Mr. Hooke and Sir Christopher Wren, and meeting with no assistance from either of them, he went to Cambridge in August (1684), to Mr. Newton, who supplied him fully with what he had so ardently sought."

A paper of Halley's in the *Philosophical Transactions* for January, 1686, professedly inserted as a preparation for Newton's work, contains some arguments against the Cartesian hypothesis of gravity, which seem to imply that Cartesian opinions had some footing among English philosophers; and we are told by Whiston, Newton's successor in his professorship at Cambridge, that Cartesianism formed a part of the studies of that place. Indeed, Rohault's *Physics* was used as a class-book at that University long after the time of which we are speaking; but the peculiar Cartesian doctrines which it contained were soon superseded by others.

With regard, then, to this part of the discovery, that the force of the sun follows the inverse duplicate proportion of the distances, we see that several other persons were on the verge of it at the same time with Newton; though he alone possessed that combination of distinctness of thought and power of mathematical invention, which enabled him to force his way across the barrier. But another, and so far as we know, an earlier train of thought, led by a different path to the same result; and it was the convergence of these two lines of reasoning that brought the conclusion to men's minds with irresistible force. I speak now of the identification of the force which retains the moon in her orbit with the force of gravity by which bodies fall at the earth's surface. In this comparison Newton had, so far as I am aware, no forerunner. We are now, therefore, arrived at the point at which the history of Newton's great discovery properly begins.

CHAPTER II.

THE INDUCTIVE EPOCH OF NEWTON.—DISCOVERY OF THE UNIVERSAL GRAVITATION OF MATTER, ACCORDING TO THE LAW OF THE INVERSE SQUARE OF THE DISTANCE.

IN order that we may the more clearly consider the bearing of this, the greatest scientific discovery ever made, we shall resolve it into the partial propositions of which it consists. Of these we may enumerate five. The doctrine of universal gravitation asserts,

1. That the force by which the *different* planets are attracted to the sun is in the inverse proportion of the squares of their distances;

2. That the force by which the *same* planet is attracted to the sun, in different parts of its orbit, is also in the inverse proportion of the squares of the distances;

3. That the *earth* also exerts such a force on the *moon*, and that this force is identical with the force of *gravity;*

4. That bodies thus act on *other* bodies, besides those which revolve round them; thus, that the sun exerts such a force on the moon and satellites, and that the planets exert such forces on *one another;*

5. That this force, thus exerted by the general masses of the sun, earth, and planets, arises from the attraction *of each particle* of these masses; which attraction follows the above law, and belongs to all matter alike.

The history of the establishment of these five truths will be given in order.

1. *Sun's Force on Different Planets.*—With regard to the first of the above five propositions, that the different planets are attracted to the sun by a force which is inversely as the square of the distance, Newton had so far been anticipated, that several persons had discovered it to be true, or nearly true; that is, they had discovered that if the orbits of the planets were circles, the proportions of the central force to the inverse square of the distance would follow from Kepler's third law, of the sesquiplicate proportion of the periodic times. As we have seen, Huyghens' theorems would have proved this, if they had been so applied; Wren knew it; Hooke not only knew it, but claimed a prior knowledge to Newton; and Halley had satisfied himself that it was at

least nearly true, before he visited Newton. Hooke was reported to Newton at Cambridge, as having applied to the Royal Society to do him justice with regard to his claims; but when Halley wrote and informed Newton (in a letter dated June 29, 1686), that Hooke's conduct "had been represented in worse colors than it ought," Newton inserted in his book a notice of these his predecessors, in order, as he said, "to compose the dispute."[1] This notice appears in a Scholium to the fourth Proposition of the *Principia*, which states the general law of revolutions in circles. "The case of the sixth corollary," Newton there says, "obtains in the celestial bodies, as has been separately inferred by our countrymen, Wren, Hooke, and Halley;" he soon after names Huyghens, "who, in his excellent treatise *De Horologio Oscillatorio*, compares the force of gravity with the centrifugal forces of revolving bodies."

The two steps requisite for this discovery were, to propose the motions of the planets as simply a mechanical problem, and to apply mathematical reasoning so as to solve this problem, with reference to Kepler's third law considered as a fact. The former step was a consequence of the mechanical discoveries of Galileo and his school; the result of the firm and clear place which these gradually obtained in men's mind, and of the utter abolition of all the notions of solid spheres by Kepler. The mathematical step required no small mathematical powers; as appears, when we consider that this was the first example of such a problem, and that the method of limits, under all its forms, was at this time in its infancy, or rather, at its birth. Accordingly, even in this step, though much the easiest in the path of deduction, no one before Newton completely executed.

2. *Force in different Points of an Orbit.*—The inference of the law of the force from Kepler's two laws concerning the elliptical motion, was a problem quite different from the preceding, and much more difficult; but the dispute with respect to priority in the two propositions was intermingled. Borelli, in 1666, had, as we have seen, endeavored to reconcile the general form of the orbit with the notion of a central attractive force, by taking centrifugal force into the account; and Hooke, in 1679, had asserted that the result of the law of the inverse square in the force of the earth would be an ellipse,[2] or a curve like an ellipse.[3] But it does not appear that this was any thing more than

---

[1] *Biog. Brit.* folio, art. *Hooke.* [2] Newton's Letter, *Biog. Brit.*, Hooke, p. 2660.
[3] Birch's *Hist. R. S.*, Wallis's Life.

a conjecture. Halley says[4] that "Hooke, in 1683, told him he had demonstrated all the laws of the celestial motions by the reciprocally duplicate proportion of the force of gravity; but that, being offered forty shillings by Sir Christopher Wren to produce such a demonstration, his answer was, that he had it, but would conceal it for some time, that others, trying and failing, might know how to value it when he should make it public." Halley, however, truly observes, that after the publication of the demonstration in the *Principia*, this reason no longer held; and adds, "I have plainly told him, that unless he produce another differing demonstration, and let the world judge of it, neither I nor any one else can believe it."

Newton allows that Hooke's assertions in 1679 gave occasion to his investigation on this point of the theory. His demonstration is contained in the second and third Sections of the *Principia*. He first treats of the general law of central forces in any curve; and then, on account, as he states, of the application to the motion of the heavenly bodies, he treats of the case of force varying inversely as the square of the distance, in a more diffuse manner.

In this, as in the former portion of his discovery, the two steps were, the proposing the heavenly motions as a mechanical problem, and the solving this problem. Borelli and Hooke had certainly made the former step, with considerable distinctness; but the mathematical solution required no common inventive power.

Newton seems to have been much ruffled by Hooke's speaking slightly of the value of this second step; and is moved in return to deny Hooke's pretensions with some asperity, and to assert his own. He says, in a letter to Halley, "Borelli did something in it, and wrote modestly; he (Hooke) has done nothing; and yet written in such a way as if he knew, and had sufficiently hinted all but what remained to be determined by the drudgery of calculations and observations; excusing himself from that labor by reason of his other business; whereas he should rather have excused himself by reason of his inability; for it is very plain, by his words, he knew not how to go about it. Now is not this very fine? Mathematicians that find out, settle, and do all the business, must content themselves with being nothing but dry calculators and drudges; and another that does nothing but pretend and grasp at all things, must carry away all the inventions, as well of those that were to follow him as of those that

---

[4] *Enc. Brit.* Hooke, p. 2660.

went before." This was written, however, under the influence of some degree of mistake; and in a subsequent letter, Newton says, "Now I understand he was in some respects misrepresented to me, I wish I had spared the postscript to my last," in which is the passage just quoted. We see, by the melting away of rival claims, the undivided honor which belongs to Newton, as the real discoverer of the proposition now under notice. We may add, that in the sequel of the third Section of the *Principia*, he has traced its consequences, and solved various problems flowing from it with his usual fertility and beauty of mathematical resource; and has there shown the necessary connection of Kepler's third law with his first and second.

3. *Moon's Gravity to the Earth.*—Though others had considered cosmical forces as governed by the general laws of motion, it does not appear that they had identified such forces with the force of terrestrial gravity. This step in Newton's discoveries has generally been the most spoken of by superficial thinkers; and a false kind of interest has been attached to it, from the story of its being suggested by the fall of an apple. The popular mind is caught by the character of an eventful narrative which the anecdote gives to this occurrence; and by the antithesis which makes a profound theory appear the result of a trivial accident. How inappropriate is such a view of the matter we shall soon see. The narrative of the progress of Newton's thoughts, is given by Pemberton (who had it from Newton himself) in his preface to his *View of Newton's Philosophy*, and by Voltaire, who had it from Mrs. Conduit, Newton's niece.[5] "The first thoughts," we are told, "which gave rise to his *Principia*, he had when he retired from Cambridge, in 1666, on account of the plague (he was then twenty-four years of age). As he sat alone in a garden, he fell into a speculation on the power of gravity; that as this power is not found sensibly diminished at the remotest distance from the centre of the earth to which we can rise, neither at the tops of the loftiest buildings, nor even on the summits of the highest mountains, it appeared to him reasonable to conclude that this power must extend much further than was usually thought: Why not as high as the moon? said he to himself; and if so, her motion must be influenced by it; perhaps she is retained in her orbit thereby."

The thought of cosmical gravitation was thus distinctly brought into being; and Newton's superiority here was, that he conceived the

[5] *Elémens de Phil. de Newton*, 3me partie, chap. iii.

celestial motions as distinctly as the motions which took place close to him;—considered them as of the same kind, and applied the same rules to each, without hesitation or obscurity. But so far, this thought was merely a guess: its occurrence showed the activity of the thinker; but to give it any value, it required much more than a "why not?"—a "perhaps." Accordingly, Newton's "why not?" was immediately succeeded by his "if so, what then?" His reasoning was, that if gravity reach to the moon, it is probably of the same kind as the central force of the sun, and follows the same rule with respect to the distance. What is this rule? We have already seen that, by calculating from Kepler's laws, and supposing the orbits to be circles, the rule of the force appears to be the inverse duplicate proportion of the distance; and this, which had been current as a conjecture among the previous generation of mathematicians, Newton had already proved by indisputable reasonings, and was thus prepared to proceed in his train of inquiry. If, then, he went on, pursuing his train of thought, the earth's gravity extend to the moon, diminishing according to the inverse square of the distance, will it, at the moon's orbit, be of the proper magnitude for retaining her in her path? Here again came in calculation, and a calculation of extreme interest; for how important and how critical was the decision which depended on the resulting numbers? According to Newton's calculations, made at this time, the moon by her motion in her orbit, was deflected from the tangent every minute through a space of thirteen feet. But by noticing the space through which bodies would fall in one minute at the earth's surface, and supposing this to be diminished in the ratio of the inverse square, it appeared that gravity would, at the moon's orbit, draw a body through more than fifteen feet. The difference seems small, the approximation encouraging, the theory plausible; a man in love with his own fancies would readily have discovered or invented some probable cause of this difference. But Newton acquiesced in it as a disproof of his conjecture, and "laid aside at that time any further thoughts of this matter; thus resigning a favorite hypothesis, with a candor and openness to conviction not inferior to Kepler, though his notion had been taken up on far stronger and sounder grounds than Kepler dealt in; and without even, so far as we know, Kepler's regrets and struggles. Nor was this levity or indifference; the idea, though thus laid aside, was not finally condemned and abandoned. When Hooke, in 1679, contradicted Newton on the subject of the curve described by a falling body, and asserted it to be an ellipse, Newton was led to investigate the subject, and was then again conducted, by another road, to the same law of the inverse square of the distance. This naturally turned his thoughts to his former speculations. Was there really no way of explaining the discrepancy which this law gave, when he attempted to reduce the moon's motion to the action of gravity? A scientific operation then recently completed, gave the explanation at once. He had been mistaken in the magnitude of the earth, and consequently in the distance of the moon, which is determined by measurements of which the earth's radius is the base. He had taken the common estimate, current among geographers and seamen, that sixty English miles are contained in one degree of latitude. But Picard, in 1670, had measured the length of a certain portion of the meridian in France, with far greater accuracy than had yet been attained; and this measure enabled Newton to repeat his calculations with these amended data. We may imagine the strong curiosity which he must have felt as to the result of these calculations. His former conjecture was now found to agree with the phenomena to a remarkable degree of precision. This conclusion, thus coming after long doubts and delays, and falling in with the other results of mechanical calculation for the solar system, gave a stamp from that moment to his opinions, and through him to those of the whole philosophical world.

[2d Ed.] [Dr. Robison (*Mechanical Philosophy*, p. 288) says that Newton having become a member of the Royal Society, there learned the accurate measurement of the earth by Picard, differing very much from the estimation by which he had made his calculations in 1666. And M. Biot, in his Life of Newton, published in the *Biographie Universelle*, says, "*According to conjecture*, about the month of June, 1682, Newton being in London at a meeting of the Royal Society, mention was made of the new measure of a degree of the earth's surface, recently executed in France by Picard; and great praise was given to the care which had been employed in making this measure exact."

I had adopted this conjecture as a fact in my first edition; but it has been pointed out by Prof. Rigaud (*Historical Essay on the First Publication of the Principia*, 1838), that Picard's measurement was probably well known to the Fellows of the Royal Society as early as 1675, there being an account of the results of it given in the *Philosophical Transactions* for that year. Newton appears to have discovered the method of determining that a body might describe an ellipse when acted upon by a force residing in the focus, and varying inversely as the square of the distance, in 1679, upon occasion of his correspondence with Hooke. In 1684, at Halley's request, he returned to the subject, and in February, 1685, there was inserted in the Register of the Royal Society a paper of Newton's (*Isaaci Newtoni Propositiones de Motu*) which contained some of the principal Propositions of the first two Books of the *Principia*. This paper, however, does not contain the Proposition "Lunam gravitare in terram," nor any of the other propositions of the third Book. The *Principia* was printed in 1686 and 7, apparently at the expense of Halley. On the 6th of April, 1687, the third Book was presented to the Royal Society.]

It does not appear, I think, that before Newton, philosophers in general had supposed that terrestrial gravity was the very force by which the moon's motions are produced. Men had, as we have seen, taken up the conception of such forces, and had probably called them gravity: but this was done only to explain, by analogy, what *kind* of forces they were, just as at other times they compared them with magnetism; and it did not imply that terrestrial gravity was a force which acted in the celestial spaces. After Newton had discovered that this was so, the application of the term "gravity" did undoubtedly convey such a suggestion; but we should err if we inferred from this coincidence of expression that the notion was commonly entertained before him. Thus Huyghens appears to use language which may be mistaken, when he says,[6] that Borelli was of opinion that the primary planets were urged by "gravity" towards the sun, and the satellites towards the primaries. The notion of terrestrial gravity, as being actually a cosmical force, is foreign to all Borelli's speculations.[7] But Horrox, as early as 1635, appears to have entertained the true view on this subject, although vitiated by Keplerian errors concerning the connection between the rotation of the central body and its effect on the body which revolves about it. Thus he says,[8] that the emanation of the earth carries a projected stone along with the motion of the earth, just in the same way as it carries the moon in her orbit; and that this force is greater on the stone than on the moon, because the distance is less.

The Proposition in which Newton has stated the discovery of which we are now speaking, is the fourth of his third Book: "That the moon gravitates to the earth, and by the force of gravity is perpetually de-

---

[6] *Cosmotheros*, l. 2. p. 720.

[7] I have found no instance in which the word is so used by him.

[8] *Astronomia Kepleriana defensa et promota*, cap. 2. See further on this subject in the *Additions* to this volume.

flected from a rectilinear motion, and retained in her orbit." The proof consists in the numerical calculation, of which he only gives the elements, and points out the method; but we may observe, that no small degree of knowledge of the way in which astronomers had obtained these elements, and judgment in selecting among them, were necessary: thus, the mean distance of the moon had been made as little as fifty-six and a half semidiameters of the earth by Tycho, and as much as sixty-two and a half by Kircher: Newton gives good reasons for adopting sixty-one.

The term "gravity," and the expression "to gravitate," which, as we have just seen, Newton uses of the moon, were to receive a still wider application in consequence of his discoveries; but in order to make this extension clearer, we consider it as a separate step.

4. *Mutual Attraction of all the Celestial Bodies.*—If the preceding parts of the discovery of gravitation were comparatively easy to conjecture, and difficult to prove, this was much more the case with the part of which we have now to speak, the attraction of other bodies, besides the central ones, upon the planets and satellites. If the mathematical calculation of the unmixed effect of a central force required transcendent talents, how much must the difficulty be increased, when other influences prevented those first results from being accurately verified, while the deviations from accuracy were far more complex than the original action! If it had not been that these deviations, though surprisingly numerous and complicated in their nature, were very small in their quantity, it would have been impossible for the intellect of man to deal with the subject; as it was, the struggle with its difficulties is even now a matter of wonder.

The conjecture that there is some mutual action of the planets, had been put forth by Hooke in his *Attempt to prove the Motion of the Earth* (1674). It followed, he said, from his doctrine, that not only the sun and moon act upon the course and motion of the earth, but that Mercury, Venus, Mars, Jupiter, and Saturn, have also, by their attractive power, a considerable influence upon the motion of the earth, and the earth in like manner powerfully affects the motions of those bodies. And Borelli, in attempting to form "theories" of the satellites of Jupiter, had seen, though dimly and confusedly, the probability that the sun would disturb the motions of these bodies. Thus he says (cap. 14), "How can we believe that the Medicean globes are not, like other planets, impelled with a greater velocity when they approach the sun: and thus they are acted upon by two moving forces, one of which produces their proper revolution about Jupiter, the other regulates their motion round the sun." And in another place (cap. 20), he attempts to show an effect of this principle upon the inclination of the orbit; though, as might be expected, without any real result.

The case which most obviously suggests the notion that the sun exerts a power to disturb the motions of secondary planets about primary ones, might seem to be our own moon; for the great inequalities which had hitherto been discovered, had all, except the first, or elliptical anomaly, a reference to the position of the sun. Nevertheless, I do not know that any one had attempted thus to explain the curiously irregular course of the earth's attendant. To calculate, from the disturbing agency, the amount of the irregularities, was a problem which could not, at any former period, have been dreamt of as likely to be at any time within the verge of human power.

Newton both made the step of inferring that there were such forces, and, to a very great extent, calculated the effects of them. The inference is made on mechanical principles, in the sixth Theorem of the third Book of the *Principia;*—that the moon is attracted by the sun, as the earth is;—that the satellites of Jupiter and Saturn are attracted as the primaries are; in the same manner, and with the same forces. If this were not so, it is shown that these attendant bodies could not accompany the principal ones in the regular manner in which they do. All those bodies at equal distances from the sun would be equally attracted.

But the complexity which must occur in tracing the results of this principle will easily be seen. The satellite and the primary, though nearly at the same distance, and in the same direction, from the sun, are not exactly so. Moreover the difference of the distances and of the directions is perpetually changing; and if the motion of the satellite be elliptical, the cycle of change is long and intricate: on this account alone the effects of the sun's action will inevitably follow cycles as long and as perplexed as those of the positions. But on another account they will be still more complicated; for in the continued action of a force, the effect which takes place at first, modifies and alters the effect afterwards. The result at any moment is the sum of the results in preceding instants: and since the terms, in this series of instantaneous effects, follow very complex rules, the sums of such series will be, it might be expected, utterly incapable of being reduced to any manageable degree of simplicity.

It certainly does not appear that any one but Newton could make any impression on this problem, or course of problems. No one for sixty years after the publication of the *Principia*, and, with Newton's methods, no one up to the present day, had added any thing of any value to his deductions. We know that he calculated all the principal lunar inequalities; in many of the cases, he has given us his processes; in others, only his results. But who has presented, in his beautiful geometry, or deduced from his simple principles, any of the inequalities which he left untouched? The ponderous instrument of synthesis, so effective in his hands, has never since been grasped by one who could use it for such purposes; and we gaze at it with admiring curiosity, as on some gigantic implement of war, which stands idle among the memorials of ancient days, and makes us wonder what manner of man he was who could wield as a weapon what we can hardly lift as a burden.

It is not necessary to point out in detail the sagacity and skill which mark this part of the *Principia*. The mode in which the author obtains the effect of a disturbing force in producing a motion of the apse of an elliptical orbit (the ninth Section of the first Book), has always been admired for its ingenuity and elegance. The general statement of the nature of the principal inequalities produced by the sun in the motion of a satellite, given in the sixty-sixth Proposition, is, even yet, one of the best explanations of such action; and the calculations of the quantity of the effects in the third Book, for instance, the *variation* of the moon, the *motion of the nodes* and its inequalities, the *change of inclination* of the orbit,—are full of beautiful and efficacious artifices. But Newton's inventive faculty was exercised to an extent greater than these published investigations show. In several cases he has suppressed the demonstration of his method, and given us the result only; either from haste or from mere weariness, which might well overtake one who, while he was struggling with facts and numbers, with difficulties of conception and practice, was aiming also at that geometrical elegance of exposition, which he considered as alone fit for the public eye. Thus, in stating the effect of the eccentricity of the moon's orbit upon the motion of the apogee, he says,[9] "The computations, as too intricate and embarrassed with approximations, I do not choose to introduce."

The computations of the theoretical motion of the moon being thus difficult, and its irregularities numerous and complex, we may ask,

[9] Schol. to Prop. 35, first edit.

whether Newton's reasoning was sufficient to establish this part of his theory; namely, that her actual motions arise from her gravitation to the sun. And to this we may reply, that it was sufficient for that purpose,—since it showed that, from Newton's hypothesis, inequalities must result, following the laws which the moon's inequalities were known to follow;—since the amount of the inequalities given by the theory agreed nearly with the rules which astronomers had collected from observation;—and since, by the very intricacy of the calculation, it was rendered probable, that the first results might be somewhat inaccurate, and thus might give rise to the still remaining differences between the calculations and the facts. A *Progression of the Apogee;* a *Regression of the Nodes;* and, besides the Elliptical, or first Inequality, an inequality, following the law of the *Evection*, or second inequality discovered by Ptolemy; another, following the law of the *Variation* discovered by Tycho;—were pointed out in the first edition of the *Principia*, as the consequences of the theory. Moreover, the quantities of these inequalities were calculated and compared with observation with the utmost confidence, and the agreement in most instances was striking. The Variation agreed with Halley's recent observations within a minute of a degree.[10] The Mean Motion of the Nodes in a year agreed within less than one-hundredth of the whole.[11] The Equation of the Motion of the Nodes also agreed well.[12] The Inclination of the Plane of the Orbit to the ecliptic, and its changes, according to the different situations of the nodes, likewise agreed.[13] The Evection has been already noticed as encumbered with peculiar difficulties: here the accordance was less close. The Difference of the daily progress of the Apogee in syzygy, and its daily Regress in Quadratures, is, Newton says, "$4\frac{1}{4}$ minutes by the Tables, $6\frac{2}{3}$ by our calculation." He boldly adds, "I suspect this difference to be due to the fault of the Tables." In the second edition (1711) he added the calculation of several other inequalities, as the *Annual Equation*, also discovered by Tycho; and he compared them with more recent observations made by Flamsteed at Greenwich; but even in what has already been stated, it must be allowed that there is a wonderful accordance of theory with phenomena, both being very complex in the rules which they educe.

The same theory which gave these Inequalities in the motion of the Moon produced by the disturbing force of the sun, gave also corres-

---

[10] B. iii. Prop. 29. [11] Prop. 32. [12] Prop. 33. [13] Prop. 35.

ponding Inequalities in the motions of the Satellites of other planets, arising from the same cause; and likewise pointed out the necessary existence of irregularities in the motions of the Planets arising from their mutual attraction. Newton gave propositions by which the Irregularities of the motion of Jupiter's moons might be deduced from those of our own;[14] and it was shown that the motions of their nodes would be slow by theory, as Flamsteed had found it to be by observation.[15] But Newton did not attempt to calculate the effect of the mutual action of the planets, though he observes, that in the case of Jupiter and Saturn this effect is too considerable to be neglected;[16] and he notices in the second edition,[17] that it follows from the theory of gravity, that the aphelia of Mercury, Venus, the Earth, and Mars, slightly progress.

In one celebrated instance, indeed, the deviation of the theory of the *Principia* from observation was wider, and more difficult to explain; and as this deviation for a time resisted the analysis of Euler and Clairaut, as it had resisted the synthesis of Newton, it at one period staggered the faith of mathematicians in the exactness of the law of the inverse square of the distance. I speak of the Motion of the Moon's Apogee, a problem which has already been referred to; and in which Newton's method, and all the methods which could be devised for some time afterwards, gave only half the observed motion; a circumstance which arose, as was discovered by Clairaut in 1750, from the insufficiency of the method of approximation. Newton does not attempt to conceal this discrepancy. After calculating what the motion of apse would be, upon the assumption of a disturbing force of the same amount as that which the sun exerts on the moon, he simply says,[18] "the apse of the moon moves about twice as fast."

The difficulty of doing what Newton did in this branch of the subject, and the powers it must have required, may be judged of from what has already been stated;—that no one, with his methods, has yet been able to add any thing to his labors: few have undertaken to illustrate what he has written, and no great number have understood it throughout. The extreme complication of the forces, and of the conditions under which they act, makes the subject by far the most thorny walk of mathematics. It is necessary to resolve the action

---

[14] B. i. Prop. 66. [15] B. iii. Prop. 23. [16] B. iii. Prop. 13.

[17] Scholium to Prop. 14. B. iii.

[18] B. i. Prop. 44, second edit. There is reason to believe, however, that Newton had, in his unpublished calculations, rectified this discrepancy.

into many elements, such as can be separated; to invent artifices for dealing with each of these; and then to recompound the laws thus obtained into one common conception. The moon's motion cannot be conceived without comprehending a scheme more complex than the Ptolemaic epicycles and eccentrics in their worst form; and the component parts of the system are not, in this instance, mere geometrical ideas, requiring only a distinct apprehension of relations of space in order to hold them securely; they are the foundations of mechanical notions, and require to be grasped so that we can apply to them sound mechanical reasonings. Newton's successors, in the next generation, abandoned the hope of imitating him in this intense mental effort; they gave the subject over to the operation of algebraical reasoning, in which symbols think for us, without our dwelling constantly upon their meaning, and obtain for us the consequences which result from the relations of space and the laws of force, however complicated be the conditions under which they are combined. Even Newton's countrymen, though they were long before they applied themselves to the method thus opposed to his, did not produce any thing which showed that they had mastered, or could retrace, the Newtonian investigations.

Thus the Problem of Three Bodies,[19] treated geometrically, belongs exclusively to Newton; and the proofs of the mutual action of the sun, planets, and satellites, which depend upon such reasoning, could not be discovered by any one but him.

But we have not yet done with his achievements on this subject; for some of the most remarkable and beautiful of the reasonings which he connected with this problem, belong to the next step of his generalization.

5. *Mutual Attraction of all Particles of Matter.*—That all the parts of the universe are drawn and held together by love, or harmony, or some affection to which, among other names, that of *attraction* may have been given, is an assertion which may very possibly have been made at various times, by speculators writing at random, and taking their chance of meaning and truth. The authors of such casual dogmas have generally nothing accurate or substantial, either in their conception of the general proposition, or in their reference to examples of it; and, therefore, their doctrines are no concern of ours at present. But among those who were really the first to think of the mutual at-

---

[19] See the history of the *Problem of Three Bodies, ante*, in Book vi. Chap. vi. Sect. 7.

traction of matter, we cannot help noticing Francis Bacon; for his notions were so far from being chargeable with the looseness and indistinctness to which we have alluded, that he proposed an experiment[20] which was to decide whether the facts were so or not;—whether the gravity of bodies to the earth arose from an attraction of the parts of matter towards each other, or was a tendency towards the centre of the earth. And this experiment is, even to this day, one of the best which can be devised, in order to exhibit the universal gravitation of matter: it consists in the comparison of the rate of going of a clock in a deep mine, and on a high place. Huyghens, in his book *De Causâ Gravitatis*, published in 1690, showed that the earth would have an oblate form, in consequence of the action of the centrifugal force; but his reasoning does not suppose gravity to arise from the mutual attraction of the parts of the earth. The apparent influence of the moon upon the tides had long been remarked; but no one had made any progress in truly explaining the mechanism of this influence; and all the analogies to which reference had been made, on this and similar subjects, as magnetic and other attractions, were rather delusive than illustrative, since they represented the attraction as something peculiar in particular bodies, depending upon the nature of each body.

That all such forces, cosmical and terrestrial, were the same single force, and that this was nothing more than the insensible attraction which subsists between one stone and another, was a conception equally bold and grand; and would have been an incomprehensible thought, if the views which we have already explained had not prepared the mind for it. But the preceding steps having disclosed, between all the bodies of the universe, forces of the same kind as those which produce the weight of bodies at the earth, and, therefore, such as exist in every particle of terrestrial matter; it became an obvious question, whether such forces did not also belong to all particles of planetary matter, and whether this was not, in fact, the whole account of the forces of the solar system. But, supposing this conjecture to be thus suggested, how formidable, on first appearance at least, was the undertaking of verifying it! For if this be so, every finite mass of matter exerts forces which are the result of the infinitely numerous forces of its particles, these forces acting in different directions. It does not appear, at first sight, that the law by which the force is related to the distance, will be the same for the particles as it is for the masses; and, in reality, it

---

[20] *Nov. Org.* Lib. ii. Aph. 36.

is not so, except in special cases. And, again, in the instance of any effect produced by the force of a body, how are we to know whether the force resides in the whole mass as a unit, or in the separate particles? We may reason, as Newton does,[21] that the rule which proves gravity to belong universally to the planets, proves it also to belong to their parts; but the mind will not be satisfied with this extension of the rule, except we can find decisive instances, and calculate the effects of both suppositions, under the appropriate conditions. Accordingly, Newton had to solve a new series of problems suggested by this inquiry; and this he did.

These solutions are no less remarkable for the mathematical power which they exhibit, than the other parts of the *Principia*. The propositions in which it is shown that the law of the inverse square for the particles gives the same law for spherical masses, have that kind of beauty which might well have justified their being published for their mathematical elegance alone, even if they had not applied to any real case. Great ingenuity is also employed in other instances, as in the case of spheroids of small eccentricity. And when the amount of the mechanical action of masses of various forms has thus been assigned, the sagacity shown in tracing the results of such action in the solar system is truly admirable; not only the general nature of the effect being pointed out, but its quantity calculated. I speak in particular of the reasonings concerning the Figure of the Earth, the Tides, the Precession of the Equinoxes, the Regression of the Nodes of a ring such as Saturn's; and of some effects which, at that time, had not been ascertained even as facts of observation; for instance, the difference of gravity in different latitudes, and the Nutation of the earth's axis. It is true, that in most of these cases, Newton's process could be considered only as a rude approximation. In one (the Precession) he committed an error, and in all, his means of calculation were insufficient. Indeed these are much more difficult investigations than the Problem of Three Bodies, in which three points act on each other by explicit laws. Up to this day, the resources of modern analysis have been employed upon some of them with very partial success; and the facts, in all of them, required to be accurately ascertained and measured, a process which is not completed even now. Nevertheless the form and nature of the conclusions which Newton did obtain, were such as to inspire a strong confidence in the competency of his theory to explain

---

[21] *Princip.* B. iii. Prop. 7.

all such phenomena as have been spoken of. We shall afterwards have to speak of the labors, undertaken in order to examine the phenomena more exactly, to which the theory gave occasion.

Thus, then, the theory of the universal mutual gravitation of all the particles of matter, according to the law of the inverse square of the distances, was conceived, its consequences calculated, and its results shown to agree with phenomena. It was found that this theory took up all the facts of astronomy as far as they had hitherto been ascertained; while it pointed out an interminable vista of new facts, too minute or too complex for observation alone to disentangle, but capable of being detected when theory had pointed out their laws, and of being used as criteria or confirmations of the truth of the doctrine. For the same reasoning which explained the evection, variation, and annual equation of the moon, showed that there must be many other inequalities besides these; since these resulted from approximate methods of calculation, in which small quantities were neglected. And it was known that, in fact, the inequalities hitherto detected by astronomers did not give the place of the moon with satisfactory accuracy; so that there was room, among these hitherto untractable irregularities, for the additional results of the theory. To work out this comparison was the employment of the succeeding century; but Newton began it. Thus, at the end of the proposition in which he asserts,[22] that "all the lunar motions and their irregularities follow from the principles here stated," he makes the observation which we have just made; and gives, as examples, the different motions of the apogee and nodes, the difference of the change of the eccentricity, and the difference of the moon's variation, according to the different distances of the sun. "But this inequality," he says, "in astronomical calculations, is usually referred to the prosthaphæresis of the moon, and confounded with it."

Reflections on the Discovery.—Such, then, is the great Newtonian Induction of Universal Gravitation, and such its history. It is indisputably and incomparably the greatest scientific discovery ever made, whether we look at the advance which it involved, the extent of the truth disclosed, or the fundamental and satisfactory nature of this truth. As to the first point, we may observe that any one of the five steps into which we have separated the doctrine, would, of itself, have been considered as an important advance;—would have conferred distinction on the persons who made it, and the time to which it belonged. All

[22] B. iii. Prop. 22.

the five steps made at once, formed not a leap, but a flight,—not an improvement merely, but a metamorphosis,—not an epoch, but a termination. Astronomy passed at once from its boyhood to mature manhood. Again, with regard to the extent of the truth, we obtain as wide a generalization as our physical knowledge admits, when we learn that every particle of matter, in all times, places, and circumstances, attracts every other particle in the universe by one common law of action. And by saying that the truth was of a fundamental and satisfactory nature, I mean that it assigned, not a rule merely, but a cause, for the heavenly motions; and that kind of cause which most eminently and peculiarly we distinctly and thoroughly conceive, namely, mechanical force. Kepler's laws were merely *formal* rules, governing the celestial motions according to the relations of space, time, and number; Newton's was a *casual* law, referring these motions to mechanical reasons. It is no doubt conceivable that future discoveries may both extend and further explain Newton's doctrines;—may make gravitation a case of some wider law, and may disclose something of the mode in which it operates; questions with which Newton himself struggled. But, in the mean time, few persons will dispute, that both in generality and profundity, both in width and depth, Newton's theory is altogether without a rival or neighbor.[23]

The requisite conditions of such a discovery in the mind of its author were, in this as in other cases, the idea, and its comparison with facts; —the conception of the law, and the moulding this conception in such a form as to correspond with known realities. The idea of mechanical

---

[23] The value and nature of this step have long been generally acknowledged wherever science is cultivated. Yet it would appear that there is, in one part of Europe, a school of philosophers who contest the merit of this part of Newton's discoveries. "Kepler," says a celebrated German metaphysician,* "discovered the laws of free motion; a discovery of immortal glory. It has since been the fashion to say that Newton first found out the proof of these rules. It has seldom happened that the glory of the first discoverer has been more unjustly transferred to another person." It may appear strange that any one in the present day should hold such language; but if we examine the reasons which this author gives, they will be found, I think, to amount to this: that his mind is in the condition in which Kepler's was; and that the whole range of mechanical ideas and modes of conception which made the transition from Kepler and Newton possible, are extraneous to the domain of his philosophy. Even this author, however, if I understand him rightly, recognizes Newton as the author of the doctrine of Perturbations.

I have given a further account of these views, in a Memoir *On Hegel's Criticism of Newton's Principia.* Cambridge Transactions, 1849.

* Hegel, *Encyclopædia*, § 270.

force as the cause of the celestial motions, had, as we have seen, been for some time growing up in men's minds; had gone on becoming more distinct and more general; and had, in some persons, approached the form in which it was entertained by Newton. Still, in the mere conception of universal gravitation, Newton must have gone far beyond his predecessors and contemporaries, both in generality and distinctness; and in the inventiveness and sagacity with which he traced the consequences of this conception, he was, as we have shown, without a rival, and almost without a second. As to the facts which he had to include in his law, they had been accumulating from the very birth of astronomy; but those which he had more peculiarly to take hold of, were the facts of the planetary motions as given by Kepler, and those of the moon's motions as given by Tycho Brahe and Jeremy Horrox.

We find here occasion to make a remark which is important in its bearing on the nature of progressive science. What Newton thus used and referred to as *facts*, were the *laws* which his predecessors had established. What Kepler and Horrox had put forth as "theories," were now established truths, fit to be used in the construction of other theories. It is in this manner that one theory is built upon another;—that we rise from particulars to generals, and from one generalization to another;—that we have, in short, successive steps of induction. As Newton's laws assumed Kepler's, Kepler's laws assumed as facts the results of the planetary theory of Ptolemy; and thus the theories of each generation in the scientific world are (when thoroughly verified and established, the facts of the next generation. Newton's theory is the circle of generalization which includes all the others;—the highest point of the inductive ascent;—the catastrophe of the philosophic drama to which Plato had prologized;—the point to which men's minds had been journeying for two thousand years.

Character of Newton.—It is not easy to anatomize the constitution and the operations of the mind which makes such an advance in knowledge. Yet we may observe that there must exist in it, in an eminent degree, the elements which compose the mathematical talent. It must possess distinctness of intuition, tenacity and facility in tracing logical connection, fertility of invention, and a strong tendency to generalization. It is easy to discover indications of these characteristics in Newton. The distinctness of his intuitions of space, and we may add of force also, was seen in the amusements of his youth; in his constructing clocks and mills, carts and dials, as well as the facility with which he mastered geometry. This fondness for handicraft employments, and for making models and machines, appears to be a common prelude of excellence in physical science;[24] probably on this very account, that it arises from the distinctness of intuitive power with which the child conceives the shapes and the working of such material combinations. Newton's inventive power appears in the number and variety of the mathematical artifices and combinations which he devised, and of which his books are full. If we conceive the operation of the inventive faculty in the only way in which it appears possible to conceive it; —that while some hidden source supplies a rapid stream of possible suggestions, the mind is on the watch to seize and detain any one of these which will suit the case in hand, allowing the rest to pass by and be forgotten;—we shall see what extraordinary fertility of mind is implied by so many successful efforts; what an innumerable host of thoughts must have been produced, to supply so many that deserved to be selected. And since the selection is performed by tracing the consequences of each suggestion, so as to compare them with the requisite conditions, we see also what rapidity and certainty in drawing conclusions the mind must possess as a talent, and what watchfulness and patience as a habit.

The hidden fountain of our unbidden thoughts is for us a mystery; and we have, in our consciousness, no standard by which we can measure our own talents; but our acts and habits are something of which we are conscious; and we can understand, therefore, how it was that Newton could not admit that there was any difference between himself and other men, except in his possession of such habits as we have mentioned, perseverance and vigilance. When he was asked how he made his discoveries, he answered, "by always thinking about them;" and at another time he declared that if he had done any thing, it was due to nothing but industry and patient thought: "I keep the subject of my inquiry constantly before me, and wait till the first dawning opens gradually, by little and little, into a full and clear light." No better account can be given of the nature of the mental *effort* which gives to the philosopher the full benefit of his powers; but the natural *powers* of men's minds are not on that account the less different. There are many who might wait through ages of darkness without being visited by any dawn.

The habit to which Newton thus, in some sense, owed his discover-

---

[24] As in Galileo, Hooke, Huyghens, and others.

ies, this constant attention to the rising thought, and development of its results in every direction, necessarily engaged and absorbed his spirit, and made him inattentive and almost insensible to external impressions and common impulses. The stories which are told of his extreme absence of mind, probably refer to the two years during which he was composing his *Principia*, and thus following out a train of reasoning the most fertile, the most complex, and the most important, which any philosopher had ever had to deal with. The magnificent and striking questions which, during this period, he must have had daily rising before him; the perpetual succession of difficult problems of which the solution was neçessary to his great object; may well have entirely occupied and possessed him. "He existed only to calculate and to think."[25] Often, lost in meditation, he knew not what he did, and his mind appeared to have quite forgotten its connection with the body. His servant reported that, on rising in a morning, he frequently sat a large portion of the day, half-dressed, on the side of his bed; and that his meals waited on the table for hours before he came to take them. Even with his transcendent powers, to do what he did was almost irreconcilable with the common conditions of human life; and required the utmost devotion of thought, energy of effort, and steadiness of will—the strongest character, as well as the highest endowments, which belong to man.

Newton has been so universally considered as the greatest example of a natural philosopher, that his moral qualities, as well as his intellect, have been referred to as models of the philosophical character; and those who love to think that great talents are naturally associated with virtue, have always dwelt with pleasure upon the views given of Newton by his contemporaries; for they have uniformly represented him as candid and humble, mild and good. We may take as an example of the impressions prevalent about him in his own time, the expressions of Thomson, in the Poem on his Death.[26]

---

[25] Biot.

[26] In the same strain we find the general voice of the time. For instance, one of Loggan's "Views of Cambridge" is dedicated "Isaaco Newtono . . Mathematico, Physico, Chymico consummatissimo; nec minus suavitate morum et candore animi . . . spectabili."

In opposition to the general current of such testimony, we have the complaints of Flamsteed, who ascribes to Newton angry language and harsh conduct in the matter of the publication of the Greenwich Observations, and of Whiston. Yet even Flamsteed speaks well of his general disposition. Whiston was himself so weak and prejudiced that his testimony is worth very little.

Say ye who best can tell, ye happy few,
Who saw him in the softest lights of life,
All unwithheld, indulging to his friends
The vast unborrowed treasures of his mind,
Oh, speak the wondrous man! how mild, how calm,
How greatly humble, how divinely good,
How firm established on eternal truth!
Fervent in doing well, with every nerve
Still pressing on, forgetful of the past,
And panting for perfection; far above
Those little cares and visionary joys
That so perplex the fond impassioned heart
Of ever-cheated, ever-trusting man.

[2d Ed.] [In the first edition of the *Principia*, published in 1687, Newton showed that the nature of all the then known inequalities of the moon, and in some cases, their quantities, might be deduced from the principles which he laid down: but the determination of the amount and law of most of the inequalities was deferred to a more favorable opportunity, when he might be furnished with better astronomical observations. Such observations as he needed for this purpose had been made by Flamsteed, and for these he applied, representing how much value their use would add to the observations. "If," he says, in 1694, "you publish them without such a theory to recommend them, they will only be thrown into the heap of the observations of former astronomers, till somebody shall arise that by perfecting the theory of the moon shall discover your observations to be exacter than the rest; but when that shall be, God knows: I fear, not in your lifetime, if I should die before it is done. For I find this theory so very intricate, and the theory of gravity so necessary to it, that I am satisfied it will never be perfected but by somebody who understands the theory of gravity as well, or better than I do." He obtained from Flamsteed the lunar observations for which he applied, and by using these he framed the Theory of the Moon which is given as his in David Gregory's *Astronomiæ Elementa*.[27] He also obtained from Flamsteed the diameters of the planets as observed at various times, and the greatest elongation of Jupiter's Satellites, both of which, Flamsteed says, he made use of in his *Principia*.

Newton, in his letters to Flamsteed in 1694 and 5, acknowledges this service.[28]

---

[27] In the Preface to a *Treatise on Dynamics*, Part i., published in 1836, I have endeavored to show that Newton's modes of determining several of the lunar inequalities admitted of an accuracy not very inferior to the modern analytical methods.

[28] The quarrel on the subject of the publication of Flamsteed's Observations took

CHAPTER III.

Sequel to the Epoch of Newton.—Reception of the Newtonian Theory.

Sect. 1.—*General Remarks.*

THE doctrine of universal gravitation, like other great steps in science, required a certain time to make its way into men's minds; and had to be confirmed, illustrated, and completed, by the labors of succeeding philosophers. As the discovery itself was great beyond former example, the features of the natural sequel to the discovery were also on a gigantic scale; and many vast and laborious trains of research, each of which might, in itself, be considered as forming a wide science, and several of which have occupied many profound and zealous inquirers from that time to our own day, come before us as parts only of the verification of Newton's Theory. Almost every thing that has been done, and is doing, in astronomy, falls inevitably under this description; and it is only when the astronomer travels to the very limits of his vast field of labor, that he falls in with phenomena which do not acknowledge the jurisdiction of the Newtonian legislation. We must give some account of the events of this part of the history of astronomy; but our narrative must necessarily be extremely brief and imperfect; for the subject is most large and copious, and our limits are fixed and narrow. We have here to do with the history of discoveries, only so far as it illustrates their philosophy. And though the

---

place at a later period. Flamsteed wished to have his Observations printed complete and entire. Halley, who, under the authority of Newton and others, had the management of the printing, made many alterations and omissions, which Flamsteed considered as deforming and spoiling the work. The advantages of publishing a *complete* series of observations, now generally understood, were not then known to astronomers in general, though well known to Flamsteed, and earnestly insisted upon in his remonstrances. The result was that Flamsteed published his Observations at his own expense, and finally obtained permission to destroy the copies printed by Halley, which he did. In 1726, after Flamsteed's death, his widow applied to the Vice-Chancellor of Oxford, requesting that the volume printed by Halley might be removed out of the Bodleian Library, where it exists, as being "nothing more than an erroneous abridgment of Mr. Flamsteed's works," and unfit to see the light.

astronomical discoveries of the last century are by no means poor, even in interest of this kind, the generalizations which they involve are far less important for our object, in consequence of being included in a previous generalization. Newton shines out so brightly, that all who follow seem faint and dim. It is not precisely the case which the poet describes—

> As in a theatre the eyes of men,
> After some well-graced actor leaves the stage,
> Are idly bent on him that enters next,
> Thinking his prattle to be tedious :

but our eyes are at least less intently bent on the astronomers who succeeded, and we attend to their communications with less curiosity, because we know the end, if not the course of their story; we know that their speeches have all closed with Newton's sublime declaration, asserted in some new form.

Still, however, the account of the verification and extension of any great discovery is a highly important part of its history. In this instance it is most important; both from the weight and dignity of the theory concerned, and the ingenuity and extent of the methods employed: and, of course, so long as the Newtonian theory still required verification, the question of the truth or falsehood of such a grand system of doctrines could not but excite the most intense curiosity. In what I have said, I am very far from wishing to depreciate the value of the achievements of modern astronomers, but it is essential to my purpose to mark the subordination of narrower to wider truths—the different character and import of the labors of those who come before and after the promulgation of a master-truth. With this warning I now proceed to my narrative.

Sect. 2.—Reception of the Newtonian Theory in England.

There appears to be a popular persuasion that great discoveries are usually received with a prejudiced and contentious opposition, and the authors of them neglected or persecuted. The reverse of this was certainly the case in England with regard to the discoveries of Newton. As we have already seen, even before they were published, they were proclaimed by Halley to be something of transcendent value; and from the moment of their appearance, they rapidly made their way from one class of thinkers to another, nearly as fast as the nature of men's intellectual capacity allows. Halley, Wren, and all the leading members of the Royal Society, appear to have embraced the system immediately and zealously. Men whose pursuits had lain rather in literature than in science, and who had not the knowledge and habits of mind which the strict study of the system required, adopted, on the credit of their mathematical friends, the highest estimation of the *Principia*, and a strong regard for its author, as Evelyn, Locke, and Pepys. Only five years after the publication, the principles of the work were referred to from the pulpit, as so incontestably proved that they might be made the basis of a theological argument. This was done by Dr. Bentley, when he preached the Boyle's Lectures in London, in 1692. Newton himself, from the time when his work appeared, is never mentioned except in terms of profound admiration; as, for instance, when he is called by Dr. Bentley, in his sermon,[1] "That very excellent and divine theorist, Mr. Isaac Newton." It appears to have been soon suggested, that the Government ought to provide in some way for a person who was so great an honor to the nation. Some delay took place with regard to this; but, in 1695, his friend Mr. Montague, afterwards Earl of Halifax, at that time Chancellor of the Exchequer, made him Warden of the Mint; and in 1699, he succeeded to the higher office of Master of the Mint, a situation worth £1200 or £1500 a year, which he filled to the end of his life. In 1703, he became President of the Royal Society, and was annually re-elected to this office during the remaining twenty-five years of his life. In 1705, he was knighted in the Master's Lodge, at Trinity College, by Queen Anne, then on a visit to the University of Cambridge. After the accession of George the First, Newton's conversation was frequently sought by the Princess, afterwards Queen Caroline, who had a taste for speculative studies, and was often heard to declare in public, that she thought herself fortunate in living at a time which enabled her to enjoy the society of so great a genius. His fame, and the respect paid him, went on increasing to the end of his life; and when, in 1727, full of years and glory, his earthly career was ended, his death was mourned as a national calamity, with the forms usually confined to royalty. His body lay in state in the Jerusalem chamber; his pall was borne by the first nobles of the land; and his earthly remains were deposited in the centre of Westminster Abbey, in the midst of the memorials of the greatest and wisest men whom England has produced.

It cannot be superfluous to say a word or two on the reception of

---

[1] Serm. vii. 221.

his philosophy in the universities of England. These are often represented as places where bigotry and ignorance resist, as long as it is possible to resist, the invasion of new truths. We cannot doubt that such opinions have prevailed extensively, when we find an intelligent and generally temperate writer, like the late Professor Playfair of Edinburgh, so far possessed by them, as to be incapable of seeing, or interpreting, in any other way, any facts respecting Oxford and Cambridge. Yet, notwithstanding these opinions, it will be found that, in the English universities, new views, whether in science or in other subjects, have been introduced as soon as they were clearly established;—that they have been diffused from the few to the many more rapidly there than elsewhere occurs;—and that from these points, the light of newly-discovered truths has most usually spread over the land. In most instances undoubtedly there has been something of a struggle, on such occasions, between the old and the new opinions. Few men's minds can at once shake off a familiar and consistent system of doctrines, and adopt a novel and strange set of principles as soon as presented; but all can see that one change produces many, and that change, in itself, is a source of inconvenience and danger. In the case of the admission of the Newtonian opinions into Cambridge and Oxford, however, there are no traces even of a struggle. Cartesianism had never struck its roots deep in this country; that is, the peculiar hypotheses of Descartes. The Cartesian books, such, for instance, as that of Rohault, were indeed in use; and with good reason, for they contained by far the best treatises on most of the physical sciences, such as Mechanics, Hydrostatics, Optics, and Formal Astronomy, which could then be found. But I do not conceive that the Vortices were ever dwelt upon as a matter of importance in our academic teaching. At any rate, if they were brought among us, they were soon dissipated. Newton's College, and his University, exulted in his fame, and did their utmost to honor and aid him. He was exempted by the king from the obligation of taking orders, under which the fellows of Trinity College in general are; by his college he was relieved from all offices which might interfere, however slightly, with his studious employments, though he resided within the walls of the society thirty-five years, almost without the interruption of a month.[2] By the University he was elected their representative in parliament in 1688,

[2] His name is nowhere found on the college-books, as appointed to any of the offices which usually pass down the list of resident fellows in rotation. This might be owing in part, however, to his being Lucasian Professor. The constancy of his

and again in 1701; and though he was rejected in the dissolution of 1705, those who opposed him acknowledged him[3] to be "the glory of the University and nation," but considered the question as a political one, and Newton as sent "to tempt them from their duty, by the great and just veneration they had for him." Instruments and other memorials, valued because they belonged to him, are still preserved in his college, along with the tradition of the chambers which he occupied.

The most active and powerful minds at Cambridge became at once disciples and followers of Newton. Samuel Clarke, afterwards his friend, defended in the public schools a thesis taken from his philosophy, as early as 1694; and in 1697 published an edition of Rohault's *Physics*, with notes, in which Newton is frequently referred to with expressions of profound respect, though the leading doctrines of the *Principia* are not introduced till a later edition, in 1703. In 1699, Bentley, whom we have already mentioned as a Newtonian, became Master of Trinity College; and in the same year, Whiston, another of Newton's disciples, was appointed his deputy as professor of mathematics. Whiston delivered the Newtonian doctrines, both from the professor's chair, and in works written for the use of the University; yet it is remarkable that a taunt respecting the late introduction of the Newtonian system into the Cambridge course of education, has been founded on some peevish expressions which he uses in his Memoirs, written at a period when, having incurred expulsion from his professorship and the University, he was naturally querulous and jaundiced in his views. In 1709–10, Dr. Laughton, who was tutor in Clare Hall, procured himself to be appointed moderator of the University disputations, in order to promote the diffusion of the new mathematical doctrines. By this time the first edition of the *Principia* was become rare, and fetched a great price. Bentley urged Newton to publish a new one; and Cotes, by far the first, at that time, of the mathematicians of Cambridge, undertook to superintend the printing, and the edition was accordingly published in 1713.

[2d Ed.] [I perceive that my accomplished German translator, Littrow, has incautiously copied the insinuations of some modern writers to the effect that Clarke's reference to Newton, in his Edition of Rohault's *Physics*, was a mode of introducing Newtonian doctrines covertly, when it was not allowed him to introduce such novelties

---

residence in college appears from the *exit* and *redit* book of that time, which is still preserved.

[3] A pamphlet by Styan Thurlby.

openly. I am quite sure that any one who looks into this matter will see that this supposition of any unwillingness at Cambridge to receive Newton's doctrine is quite absurd, and can prove nothing but the intense prejudices of those who maintain such an opinion. Newton received and held his professorship amid the unexampled admiration of all contemporary members of the University. Whiston, who is sometimes brought as an evidence against Cambridge on this point, says, "I with immense pains set myself with the utmost zeal to the study of Sir Isaac Newton's wonderful discoveries in his *Philosophiæ Naturalis Principia Mathematica*, one or two of which *lectures I had heard him read in the public schools*, though I understood them not at the time." As to Rohault's *Physics*, it really did contain the best mechanical philosophy of the time;—the doctrines which were held by Descartes in common with Galileo, and with all the sound mathematicians who succeeded them. Nor does it look like any great antipathy to novelty in the University of Cambridge, that this book, which was quite as novel in its doctrines as Newton's *Principia*, and which had only been published at Paris in 1671, had obtained a firm hold on the University in less than twenty years. Nor is there any attempt made in Clarke's notes to conceal the novelty of Newton's discoveries, but on the contrary, admiration is claimed for them as new.

The promptitude with which the Mathematicians of the University of Cambridge adopted the best parts of the mechanical philosophy of Descartes, and the greater philosophy of Newton, in the seventeenth century, has been paralleled in our own times, in the promptitude with which they have adopted and followed into their consequences the Mathematical Theory of Heat of Fourier and Laplace, and the Undulatory Theory of Light of Young and Fresnel.

In Newton's College, we possess, besides the memorials of him mentioned above (which include two locks of his silver-white hair), a paper in his own handwriting, describing the preparatory reading which was necessary in order that our College students might be able to read the *Principia*. I have printed this paper in the Preface to my Edition of the First Three Sections of the *Principia* in the original Latin (1846).

Bentley, who had expressed his admiration for Newton in his Boyle's Lectures in 1692, was made Master of the College in 1699, as I have stated; and partly, no doubt, in consequence of the Newtonian sermons which he had preached. In his administration of the College, he zealously stimulated and assisted the exertions of Cotes, Whiston, and other disciples of Newton. Smith, Bentley's successor as Master of the College, erected a statue of Newton in the College Chapel (a noble work of Roubiliac), with the inscription, *Qui genus humanum ingenio superavit.*]

At Oxford, David Gregory and Halley, both zealous and distinguished disciples of Newton, obtained the Savilian professorships of astronomy and geometry in 1691 and 1703.

David Gregory's *Astronomiæ Physicæ et Geometricæ Elementa* issued from the Oxford Press in 1702. The author, in the first sentence of the Preface, states his object to be to explain the mechanics of the universe (Physica Cœlestis), which Isaac Newton, the Prince of Geometers, has carried to a point of elevation which all look up to with admiration. And this design is executed by a full exposition of the Newtonian doctrines and their results. Keill, a pupil of Gregory, followed his tutor to Oxford, and taught the Newtonian philosophy there in 1700, being then Deputy Sedleian Professor. He illustrated his lectures by experiments, and published an Introduction to the *Principia* which is not out of use even yet.

In Scotland, the Newtonian philosophy was accepted with great alacrity, as appears by the instances of David Gregory and Keill. David Gregory was professor at Edinburgh before he removed to Oxford, and was succeeded there by his brother James. The latter had, as early as 1690, printed a thesis, containing in twenty-two propositions, a compend of Newton's *Principia.*[4] Probably these were intended as theses for academical disputations; as Laughton at Cambridge introduced the Newtonian philosophy into these exercises. The formula at Cambridge, in use till very recently in these disputations, was "*Rectè statuit Newtonus de Motu Lunæ;*" or the like.

The general diffusion of these opinions in England took place, not only by means of books, but through the labors of various experimental lecturers, like Desaguliers, who removed from Oxford to London in 1713; when he informs us,[5] that "he found the Newtonian philosophy generally received among persons of all ranks and professions, and even among the ladies by the help of experiments."

---

[4] See Hutton's *Math. Dict.*, art. *James Gregory.* If it fell in with my plan to notice derivative works, I might speak of Maclaurin's admirable *Account of Sir Isaac Newton's Discoveries*, published in 1748. This is still one of the best books on the subject. The late Professor Rigaud's *Historical Essay on the First Publication of Sir Isaac Newton's "Principia"* (Oxf. 1838) contains a careful and candid view of the circumstances of that event.

[5] Desag. *Pref.*

We might easily trace in our literature indications of the gradual progress of the Newtonian doctrines. For instance, in the earlier editions of Pope's *Dunciad*, this couplet occurred, in the description of the effects of the reign of Dulness:

Philosophy, that reached the heavens before,
Shrinks to her hidden cause, and is no more.

"And this," says his editor, Warburton, "was intended as a censure on the Newtonian philosophy. For the poet had been misled by the prejudices of foreigners, as if that philosophy had recurred to the occult qualities of Aristotle. This was the idea he received of it from a man educated much abroad, who had read every thing, but every thing superficially.[6] When I hinted to him how he had been imposed upon, he changed the lines with great pleasure into a compliment (as they now stand) on that divine genius, and a satire on that very folly by which he himself had been misled." In 1743 it was printed,

Philosophy, that leaned on heaven before,
Shrinks to her second cause, and is no more.

The Newtonians repelled the charge of dealing in occult causes;[7] and, referring gravity to the will of the Deity, as the First Cause, assumed a superiority over those whose philosophy rested in second causes.

To the cordial reception of the Newtonian theory by the English astronomers, there is only one conspicuous exception; which is, however, one of some note, being no other than Flamsteed, the Astronomer Royal, a most laborious and exact observer. Flamsteed at first listened with complacency to the promises of improvements in the Lunar Tables, which the new doctrines held forth, and was willing to assist Newton, and to receive assistance from him. But after a time, he lost his respect for Newton's theory, and ceased to take any interest in it. He then declared to one of his correspondents,[8] "I have determined to lay these crotchets of Sir Isaac Newton's wholly aside." We need not, however, find any difficulty in this, if we recollect that Flamsteed, though a good observer, was no philosopher;—never understood by a Theory any thing more than a Formula which should predict results;—and was incapable of comprehending the object of Newton's theory, which was to assign causes as well as rules, and to satisfy the conditions of Mechanics as well as of Geometry.

---

6 I presume Bolingbroke is here meant. 7 See Cotes's Pref. to the *Principia*.

8 Baily's *Account of Flamsteed, &c.*, p. 309.

[2d Ed.] [I do not see any reason to retract what was thus said; but it ought perhaps to be distinctly said that on these very accounts Flamsteed's rejection of Newton's rules did not imply a denial of the doctrine of gravitation. In the letter above quoted, Flamsteed says that he has been employed upon the Moon, and that "the heavens reject that equation of Sir I. Newton which Gregory and Newton called his sixth: I had then [when he wrote before] compared but 72 of my observations with the tables, now I have examined above 100 more. I find them all firm in the same, and the seventh [equation] too." And thereupon he comes to the determination above stated.

At an earlier period Flamsteed, as I have said, had received Newton's suggestions with great deference, and had regulated his own observations and theories with reference to them. The calculation of the lunar inequalities upon the theory of gravitation was found by Newton and his successors to be a more difficult and laborious task than he had anticipated, and was not performed without several trials and errors. One of the equations was at first published (in Gregory's *Astronomiæ Elementa*) with a wrong sign. And when Newton had done all, Flamsteed found that the rules were far from coming up to the degree of accuracy which had been claimed for them, that they could give the moon's place true to 2 or 3 minutes. It was not till considerably later that this amount of exactness was attained.

The late Mr. Baily, to whom astronomy and astronomical literature are so deeply indebted, in his *Supplement to the Account of Flamsteed*, has examined with great care and great candor the assertion that Flamsteed did not understand Newton's Theory. He remarks, very justly, that what Newton himself at first presented as his Theory, might more properly be called *Rules* for computing lunar tables, than a physical *Theory* in the modorn acceptation of the term. He shows, too, that Flamsteed had read the *Principia* with attention.[9] Nor do I doubt that many considerable mathematicians gave the same imperfect assent to Newton's doctrine which Flamsteed did. But when we find that others, as Halley, David Gregory, and Cotes, at once not only saw in the doctrine a source of true formulæ, but also a magnificent physical discovery, we are obliged, I think, to make Flamsteed, in this respect, an exception to the first class of astronomers of his own time.

Mr. Baily's suggestion that the annual equations for the corrections of the lunar apogee and node were collected from Flamsteed's tables

[9] *Supp*. p. 691.

and observations independently of their suggestion by Newton as the results of Theory (*Supp.* p. 692, Note, and p. 698), appears to me not to be adequately supported by the evidence given.]

Sect. 3.—Reception of the Newtonian Theory abroad.

The reception of the Newtonian theory on the Continent, was much more tardy and unwilling than in its native island. Even those whose mathematical attainments most fitted them to appreciate its proofs, were prevented by some peculiarity of view from adopting it as a system; as Leibnitz, Bernoulli, Huyghens; who all clung to one modification or other of the system of vortices. In France, the Cartesian system had obtained a wide and popular reception, having been recommended by Fontenelle with the graces of his style; and its empire was so firm and well established in that country, that it resisted for a long time the pressure of Newtonian arguments. Indeed, the Newtonian opinions had scarcely any disciples in France, till Voltaire asserted their claims, on his return from England in 1728: until then, as he himself says, there were not twenty Newtonians out of England.

The hold which the Philosophy of Descartes had upon the minds of his countrymen is, perhaps, not surprising. He really had the merit, a great one in the history of science, of having completely overturned the Aristotelian system, and introduced the philosophy of matter and motion. In all branches of mixed mathematics, as we have already said, his followers were the best guides who had yet appeared. His hypothesis of vortices, as an explanation of the celestial motions, had an apparent advantage over the Newtonian doctrine, in this respect;—that it referred effects to the most intelligible, or at least most familiar kinds of mechanical causation, namely, pressure and impulse. And above all, the system was acceptable to most minds, in consequence of being, as was pretended, deduced from a few simple principles by necessary consequences; and of being also directly connected with metaphysical and theological speculations. We may add, that it was modified by its mathematical adherents in such a way as to remove most of the objections to it. A vortex revolving about a centre could be constructed, or at least it was supposed that it could be constructed, so as to produce a tendency of bodies to the centre. In all cases, therefore, where a central force acted, a vortex was supposed; but in reasoning to the results of this hypothesis, it was easy to leave out of sight all other effects of the vortex, and to consider only the central force; and when this was done, the Cartesian mathematician could apply to his problems a mechanical principle of some degree of consistency. This reflection will, in some degree, account for what at first seems so strange;—the fact that the language of the French mathematicians is Cartesian, for almost half a century after the publication of the *Principia* of Newton.

There was, however, a controversy between the two opinions going on all this time, and every day showed the insurmountable difficulties under which the Cartesians labored. Newton, in the *Principia*, had inserted a series of propositions, the object of which was to prove, that the machinery of vortices could not be accommodated to one part of the celestial phenomena, without contradicting another part. A more obvious difficulty was the case of gravity of the earth; if this force arose, as Descartes asserted, from the rotation of the earth's vortex about its axis, it ought to tend directly to the axis, and not to the centre. The asserters of vortices often tried their skill in remedying this vice in the hypothesis, but never with much success. Huyghens supposed the ethereal matter of the vortices to revolve about the centre in all directions; Perrault made the strata of the vortex increase in velocity of rotation as they recede from the centre; Saurin maintained that the circumambient resistance which comprises the vortex will produce a pressure passing through the centre. The elliptic form of the orbits of the planets was another difficulty. Descartes had supposed the vortices themselves to be oval; but others, as John Bernoulli, contrived ways of having elliptical motion in a circular vortex.

The mathematical prize-questions proposed by the French Academy, naturally brought the two sets of opinions into conflict. The Cartesian memoir of John Bernoulli, to which we have just referred, was the one which gained the prize in 1730. It not unfrequently happened that the Academy, as if desirous to show its impartiality, divided the prize between the Cartesians and Newtonians. Thus in 1734, the question being, the cause of the inclination of the orbits of the planets, the prize was shared between John Bernoulli, whose Memoir was founded on the system of vortices, and his son Daniel, who was a Newtonian. The last act of homage of this kind to the Cartesian system was performed in 1740, when the prize on the question of the Tides was distributed between Daniel Bernoulli, Euler, Maclaurin, and Cavallieri; the last of whom had tried to patch up and amend the Cartesian hypothesis on this subject.

Thus the Newtonian system was not adopted in France till the Cartesian generation had died off; Fontenelle, who was secretary to the Academy of Sciences, and who lived till 1756, died a Cartesian. There were exceptions; for instance, Delisle, an astronomer who was selected by Peter the Great of Russia, to found the Academy of St. Petersburg; who visited England in 1724, and to whom Newton then gave his picture, and Halley his Tables. But in general, during the interval, that country and this had a national difference of creed on physical subjects. Voltaire, who visited England in 1727, notices this difference in his lively manner. "A Frenchman who arrives in London, finds a great alteration in philosophy, as in other things. He left the world full [*a plenum*], he finds it empty. At Paris you see the universe composed of vortices of subtle matter, in London we see nothing of the kind. With you it is the pressure of the moon which causes the tides of the sea, in England it is the sea which gravitates towards the moon; so that when you think the moon ought to give us high water, these gentlemen believe that you ought to have low water; which unfortunately we cannot test by experience; for in order to do that, we should have examined the Moon and the Tides at the moment of the creation. You will observe also that the sun, which in France has nothing to do with the business, here comes in for a quarter of it. Among you Cartesians, all is done by an impulsion which one does not well understand; with the Newtonians, it is done by an attraction of which we know the cause no better. At Paris you fancy the earth shaped like a melon, at London it is flattened on the two sides."

It was Voltaire himself, as we have said, who was mainly instrumental in giving the Newtonian doctrines currency in France. He was at first refused permission to print his *Elements of the Newtonian Philosophy*, by the Chancellor, D'Aguesseaux, who was a Cartesian; but after the appearance of this work in 1738, and of other writings by him on the same subject, the Cartesian edifice, already without real support or consistency, crumbled to pieces and disappeared. The first Memoir in the *Transactions of the French Academy* in which the doctrine of central force is applied to the solar system, is one by the Chevalier de Louville in 1720, *On the Construction and Theory of Tables of the Sun*. In this, however, the mode of explaining the motions of the planets by means of an original impulse and an attractive force is attributed to Kepler, not to Newton. The first Memoir which refers to the universal gravitation of matter is by Maupertuis, in

1736. But Newton was not unknown or despised in France till this time. In 1699 he was admitted one of the very small number of foreign associates of the French Academy of Sciences. Even Fontenelle, who, as we have said, never adopted his opinions, spoke of him in a worthy manner, in the *Eloge* which he composed on the occasion of his death. At a much earlier period too, Fontenelle did homage to his fame. The following passage refers, I presume, to Newton. In the *History* of the Academy for 1708, which is written by the secretary, he says,[10] in referring to the difficulty which the comets occasion in the Cartesian hypothesis: "We might relieve ourselves at once from all the embarrassment which arises from the directions of these motions, by suppressing, as has been done *by one of the greatest geniuses of the age*, all this immense fluid matter, which we commonly suppose between the planets, and conceiving them suspended in a perfect void."

Comets, as the above passage implies, were a kind of artillery which the Cartesian *plenum* could not resist. When it appeared that the paths of such wanderers traversed the vortices in all directions, it was impossible to maintain that these imaginary currents governed the movements of bodies immersed in them; and the mechanism ceased to have any real efficacy. Both these phenomena of comets, and many others, became objects of a stronger and more general interest, in consequence of the controversy between the rival parties; and thus the prevalence of the Cartesian system did not seriously impede the progress of sound knowledge. In some cases, no doubt, it made men unwilling to receive the truth, as in the instance of the deviation of the comets from the zodiacal motion; and again, when Römer discovered that light was not instantaneously propagated. But it encouraged observation and calculation, and thus forwarded the verification and extension of the Newtonian system; of which process we must now consider some of the incidents.

---

[10] *Hist. Ac. Sc.* 1708. p. 103.

CHAPTER IV.

SEQUEL TO THE EPOCH OF NEWTON, CONTINUED.—VERIFICATION AND COMPLETION OF THE NEWTONIAN THEORY.

Sect. 1.—*Division of the Subject.*

THE verification of the Law of Universal Gravitation as the governing principle of all cosmical phenomena, led, as we have already stated, to a number of different lines of research, all long and difficult. Of these we may treat successively, the motions of the Moon, of the Sun, of the Planets, of the Satellites, of Comets; we may also consider separately the Secular Inequalities, which at first sight appear to follow a different law from the other changes; we may then speak of the results of the principle as they affect this Earth, in its Figure, in the amount of Gravity at different places, and in the phenomena of the Tides. Each of these subjects has lent its aid to confirm the general law: but in each the confirmation has had its peculiar difficulties, and has its separate history. Our sketch of this history must be very rapid, for our aim is only to show what is the kind and course of the confirmation which such a theory demands and receives.

For the same reason we pass over many events of this period which are highly important in the history of astronomy. They have lost much of their interest for us, and even for common readers, because they are of a class with which we are already familiar, truths included in more general truths to which our eyes now most readily turn. Thus, the discovery of new satellites and planets is but a repetition of what was done by Galileo: the determination of their nodes and apses, the reduction of their motions to the law of the ellipse, is but a fresh exemplification of the discoveries of Kepler. Otherwise, the formation of Tables of the satellites of Jupiter and Saturn, the discovery of the eccentricities of the orbits, and of the motions of the nodes and apses, by Cassini, Halley, and others, would rank with the great achievements in astronomy. Newton's peculiar advance in the *Tables* of the celestial motions is the introduction of Perturbations. To these motions, so affected, we now proceed.

Sect. 2.—*Application of the Newtonian Theory to the Moon.*

The Motions of the Moon may be first spoken of, as the most obvious and the most important of the applications of the Newtonian Theory. The verification of such a theory consists, as we have seen in previous cases, in the construction of Tables derived from the theory, and the comparison of these with observation. The advancement of astronomy would alone have been a sufficient motive for this labor; but there were other reasons which urged it on with a stronger impulse. A perfect Lunar Theory, if the theory could be perfected, promised to supply a method of finding the Longitude of any place on the earth's surface; and thus the verification of a theory which professed to be complete in its foundations, was identified with an object of immediate practical use to navigators and geographers, and of vast acknowledged value. A good method for the near discovery of the longitude had been estimated by nations and princes at large sums of money. The Dutch were willing to tempt Galileo to this task by the offer of a chain of gold: Philip the Third of Spain had promised a reward for this object still earlier;[1] the parliament of England, in 1714, proposed a recompense of 20,000*l.* sterling; the Regent Duke of Orléans, two years afterwards, offered 100,000 francs for the same purpose. These prizes, added to the love of truth and of fame, kept this object constantly before the eyes of mathematicians, during the first half of the last century.

If the Tables could be so constructed as to represent the moon's real place in the heavens with extreme precision, as it would be seen from a *standard* observatory, the observation of her apparent place, as seen from any other point of the earth's surface, would enable the observer to find his longitude from the standard point. The motions of the moon had hitherto so ill agreed with the best Tables, that this method failed altogether. Newton had discovered the ground of this want of agreement. He had shown that the same force which produces the Evection, Variation, and Annual Equation, must produce also a long series of other Inequalities, of various magnitudes and cycles, which perpetually drag the moon before or behind the place where she would be sought by an astronomer who knew only of those principal and notorious inequalities. But to calculate and apply the new inequalities, was no slight undertaking.

---

[1] Del. *A. M.* i. 39, 66.

In the first edition of the *Principia* in 1687, Newton had not given any calculations of new inequalities affecting the longitude of the moon. But in David Gregory's *Elements of Physical and Geometrical Astronomy*, published in 1702, is inserted[2] "Newton's Lunar Theory as applied by him to Practice;" in which the great discoverer has given the results of his calculations of eight of the lunar Equations, their quantities, epochs, and periods. These calculations were for a long period the basis of new Tables of the Moon, which were published by various persons;[3] as by Delisle in 1715 or 1716, Grammatici at Ingoldstadt in 1726, Wright in 1732, Angelo Capelli at Venice in 1733, Dunthorne at Cambridge in 1739.

Flamsteed had given Tables of the Moon upon Horrox's theory in 1681, and wished to improve them; and though, as we have seen, he would not, or could not, accept Newton's doctrines in their whole extent, Newton communicated his theory to the observer in the shape in which he could understand it and use it:[4] and Flamsteed employed these directions in constructing new Lunar Tables, which he called his *Theory*.[5] These Tables were not published till long after his death, by Le Monnier at Paris in 1746. They are said, by Lalande,[6] not to differ much from Halley's. Halley's Tables of the Moon were printed in 1719 or 1720, but not published till after his death in 1749. They had been founded on Flamsteed's observations and his own; and when, in 1720, Halley succeeded Flamsteed in the post of Astronomer Royal at Greenwich, and conceived that he had the means of much improving what he had done before, he began by printing what he had already executed.[7]

But Halley had long proposed a method, different from that of Newton, but marked by great ingenuity, for amending the Lunar Tables. He proposed to do this by the use of a cycle, which we have mentioned as one of the earliest discoveries in astronomy;—the Period of 223 lunations, or eighteen years and eleven days, the Chaldean

---

[2] P. 332. [3] Lalande, 1457. [4] Baily. *Account of Flamsteed*, p. 72.
[5] P. 211. [6] Lal. 1459.

[7] Mr. Baily* says that Mayer's *Nouvelles Tables de la Lune* in 1453, published upwards of fifty years after Gregory's *Astronomy*, may be considered as the first lunar tables formed *solely* on Newton's principles. Though Wright in 1732 published *New and Correct Tables of the Lunar Motions according to the Newtonian Theory*, Newton's rules were in them only partially adopted. In 1735 Leadbetter published his *Uranoscopia*, in which those rules were more fully followed. But these *Newtonian Tables* did not supersede Flamsteed's Horroxian Tables, till both were supplanted by those of Mayer.

* *Supp.* p. 702.

Saros. This period was anciently used for predicting the eclipses of the sun and moon; for those eclipses which happen during this period, are repeated again in the same order, and with nearly the same circumstances, after the expiration of one such period and the commencement of a second. The reason of this is, that at the end of such a cycle, the moon is in nearly the same position with respect to the sun, her nodes, and her apogee, as she was at first; and is only a few degrees distant from the same part of the heavens. But on the strength of this consideration, Halley conjectured that all the irregularities of the moon's motion, however complex they may be, would recur after such an interval; and that, therefore, if the requisite corrections were determined by observation for one such period, we might by means of them give accuracy to the Tables for all succeeding periods. This idea occurred to him before he was acquainted with Newton's views.[8] After the lunar theory of the *Principia* had appeared, he could not help seeing that the idea was confirmed; for the inequalities of the moon's motion, which arise from the attraction of the sun, will depend on her positions with regard to the sun, the apogee, and the node; and therefore, however numerous, will recur when these positions recur.

Halley announced, in 1691,[9] his intention of following this idea into practice; in a paper in which he corrected the text of three passages in Pliny, in which this period is mentioned, and from which it is sometimes called the Plinian period. In 1710, in the preface to a new edition to Street's *Caroline Tables*, he stated that he had already confirmed it to a considerable extent.[10] And even after Newton's theory had been applied, he still resolved to use his cycle as a means of obtaining further accuracy. On succeeding to the Observatory at Greenwich in 1720, he was further delayed by finding that the instruments had belonged to Flamsteed, and were removed by his executors. "And this," he says,[11] "was the more grievous to me, on account of my advanced age, being then in my sixty-fourth year; which put me past all hopes of ever living to see a complete period of eighteen years' observation. But, thanks to God, he has been pleased hitherto (in 1731) to afford me sufficient health and strength to execute my office, in all its parts, with my own hands and eyes, without any assistance or interruption, during one whole period of the moon's

---

[8] *Phil Trans.* 1731, p. 188.

[9] Ib. p. 536.

[10] Ib. 1731, p. 187.

[11] Ib. p. 193.

apogee, which period is performed in somewhat less than nine years." He found the agreement very remarkable, and conceived hopes of attaining the great object, of finding the Longitude with the requisite degree of exactness; nor did he give up his labors on this subject till he had completed his Plinian period in 1739.

The accuracy with which Halley conceived himself able to predict the moon's place[12] was within two minutes of space, or one fifteenth of the breadth of the moon herself. The accuracy required for obtaining the national reward was considerably greater. Le Monnier pursued the idea of Halley.[13] But before Halley's method had been completed, it was superseded by the more direct prosecution of Newton's views.

We have already remarked, in the history of analytical mechanics, that in the Lunar Theory, considered as one of the cases of the Problem of Three Bodies, no advance was made beyond what Newton had done, till mathematicians threw aside the Newtonian artifices, and applied the newly developed generalizations of the analytical method. The first great apparent deficiency in the agreement of the law of universal gravitation with astronomical observation, was removed by Clairaut's improved approximation to the theoretical Motion of the Moon's Apogee, in 1750; yet not till it had caused so much disquietude, that Clairaut himself had suggested a modification of the law of attraction; and it was only in tracing the consequences of this suggestion, that he found the Newtonian law of the inverse square to be that which, when rightly developed, agreed with the facts. Euler solved the problem by the aid of his analysis in 1745,[14] and published Tables of the Moon in 1746. His tables were not very accurate at first;[15] but he, D'Alembert, and Clairaut, continued to labor at this object, and the two latter published Tables of the Moon in 1754.[16] Finally, Tobias Mayer, an astronomer of Göttingen, having compared Euler's tables with observations, corrected them so successfully, that in 1753 he published Tables of the Moon, which really did possess the accuracy which Halley only flattered himself that he had attained. Mayer's success in his first Tables encouraged him to make them still more perfect. He applied himself to the mechanical theory of the moon's orbit; corrected all the coefficients of the series by a great number of observations; and in 1755, sent his new Tables to London as worthy to claim the prize offered for the discovery of longitude. He died soon after

---

[12] *Phil. Trans.* 1731, p. 195. [13] Bailly, *A. M.* c. 131.
[14] Lal. 1460. [15] Bradley's Correspondence. [16] Lal. 1460.

(in 1762), at the early age of thirty-nine, worn out by his incessant labors; and his widow sent to London a copy of his Tables with additional corrections. These Tables were committed to Bradley, then Astronomer Royal, in order to be compared with observation. Bradley labored at this task with unremitting zeal and industry, having himself long entertained hopes that the Lunar Method of finding the Longitude might be brought into general use. He and his assistant, Gael Morris, introduced corrections into Mayer's Tables of 1755. In his report of 1756, he says,[17] that he did not find any difference so great as a minute and a quarter; and in 1760, he adds, that this deviation had been further diminished by his corrections. It is not foreign to our purpose to observe the great labor which this verification required. Not less than 1220 observations, and long calculations founded upon each, were employed. The accuracy which Mayer's Tables possessed was considered to entitle them to a part of the parliamentary reward; they were printed in 1770, and his widow received 3000*l.* from the English nation. At the same time, Euler, whose Tables had been the origin and foundation of Mayer's, also had a recompense of the same amount.

This public national acknowledgment of the practical accuracy of these Tables is, it will be observed, also a solemn recognition of the truth of the Newtonian theory, as far as truth can be judged of by men acting under the highest official responsibility, and aided by the most complete command of the resources of the skill and talents of others. The finding the Longitude is thus the seal of the moon's gravitation to the sun and earth; and with this occurrence, therefore, our main concern with the history of the Lunar Theory ends. Various improvements have been since introduced into this research; but on these we, with so many other subjects before us, must forbear to enter.

Sect. 3.—Application of the Newtonian Theory to the Planets, Satellites, and Earth.

THE theories of the Planets and Satellites, as affected by the law of universal gravitation, and therefore by perturbations, were naturally subjects of interest, after the promulgation of that law. Some of the effects of the mutual attraction of the planets had, indeed, already attracted notice. The inequality produced by the mutual attraction of Jupiter and Saturn cannot be overlooked by a good observer. In the

[17] Bradley's *Mem.* p. xcviii.

preface to the second edition of the *Principia*, Cotes remarks,[18] that the perturbation of Jupiter and Saturn is not unknown to astronomers. In Halley's Tables it was noticed[19] that there are very great deviations from regularity in these two planets, and these deviations are ascribed to the perturbing force of the planets on each other; but the correction of these by a suitable equation is left to succeeding astronomers.

The motion of the planes and apsides of the planetary orbits was one of the first results of their mutual perturbation which was observed. In 1706, La Hire and Maraldi compared Jupiter with the Rudolphine Tables, and those of Bullialdus: it appeared that his aphelion had advanced, and that his nodes had regressed. In 1728, J. Cassini found that Saturn's aphelion had in like manner travelled forwards. In 1720, when Louville refused to allow in his solar tables the motion of the aphelion of the earth, Fontenelle observed that this was a misplaced scrupulousness, since the aphelion of Mercury certainly advances. Yet this reluctance to admit change and irregularity was not yet overcome. When astronomers had found an approximate and apparent constancy and regularity, they were willing to believe it absolute and exact. In the satellites of Jupiter, for instance, they were unwilling to admit even the eccentricity of the orbits; and still more, the variation of the nodes, inclinations, and apsides. But all the fixedness of these was successively disproved. Fontenelle in 1732, on the occasion of Maraldi's discovery of the change of inclination of the fourth satellite, expresses a suspicion that all the elements might prove liable to change. "We see," says he, "the constancy of the inclination already shaken in the three first satellites, and the eccentricity in the fourth. The immobility of the nodes holds out so far, but there are strong indications that it will share the same fate."

The motions of the nodes and apsides of the satellites are a necessary part of the Newtonian theory; and even the Cartesian astronomers now required only data, in order to introduce these changes into their Tables.

The complete reformation of the Tables of the Sun, Planets, and Satellites, which followed as a natural consequence from the revolution which Newton had introduced, was rendered possible by the labors of the great constellation of mathematicians of whom we have spoken in the last book, Clairaut, Euler, D'Alembert, and their successors; and

---

[18] Preface to *Principia*, p. xxi.

[19] End of Planetary Tables.

it was carried into effect in the course of the last century. Thus Lalande applied Clairaut's theory to Mars, as did Mayer; and the inequalities in this case, says Bailly[20] in 1785, may amount to two minutes, and therefore must not be neglected. Lalande determined the inequalities of Venus, as did Father Walmesley, an English mathematician; these were found to reach only to thirty seconds.

The Planetary Tables[21] which were in highest repute, up to the end of the last century, were those of Lalande. In these, the perturbations of Jupiter and Saturn were introduced, their magnitude being such that they cannot be dispensed with; but the Tables of Mercury, Venus, and Mars, had no perturbations. Hence these latter Tables might be considered as accurate enough to enable the observer to find the object, but not to test the theory of perturbations. But when the calculation of the mutual disturbances of the planets was applied, it was always found that it enabled mathematicians to bring the theoretical places to coincide more exactly with those observed. In improving, as much as possible, this coincidence, it is necessary to determine the mass of each planet; for upon that, according to the law of universal gravitation, its disturbing power depends. Thus, in 1813, Lindenau published Tables of Mercury, and concluded, from them, that a considerable increase of the supposed mass of Venus was necessary to reconcile theory with observation.[22] He had published Tables of Venus in 1810, and of Mars in 1811. And, in proving Bouvard's Tables of Jupiter and Saturn, values were obtained of the masses of those planets. The form in which the question of the truth of the doctrine of universal gravitation now offers itself to the minds of astronomers, is this:—that it is taken for granted that it will account for the motions of the heavenly bodies, and the question is, with what supposed masses it will give the *best* account.[23] The continually increasing accuracy of the table shows the truth of the fundamental assumption.

The question of perturbation is exemplified in the satellites also.

---

[20] *Ast. Mod.* iii. 170.

[21] Airy, *Report on Ast. to Brit. Ass.* 1832.

[22] Airy, *Report on Ast. to Brit. Ast.* 1832.

[23] Among the most important corrections of the supposed masses of the planets, we may notice that of Jupiter, by Professor Airy. This determination of Jupiter's mass was founded, not on the effect as seen in perturbations, but on a much more direct datum, the time of revolution of his fourth satellite. It appeared, from this calculation, that Jupiter's mass required to be increased by about 1-80th. This result agrees with that which has been derived by German astronomers from the perturbations which the attractions of Jupiter produce in the four new planets, and has been generally adopted as an improvement of the elements of our system.

Thus the satellites of Jupiter are not only disturbed by the sun, as the moon is, but also by each other, as the planets are. This mutual action gives rise to some very curious relations among their motions; which, like most of the other leading inequalities, were forced upon the notice of astronomers by observation before they were obtained by mathematical calculation. In Bradley's remarks upon his own Tables of Jupiter's Satellites, published among Halley's Tables, he observes that the places of the three interior satellites are affected by errors which recur in a cycle of 437 days, answering to the time in which they return to the same relative position with regard to each other, and to the axis of Jupiter's shadow. Wargentin, who had noticed the same circumstance without knowledge of what Bradley had done, applied it, with all diligence, to the purpose of improving the tables of the satellites in 1746. But, at a later period, Laplace established, by mathematical reasoning, the very curious theorem on which this cycle depends, which he calls *the libration of Jupiter's satellites;* and Delambre was then able to publish Tables of Jupiter's Satellites more accurate than those of Wargentin, which he did in 1789.[24]

The progress of physical astronomy from the time of Euler and Clairaut, has consisted of a series of calculations and comparisons of the most abstruse and recondite kind. The formation of Tables of the Planets and Satellites from the theory, required the solution of problems much more complex than the original case of the Problem of Three Bodies. The real motions of the planets and their orbits are rendered still further intricate by this, that all the lines and points to which we can refer them, are themselves in motion. The task of carrying order and law into this mass of apparent confusion, has required a long series of men of transcendent intellectual powers; and a perseverance and delicacy of observation, such as we have not the smallest example of in any other subject. It is impossible here to give any detailed account of these labors; but we may mention one instance of the complex considerations which enter into them. The nodes of Jupiter's fourth satellite do not go backwards,[25] as the Newtonian theory seems to require; they advance upon Jupiter's orbit. But then, it is to be recollected that the theory requires the nodes to retrograde upon the orbit of the perturbing body, which is here the third satellite; and Lalande showed that, by the necessary relations of space, the latter motion may be retrograde though the former is direct.

---

[24] Voiron, *Hist. Ast.* p. 322.

[25] Bailly, iii. 175.

Attempts have been made, from the time of the solution of the Problem of three bodies to the present, to give the greatest possible accuracy to the Tables of the Sun, by considering the effect of the various perturbations to which the earth is subject. Thus, in 1756, Euler calculated the effect of the attractions of the planets on the earth (the prize-question of the French Academy of Sciences), and Clairaut soon after. Lacaille, making use of these results, and of his own numerous observations, published Tables of the Sun. In 1786, Delambre[26] undertook to verify and improve these tables, by comparing them with 314 observations made by Maskelyne, at Greenwich, in 1775 and 1784, and in some of the intermediate years. He corrected most of the elements; but he could not remove the uncertainty which occurred respecting the amount of the inequality produced by the reaction of the moon. He admitted also, in pursuance of Clairaut's theory, a second term of this inequality depending on the moon's latitude; but irresolutely, and half disposed to reject it on the authority of the observations. Succeeding researches of mathematicians have shown, that this term is not admissible as a result of mechanical principles. Delambre's Tables, thus improved, were exact to seven or eight seconds;[27] which was thought, and truly, a very close coincidence for the time. But astronomers were far from resting content with this. In 1806, the French Board of Longitude published Delambre's improved Solar Tables; and in the *Connaissance des Tems* for 1816, Burckhardt gave the results of a comparison of Delambre's Tables with a great number of Maskelyne's observations;—far greater than the number on which they were founded.[28] It appeared that the epoch, the perigee, and the eccentricity, required sensible alterations, and that the mass of Venus ought to be reduced about one-ninth, and that of the Moon to be sensibly diminished. In 1827, Professor Airy[29] compared Delambre's tables with 2000 Greenwich observations, made with the new transit-instrument at Cambridge, and deduced from this comparison the correction of the elements. These in general agreed closely with Burckhardt's, excepting that a diminution of Mars appeared necessary. Some discordances, however, led Professor Airy to suspect the existence of an inequality which had escaped the sagacity of Laplace and Burckhardt. And, a few weeks after this suspicion had been expressed, the same mathematician announced to the Royal Society that he had de-

[26] Voiron, *Hist.* p. 315.

[27] Montucla, iv. 42.

[28] Airy, *Report*, p. 150.

[29] *Phil. Trans.* 1828.

tected, in the planetary theory such an inequality, hitherto unnoticed, arising from the mutual attraction of Venus and the Earth. Its whole effect on the earth's longitude, would be to increase or diminish it by nearly three seconds of space, and its period is about 240 years. "This term," he adds, "accounts completely for the difference of the secular motions given by the comparison of the epochs of 1783 and 1821, and by that of the epochs of 1801 and 1821."

Many excellent Tables of the motions of the sun, moon, and planets, were published in the latter part of the last century; but the Bureau des Longitudes which was established in France in 1795, endeavored to give new or improved tables of most of these motions. Thus were produced Delambre's Tables of the Sun, Burg's Tables of the Moon, Bouvard's Tables of Jupiter, Saturn, and Uranus. The agreement between these and observation is, in general, truly marvellous.

We may notice here a difference in the mode of referring to observation when a theory is first established, and when it is afterwards to be confirmed and corrected. It was remarked as a merit in the method of Hipparchus, and an evidence of the mathematical coherence of his theory, that in order to determine the place of the sun's apogee, and the eccentricity of his orbit, he required to know nothing besides the lengths of winter and spring. But if the fewness of the requisite data is a beauty in the first fixation of a theory, the multitude of observations to which it applies is its excellence when it is established; and in correcting Tables, mathematicians take far more data than would be requisite to determine the elements. For the theory ought to account for *all* the facts: and since it will not do this with mathematical rigor (for observation is not perfect), the elements are determined, not so as to satisfy any selected observations, but so as to make the whole mass of error as small as possible. And thus, in the adaptation of theory to observation, even in its most advanced state, there is room for sagacity and skill, prudence and judgment.

In this manner, by selecting the best mean elements of the motions of the heavenly bodies, the observed motions deviate from this mean in the way the theory points out, and constantly return to it. To this general rule, of the constant return to a mean, there are, however, some apparent exceptions, of which we shall now speak.

Sect. 4.—Application of the Newtonian Theory to Secular Inequalities.

SECULAR Inequalities in the motions of the heavenly bodies occur in consequence of changes in the elements of the solar system, which go on progressively from age to age. The example of such changes which was first studied by astronomers, was the Acceleration of the Moon's Mean Motion, discovered by Halley. The observed fact was, that the moon now moves in a very small degree quicker than she did in the earlier ages of the world. When this was ascertained, the various hypotheses which appeared likely to account for the fact were reduced to calculation. The resistance of the medium in which the heavenly bodies move was the most obvious of these hypotheses. Another, which was for some time dwelt upon by Laplace, was the successive transmission of gravity, that is, the hypothesis that the gravity of the earth takes a certain finite time to reach the moon. But none of these suppositions gave satisfactory conclusions; and the strength of Euler, D'Alembert, Lagrange, and Laplace, was for a time foiled by this difficulty. At length, in 1787, Laplace announced to the Academy that he had discovered the true cause of this acceleration, and that it arose from the action of the sun upon the moon, combined with the secular variation of the eccentricity of the earth's orbit. It was found that the effects of this combination would exactly account for the changes which had hitherto so perplexed mathematicians. A very remarkable result of this investigation was, that "this Secular Inequality of the motion of the moon is periodical, but it requires millions of years to re-establish itself;" so that after an almost inconceivable time, the acceleration will become a retardation. Laplace some time after (in 1797), announced other discoveries relative to the secular motions of the apogee and the nodes of the moon's orbit. Laplace collected these researches in his "Theory of the Moon," which he published in the third volume of the *Mécanique Céleste* in 1802.

A similar case occurred with regard to an acceleration of Jupiter's mean motion, and a retardation of Saturn's, which had been observed by Cassini, Maraldi, and Horrox. After several imperfect attempts by other mathematicians, Laplace, in 1787, found that there resulted from the mutual attraction of these two planets a great Inequality, of which the period is 929 years and a half, and which has accelerated Jupiter and retarded Saturn ever since the restoration of astronomy.

Thus the secular inequalities of the celestial motions, like all the others, confirm the law of universal gravitation. They are called "secular," because ages are requisite to unfold their existence, and because they are not obviously periodical. They might, in some measure, be considered as extensions of the Newtonian theory, for though Newton's law accounts for such facts, he did not, so far as we know, foresee such a result of it. But on the other hand, they are exactly of the same nature as those which he did foresee and calculate. And when we call them *secular*, in opposition to *periodical*, it is not that there is any real difference, for they, too, have their cycle; but it is that we have assumed our *mean* motion without allowing for these long inequalities. And thus, as Laplace observes on this very occasion,[30] the lot of this great discovery of gravitation is no less than this, that every apparent exception becomes a proof, every difficulty a new occasion of a triumph. And such, as he truly adds, is the character of a true theory,—of a real representation of nature.

It is impossible for us here to enumerate even the principal objects which have thus filled the triumphal march of the Newtonian theory from its outset up to the present time. But among these secular changes, we may mention the Diminution of the Obliquity of the Ecliptic, which has been going on from the earliest times to the present. This change has been explained by theory, and shown to have, like all the other changes of the system, a limit, after which the diminution will be converted into an increase.

We may mention here some subjects of a kind somewhat different from those just spoken of. The true theoretical quantity of the Precession of the Equinoxes, which had been erroneously calculated by Newton, was shown by D'Alembert to agree with observation. The constant coincidence of the Nodes of the Moon's Equator with those of her Orbit, was proved to result from mechanical principles by Lagrange. The curious circumstance that the Time of the Moon's rotation on her axis is equal to the Time of her revolution about the earth, was shown to be consistent with the results of the laws of motion by Laplace. Laplace also, as we have seen, explained certain remarkable relations which constantly connect the longitudes of the three first satellites of Jupiter; Bailly and Lagrange analyzed and explained the curious librations of the nodes and inclinations of their orbits; and Laplace traced the effect of Jupiter's oblate figure on their motions,

---

[30] *Syst. du Monde*, 8vo, ii. 87.

which masks the other causes of inequality, by determining the direction of the motions of the *perijove* and node of each satellite.

Sect. 5.—Application of the Newtonian Theory to the New Planets.

We are now so accustomed to consider the Newtonian theory as true, that we can hardly imagine to ourselves the possibility that those planets which were not discovered when the theory was founded, should contradict its doctrines. We can scarcely conceive it possible that Uranus or Ceres should have been found to violate Kepler's laws, or to move without suffering perturbations from Jupiter and Saturn. Yet if we can suppose men to have had any doubt of the exact and universal truth of the doctrine of universal gravitation, at the period of these discoveries, they must have scrutinized the motions of these new bodies with an interest far more lively than that with which we now look for the predicted return of a comet. The solid establishment of the Newtonian theory is thus shown by the manner in which we take it for granted not only in our reasonings, but in our feelings. But though this is so, a short notice of the process by which the new planets were brought within the domain of the theory may properly find a place here.

William Herschel, a man of great energy and ingenuity, who had made material improvements in reflecting telescopes, observing at Bath on the 13th of March, 1781, discovered, in the constellation Gemini, a star larger and less luminous than the fixed stars. On the application of a more powerful telescope, it was seen magnified, and two days afterwards he perceived that it had changed its place. The attention of the astronomical world was directed to this new object, and the best astronomers in every part of Europe employed themselves in following it along the sky.[31]

The admission of an eighth planet into the long-established list, was a notion so foreign to men's thoughts at that time, that other suppositions were first tried. The orbit of the new body was at first calculated as if it had been a comet running in a parabolic path. But in a few days the star deviated from the course thus assigned it: and it was in vain that in order to represent the observations, the perihelion distance of the parabola was increased from fourteen to eighteen times the earth's distance from the sun. Saron, of the Academy of Sciences of Paris, is said[32] to have been the first person who perceived that the

[31] Voiron, *Hist. Ast.* p. 12. [32] Ibid.

places were better represented by a circle than by a parabola: and Lexell, a celebrated mathematician of Petersburg, found that a motion in a circular orbit, with a radius double of that of Saturn, would satisfy all the observations. This made its period about eighty-two years.

Lalande soon discovered that the circular motion was subject to a sensible inequality: the orbit was, in fact, an ellipse, like those of the other planets. To determine the equation of the centre of a body which revolves so slowly, would, according to the ancient methods, have required many years; but Laplace contrived methods by which the elliptical elements were determined from four observations, within little more than a year from its first discovery by Herschel. These calculations were soon followed by tables of the new planet, published by Nouet.

In order to obtain additional accuracy, it now became necessary to take account of the perturbations. The French Academy of Sciences proposed, in 1789, the construction of new Tables of this Planet as its prize-question. It is a curious illustration of the constantly accumulating evidence of the theory, that the calculation of the perturbations of the planet enabled astronomers to discover that it had been observed as a star in three different positions in former times; namely, by Flamsteed in 1690, by Mayer in 1756, and by Le Monnier in 1769. Delambre, aided by this discovery and by the theory of Laplace, calculated Tables of the planet, which, being compared with observation for three years, never deviated from it more than seven seconds. The Academy awarded its prize to these Tables, they were adopted by the astronomers of Europe, and the planet of Herschel now conforms to the laws of attraction, along with those ancient members of the known system from which the theory was inferred.

The history of the discovery of the other new planets, Ceres, Pallas, Juno, and Vesta, is nearly similar to that just related, except that their planetary character was more readily believed. The first of these was discovered on the first day of this century by Piazzi, the astronomer at Palermo; but he had only begun to suspect its nature, and had not completed his third observation, when his labors were suspended by a dangerous illness; and on his recovery the star was invisible, being lost in the rays of the sun.

He declared it to be a planet with an elliptical orbit; but the path which it followed, on emerging from the neighborhood of the sun, was not that which Piazzi had traced out for it. Its extreme smallness made it difficult to rediscover; and the whole of the year 1801 was employed in searching the sky for it in vain. At last, after many trials, Von Zach and Olbers again found it, the one on the last day of 1801, the other on the first day of 1802. Gauss and Burckhardt immediately used the new observations in determining the elements of the orbit; and the former invented a new method for the purpose. Ceres now moves in a path of which the course and inequalities are known, and can no more escape the scrutiny of astronomers.

The second year of the nineteenth century also produced its planet. This was discovered by Dr. Olbers, a physician of Bremen, while he was searching for Ceres among the stars of the constellation Virgo. He found a star which had a perceptible motion even in the space of two hours. It was soon announced as a new planet, and received from its discoverer the name of Pallas. As in the case of Ceres, Burckhardt and Gauss employed themselves in calculating its orbit. But some peculiar difficulties here occurred. Its eccentricity is greater than that of any of the old planets, and the inclination of its orbit to the ecliptic is not less than thirty-five degrees. These circumstances both made its perturbations large, and rendered them difficult to calculate. Burckhardt employed the known processes of analysis, but they were found insufficient: and the Imperial Institute (as the French Academy was termed during the reign of Napoleon) proposed the Perturbations of Pallas as a prize-question.

To these discoveries succeeded others of the same kind. The German astronomers agreed to examine the whole of the zone in which Ceres and Pallas move; in the hope of finding other planets, fragments, as Olbers conceived they might possibly be, of one original mass. In the course of this research, Mr. Harding of Lilienthal, on the first of September, 1804, found a new star, which he soon was led to consider as a planet. Gauss and Burckhardt also calculated the elements of this orbit, and the planet was named Juno.

After this discovery, Olbers sought the sky for additional fragments of his planet with extraordinary perseverance. He conceived that one of two opposite constellations, the Virgin or the Whale, was the place where its separation must have taken place; and where, therefore, all the orbits of all the portions must pass. He resolved to survey, three times a year, all the small stars in these two regions. This undertaking, so curious in its nature, was successful. The 29th of March, 1807, he discovered Vesta, which was soon found to be a planet. And to show the manner in which Olbers pursued his labors, we may state that he afterwards published a notification that he had examined the same parts of the heavens with such regularity, that he was certain no new planet had passed that way between 1808 and 1816. Gauss and Burckhardt computed the orbit of Vesta; and when Gauss compared one of his orbits with twenty-two observations of M. Bouvard, he found the errors below seventeen seconds of space in right ascension, and still less in declination.

The elements of all these orbits have been successively improved, and this has been done entirely by the German mathematicians.[33] These perturbations are calculated, and the places for some time before and after opposition are now given in the Berlin Ephemeris. "I have lately observed," says Professor Airy, "and compared with the Berlin Ephemeris, the right ascensions of Juno and Vesta, and I find that they are rather more accurate than those of Venus:" so complete is the confirmation of the theory by these new bodies; so exact are the methods of tracing the theory to its consequences.

We may observe that all these new-discovered bodies have received names taken from the ancient mythology. In the case of the first of these, astronomers were originally divided; the discoverer himself named it the *Georgium Sidus*, in honor of his patron, George the Third; Lalande and others called it *Herschel.* Nothing can be more just than this mode of perpetuating the fame of the author of a discovery; but it was felt to be ungraceful to violate the homogeneity of the ancient system of names. Astronomers tried to find for the hitherto neglected denizen of the skies, an appropriate place among the deities to whose assembly he was at last admitted; and *Uranus*, the father of Saturn, was fixed upon as best suiting the order of the course.

The mythological nomenclature of planets appeared, from this time, to be generally agreed to. Piazzi termed his *Ceres Ferdinandea.* The first term, which contains a happy allusion to Sicily, the country of the discovery in modern, and of the goddess in ancient, times, has been accepted; the attempt to pay a compliment to royalty out of the products of science, in this as in most other cases, has been set aside. Pallas, Juno, and Vesta, were named, without any peculiar propriety of selection, according to the choice of their discoverers.

Sect. 6.—*Application of the Newtonian Theory to Comets.*

A FEW words must be said upon another class of bodies, which at first seemed as lawless as the clouds and winds; and which astronomy

---

[33] Airy, *Rep.* 157.

has reduced to a regularity as complete as that of the sun;—upon *Comets.* No part of the Newtonian discoveries excited a more intense interest than this. These anomalous visitants were anciently gazed at with wonder and alarm; and might still, as in former times, be accused of "perplexing nations," though with very different fears and questionings. The conjecture that they, too, obeyed the law of universal gravitation, was to be verified by showing that they described a curve such as that force would produce. Hevelius, who was a most diligent observer of these objects, had, without reference to gravitation, satisfied himself that they moved in parabolas.[34] To determine the elements of the parabola from observations, even Newton called[35] "problema longe difficillimum." Newton determined the orbit of the comet of 1680 by certain graphical methods. His methods supposed the orbit to be a parabola, and satisfactorily represented the motion in the visible part of the comet's path. But this method did not apply to the possible return of the wandering star. Halley has the glory of having first detected a periodical comet, in the case of that which has since borne his name. But this great discovery was not made without labor. In 1705, Halley[36] explained how the parabolic orbit of a planet may be determined from three observations; and, joining example to precept, himself calculated the positions and orbits of twenty-four comets. He found, as the reward of this industry, that the comets of 1607 and of 1531 had the same orbit as that of 1682. And here the intervals are also nearly the same, namely, about seventy-five years. Are the three comets then identical? In looking back into the history of such appearances, he found comets recorded in 1456, in 1380, and in 1305; the intervals are still the same, seventy-five or seventy-six years. It was impossible now to doubt that they were the periods of a revolving body; that the comet was a planet; its orbit a long ellipse, not a parabola.[37]

But if this were so, the Comet must reappear in 1758 or 1759. Halley predicted that it would do so; and the fulfilment of this prediction was naturally looked forwards to, as an additional stamp of the truths of the theory of gravitation.

---

[34] Bailly, ii. 246. [35] *Principia*, ed. 1. p. 494. [36] Bailly, ii. 646.

[37] The importance of Halley's labors on Comets has always been acknowledged. In speaking of Halley's *Synopsis Astronomicæ Cometicæ*, Delambre says (*Ast.* xviii. *Siècle*, p. 130), "Voilà bien, depuis Kepler, ce qu'on a fait de plus grand, de plus beau, de plus neuf en astronomie." Halley, in predicting the comet of 1758, says, if it returns, "Hoc primum ab homine Anglo inventum fuisse non inficiabitur æqua posteritas."

But in all this, the Comet had been supposed to be affected only by the attraction of the sun. The planets must disturb its motion as they disturb each other. How would this disturbance affect the time and circumstances of its reappearance? Halley had proposed, but not attempted to solve, this question.

The effect of perturbations upon a comet defeats all known methods of approximation, and requires immense labor. "Clairaut," says Bailly,[38] "undertook this: with courage enough to dare the adventure, he had talent enough to obtain a memorable victory;" the difficulties, the labors, grew upon him as he advanced, but he fought his way through them, assisted by Lalande, and by a female calculator, Madame Lepaute. He predicted that the comet would reach its perihelion April 13, 1759, but claimed the license of a month for the inevitable inaccuracy of a calculation which, in addition to all other sources of error, was made in haste, that it might appear as a prediction. The comet justified his calculations and his caution together; for it arrived at its perihelion on the 13th of March.

Two other Comets, of much shorter period, have been detected of late years; Encke's, which revolves round the sun in three years and one-third, and Biela's, which describes an ellipse, not extremely eccentric, in six years and three-quarters. These bodies, apparently thin and vaporous masses, like other comets, have, since their orbits were calculated, punctually conformed to the law of gravitation. If it were still doubtful whether the more conspicuous comets do so, these bodies would tend to prove the fact, by showing it to be true in an intermediate case.

[2d Ed.] [A third Comet of short period was discovered by Faye, at the Observatory of Paris, Nov. 22, 1843. It is included between the orbits of Mars and Saturn, and its period is seven years and three-tenths.

This is commonly called *Faye's Comet*, as the two mentioned in the text are called *Encke's* and *Biela's*. In the former edition I had expressed my assent to the rule proposed by M. Arago, that the latter ought to be called *Gambart's Comet*, in honor of the astronomer who first proved it to revolve round the Sun. But astronomers in general have used the former name, considering that the discovery and observation of the object are more distinct and conspicuous merits than a calculation founded upon the observations of others. And in reality,

---

[38] Bailly, *A. M.* iii. 190.

Biela had great merit in the discovery of his Comet's periodicity, having set about his search of it from an anticipation of its return founded upon former observations.

Also a Comet was discovered by De Vico at Rome on Aug. 22, 1844, which was found to describe an elliptical orbit having its aphelion near the orbit of Jupiter, which is consequently one of those of short period. And on Feb. 26, 1846, M. Brorsen of Kiel discovered a telescopic Comet whose orbit is found to be elliptical.]

We may add to the history of Comets, that of Lexell's, which, in 1770, appeared to be revolving in a period of about five years, and whose motion was predicted accordingly. The prediction was disappointed; but the failure was sufficiently explained by the comet's having passed close to Jupiter, by which occurrence its orbit was utterly deranged.

It results from the theory of universal gravitation, that Comets are collections of extremely attenuated matter. Lexell's is supposed to have passed twice (in 1767 and 1779) through the system of Jupiter's Satellites, without disturbing their motions, though suffering itself so great a disturbance as to have its orbit entirely altered. The same result is still more decidedly proved by the last appearance of Biela's Comet. It appeared double, but the two bodies did not perceptibly affect each other's motions, as I am informed by Professor Challis of Cambridge, who observed both of them from Jan. 23 to Mar. 25, 1846. This proves the quantity of matter in each body to have been exceedingly small.

Thus, no verification of the Newtonian theory, which was possible in the motions of the stars, has yet been wanting. The return of Halley's Comet again in 1835, and the extreme exactitude with which it conformed to its predicted course, is a testimony of truth, which must appear striking even to the most incurious respecting such matters.[39]

Sect. 7.—Application of the Newtonian Theory to the Figure of the Earth.

The Heavens had thus been consulted respecting the Newtonian doctrine, and the answer given, over and over again, in a thousand

[39] M. de Humboldt (*Kosmos*, p. 116) speaks of *nine* returns of Halley's Comet, the comet observed in China in 1378 being identified with this. But whether we take 1378 or 1380 for the appearance in that century, if we begin with that, we have only *seven* appearances, namely, in 1378 or 1380, in 1456, in 1531, in 1607, in 1682, in 1759, and in 1835.

different forms, had been, that it was true; nor had the most persevering cross-examination been able to establish any thing of contradiction or prevarication. The same question was also to be put to the Earth and the Ocean, and we must briefly notice the result.

According to the Newtonian principles, the form of the earth must be a globe somewhat flattened at the poles. This conclusion, or at least the amount of the flattening, depends not only upon the existence and law of attraction, but upon its belonging to each particle of the mass separately; and thus the experimental confirmation of the form asserted from calculation, would be a verification of the theory in its widest sense. The application of such a test was the more necessary to the interests of science, inasmuch as the French astronomers had collected from their measures, and had connected with their Cartesian system, the opinion that the earth was not *oblate* but *oblong*. Dominic Cassini had measured seven degrees of latitude from Amiens to Perpignan, in 1701, and found them to decrease in going from south to north. The prolongation of this measure to Dunkirk confirmed the same result. But if the Newtonian doctrine was true, the contrary ought to be the case, and the degrees ought to increase in proceeding towards the pole.

The only answer which the Newtonians could at this time make to the difficulty thus presented, was, that an arc so short as that thus measured, was not to be depended upon for the determination of such a question; inasmuch as the inevitable errors of observation might exceed the differences which were the object of research. It would, undoubtedly, have become the English to have given a more complete answer, by executing measurements under circumstances not liable to this uncertainty. The glory of doing this, however, they for a long time abandoned to other nations. The French undertook the task with great spirit.[40] In 1733, in one of the meetings of the French Academy, when this question was discussed, De la Condamine, an ardent and eager man, proposed to settle this question by sending members of the Academy to measure a degree of the meridian near the equator, in order to compare it with the French degrees, and offered himself for the expedition. Maupertuis, in like manner, urged the necessity of another expedition to measure a degree in the neighborhood of the pole. The government received the applications favorably, and these remarkable scientific missions were sent out at the national expense.

---

[40] Bailly, iii. 11.

As soon as the result of these measurements was known, there was no longer any doubt as to the fact of the earth's oblateness, and the question only turned upon its quantity. Even before the return of the academicians, the Cassinis and Lacaille had measured the French arc, and found errors which subverted the former result, making the earth oblate to the amount of 1-168th of its diameter. The expeditions to Peru and to Lapland had to struggle with difficulties in the execution of their design, which make their narratives resemble some romantic history of irregular warfare, rather than the monotonous records of mere measurements. The equatorial degree employed the observers not less than eight years. When they did return, and the results were compared, their discrepancy, as to quantity, was considerable. The comparison of the Peruvian and French arcs gave an ellipticity of nearly 1-314th, that of the Peruvian and Swedish arcs gave 1-213th for its value.

Newton had deduced from his theory, by reasonings of singular ingenuity, an ellipticity of 1-230th; but this result had been obtained by supposing the earth homogeneous. If the earth be, as we should most readily conjecture it to be, more dense in its interior than at its exterior, its ellipticity will be less than that of a homogeneous spheroid revolving in the same time. It does not appear that Newton was aware of this; but Clairaut, in 1743, in his *Figure of the Earth*, proved this and many other important results of the attraction of the particles. Especially he established that, in proportion as the fraction expressing the Ellipticity becomes smaller, that expressing the Excess of the polar over the equatorial gravity becomes larger; and he thus connected the measures of the ellipticity obtained by means of Degrees, with those obtained by means of Pendulums in different latitudes.

The altered rate of a Pendulum when carried towards the equator, had been long ago observed by Richer and Halley, and had been quoted by Newton as confirmatory of his theory. Pendulums were swung by the academicians who measured the degrees, and confirmed the general character of the results.

But having reached this point of the verification of the Newtonian theory, any additional step becomes more difficult. Many excellent measures, both of Degrees and of Pendulums, have been made since those just mentioned. The results of the Arcs[41] is an Ellipticity of 1-298th; —of the Pendulums, an Ellipticity of about 1-285th. This difference

---

[41] Airy, *Fig. Earth*, p. 230.

is considerable, if compared with the quantities themselves; but does not throw a shadow of doubt on the truth of the theory. Indeed, the observations of each kind exhibit irregularities which we may easily account for, by ascribing them to the unknown distribution of the denser portions of the earth; but which preclude the extreme of accuracy and certainty in our result.

But the near agreement of the determination, from Degrees and from Pendulums, is not the only coincidence by which the doctrine is confirmed. We can trace the effect of the earth's Oblateness in certain minute apparent motions of the stars; for the attraction of the sun and moon on the protuberant matter of the spheroid produces the Precession of the equinoxes, and a Nutation of the earth's axis. The Precession had been known from the time of Hipparchus, and the existence of Nutation was foreseen by Newton; but the quantity is so small, that it required consummate skill and great labor in Bradley to detect it by astronomical observation. Being, however, so detected, its amount, as well as that of the Precession, gives us the means of determining the amount of Terrestrial Ellipticity, by which the effect is produced. But it is found, upon calculation, that we cannot obtain this determination without assuming some law of density in the homogeneous strata of which we suppose the earth to consist.[42] The density will certainly increase in proceeding towards the centre, and there is a simple and probable law of this increase, which will give 1-300th for the Ellipticity, from the amount of two lunar Inequalities (one in latitude and one in longitude), which are produced by the earth's oblateness. Nearly the same result follows from the quantity of Nutation. Thus every thing tends to convince us that the ellipticity cannot deviate much from this fraction.

[2d Ed.] [I ought not to omit another class of phenomena in which the effects of the Earth's Oblateness, acting according to the law of universal gravitation, have manifested themselves;—I speak of the Moon's Motion, as affected by the Earth's Ellipticity. In this case, as in most others, observation anticipated theory. Mason had inferred from lunar observations a certain Inequality in Longitude, depending upon the distance of the Moon's Node from the Equinox. Doubts were entertained by astronomers whether this inequality really existed; but Laplace showed that such an inequality would arise from the oblate form of the earth; and that its magnitude might serve to de-

[42] Airy, *Fig. Earth*, p. 235.

termine the amount of the oblateness. Laplace showed, at the same time, that along with this Inequality in Longitude there must be an Inequality in Latitude; and this assertion Burg confirmed by the discussion of observations. The two Inequalities, as shown in the observations, agree in assigning to the earth's form an Ellipticity of 1-305th.]

Sect. 8.—*Confirmation of the Newtonian Theory by Experiments on Attraction.*

The attraction of all the parts of the earth to one another was thus proved by experiments, in which the whole mass of the earth is concerned. But attempts have also been made to measure the attraction of smaller portions; as mountains, or artificial masses. This is an experiment of great difficulty; for the attraction of such masses must be compared with that of the earth, of which it is a scarcely perceptible fraction; and, moreover, in the case of mountains, the effect of the mountain will be modified or disguised by unknown or unappreciable circumstances. In many of the measurements of degrees, indications of the attraction of mountains had been perceived; but at the suggestion of Maskelyne, the experiment was carefully made, in 1774, upon the mountain Schehallien, in Scotland, the mountain being mineralogically surveyed by Playfair. The result obtained was, that the attraction of the mountain drew the plumb-line about six seconds from the vertical; and it was deduced from this, by Hutton's calculations, that the density of the earth was about once and four-fifths that of Schehallien, or four and a half times that of water.

Cavendish, who had suggested many of the artifices in this calculation, himself made the experiment in the other form, by using leaden balls, about nine inches diameter. This observation was conducted with an extreme degree of ingenuity and delicacy, which could alone make it valuable; and the result agreed very nearly with that of the Schehallien experiment, giving for the density of the earth about five and one-third times that of water. Nearly the same result was obtained by Carlini, in 1824, from observations of the pendulum, made at a point of the Alps (the Hospice, on Mount Cenis) at a considerable elevation above the average surface of the earth.

Sect. 9.—Application of the Newtonian Theory to the Tides.

We come, finally, to that result, in which most remains to be done for the verification of the general law of attraction—the subject of the Tides. Yet, even here, the verification is striking, as far as observations have been carried. Newton's theory explained, with singular felicity, all the prominent circumstances of the tides then known;—the difference of spring and neap tides; the effect of the moon's and sun's declination and parallax; even the difference of morning and evening tides, and the anomalous tides of particular places. About, and after, this time, attempts were made both by the Royal Society of England, and by the French Academy, to collect numerous observations; but these were not followed up with sufficient perseverance. Perhaps, indeed, the theory had not been at that time sufficiently developed; but the admirable prize-essays of Euler, Bernoulli, and D'Alembert, in 1740, removed, in a great measure, this deficiency. These dissertations supplied the means of bringing this subject to the same test to which all the other consequences of gravitation had been subjected;—namely, the calculation of tables, and the continued and orderly comparison of these with observation. Laplace has attempted this verification in another way, by calculating the results of the theory (which he has done with an extraordinary command of analysis), and then by comparing these, in supposed critical cases, with the Brest observations. This method has confirmed the theory as far as it could do so; but such a process cannot supersede the necessity of applying the proper criterion of truth in such cases, the construction and verification of Tables. Bernoulli's theory, on the other hand, has been used for the construction of Tide-tables; but these have not been properly compared with experiment; and when the comparison has been made, having been executed for purposes of gain rather than of science, it has not been published, and cannot be quoted as a verification of the theory.

Thus we have, as yet, no sufficient comparison of fact with theory, for Laplace's is far from a complete comparison. In this, as in other parts of physical astronomy, our theory ought not only to agree with observations selected and grouped in a particular manner, but with the whole course of observation, and with every part of the phenomena. In this, as in other cases, the true theory should be verified by its giving us the best Tables; but Tide-tables were never, I believe, calculated upon Laplace's theory, and thus it was never fairly brought to the test.

It is, perhaps, remarkable, considering all the experience which astronomy had furnished, that men should have expected to reach the completion of this branch of science by improving the mathematical theory, without, at the same time, ascertaining the laws of the facts. In all other departments of astronomy, as, for instance, in the cases of the moon and the planets, the leading features of the phenomena had been made out empirically, before the theory explained them. The course which analogy would have recommended for the cultivation of our knowledge of the tides, would have been, to ascertain, by an analysis of long series of observations, the effect of changes in the time of transit, parallax, and declination of the moon, and thus to obtain the laws of phenomena; and then proceed to investigate the laws of causation.

Though this was not the course followed by mathematical theorists, it was really pursued by those who practically calculated Tide-tables; and the application of knowledge to the useful purposes of life being thus separated from the promotion of the theory, was naturally treated as a gainful property, and preserved by secrecy. Art, in this instance, having cast off her legitimate subordination to Science, or rather, being deprived of the guidance which it was the duty of Science to afford, resumed her ancient practices of exclusiveness and mystery. Liverpool, London, and other places, had their Tide-tables, constructed by undivulged methods, which methods, in some instances at least, were handed down from father to son for several generations as a family possession; and the publication of new Tables, accompanied by a statement of the mode of calculation, was resented as an infringement of the rights of property.

The mode in which these secret methods were invented, was that which we have pointed out;—the analysis of a considerable series of observations. Probably the best example of this was afforded by the Liverpool Tide-tables. These were deduced by a clergyman named Holden, from observations made at that port by a harbor-master of the name of Hutchinson; who was led, by a love of such pursuits, to observe the tides carefully for above twenty years, day and night. Holden's Tables, founded on four years of these observations, were remarkably accurate.

At length men of science began to perceive that such calculations were part of their business; and that they were called upon, as the guardians of the established theory of the universe, to compare it in the greatest possible detail with the facts. Mr. Lubbock was the first mathematician who undertook the extensive labors which such a conviction suggested. Finding that regular tide-observations had been made at the London Docks from 1795, he took nineteen years of these (purposely selecting the length of a cycle of the motions of the lunar orbit), and caused them (in 1831) to be analyzed by Mr. Dessiou, an expert calculator. He thus obtained[43] Tables for the effect of the Moon's Declination, Parallax, and hour of Transit, on the tides; and was enabled to produce Tide-tables founded upon the data thus obtained. Some mistakes in these as first published (mistakes unimportant as to the theoretical value of the work), served to show the jealousy of the practical tide-table calculators, by the acrimony with which the oversights were dwelt upon; but in a very few years, the tables thus produced by an open and scientific process were more exact than those which resulted from any of the secrets; and thus practice was brought into its proper subordination to theory.

The theory with which Mr. Lubbock was led to compare his results, was the Equilibrium-theory of Daniel Bernoulli; and it was found that this theory, with certain modifications of its elements, represented the facts to a remarkable degree of precision. Mr. Lubbock pointed out this agreement especially in the semi-mensual inequality of the times of high water. The like agreement was afterwards (in 1833) shown by Mr. Whewell[44] to obtain still more accurately at Liverpool, both for the Times and Heights; for by this time, nineteen years of Hutchinson's Liverpool Observations had also been discussed by Mr. Lubbock. The other inequalities of the Times and Heights (depending upon the Declination and Parallax of the Moon and Sun,) were variously compared with the Equilibrium-theory by Mr. Lubbock and Mr. Whewell; and the general result was, that the facts agreed with the condition of equilibrium at a certain anterior time, but that this anterior time was different for different phenomena. In like manner it appeared to follow from these researches, that in order to explain the facts, the mass of the moon must be supposed different in the calculation at different places. A result in effect the same was obtained by M. Daussy,[45] an active French Hydrographer; for he found that observations at various stations could not be reconciled with the formulæ of Laplace's *Mécanique*

---

[43] *Phil. Trans.* 1831. *British Almanac*, 1832. [44] *Phil. Trans.* 1834.

[45] *Connaissance des Tems*, 1838.

Céleste (in which the ratio of the heights of spring-tides and neap-tides was computed on an assumed mass of the moon) without an alteration of level which was, in fact, equivalent to an alteration of the moon's mass. Thus all things appeared to tend to show that the Equilibrium-theory would give the *formulæ* for the inequalities of the tides, but that the *magnitudes* which enter into these formulæ must be sought from observation.

Whether this result is consistent with theory, is a question not so much of Physical Astronomy as of Hydrodynamics, and has not yet been solved. A Theory of the Tides which should include in its conditions the phenomena of Derivative Tides, and of their combinations, will probably require all the resources of the mathematical mechanician.

As a contribution of empirical materials to the treatment of this hydrodynamical problem, it may be allowable to mention here Mr. Whewell's attempts to trace the progress of the tide into all the seas of the globe, by drawing on maps of the ocean what he calls *Cotidal Lines*;—lines marking the contemporaneous position of the various points of the great wave which carries high water from shore to shore.[46] This is necessarily a task of labor and difficulty, since it requires us to know the time of high water on the same day in every part of the world; but in proportion as it is completed, it supplies steps between our general view of the movements of the ocean and the phenomena of particular ports.

Looking at this subject by the light which the example of the history of astronomy affords, we may venture to repeat, that it will never have justice done it till it is treated as other parts of astronomy are treated; that is, till Tables of all the phenomena which can be observed, are calculated by means of the best knowledge which we at present possess, and till these tables are constantly improved by a comparison of the predicted with the observed fact. A set of Tide-observations and Tide-ephemerides of this kind, would soon give to this subject that precision which marks the other parts of astronomy; and would leave an assemblage of unexplained *residual phenomena*, in which a careful research might find the materials of other truths as yet unsuspected.

[2d Ed.] [That there would be, in the tidal movements of the ocean, inequalities of the heights and times of high and low water *corres-*

---

[46] Essay towards a First Approximation to a Map of Cotidal Lines. *Phil. Trans.* 1833, 1836.

ponding to those which the equilibrium theory gives, could be considered only as a conjecture, till the comparison with observation was made. It was, however, a natural conjecture; since the waters of the ocean are at every moment *tending* to acquire the form assumed in the equilibrium theory: and it may be considered likely that the causes which prevent their assuming this form produce an effect nearly constant for each place. Whatever be thought of this reasoning, the conjecture is confirmed by observation with curious exactness. The laws of a great number of the tidal phenomena—namely, of the Semi-mensual Inequality of the Heights, of the Semi-mensual Inequality of the Times, of the Diurnal Inequality, of the effect of the Moon's Declination, of the effect of the Moon's Parallax—are represented very closely by formulæ derived from the equilibrium theory. The hydrodynamical mode of treating the subject has not added any thing to the knowledge of the laws of the phenomena to which the other view had conducted us.

We may add, that Laplace's assumption, that in the moving fluid the motions must have a *periodicity* corresponding to that of the forces, is also a conjecture. And though this conjecture may, in some cases of the problem, be verified, by substituting the resulting expressions in the equations of motion, this cannot be done in the actual case, where the revolving motion of the ocean is prevented by the intrusion of tracts of land running nearly from pole to pole.

Yet in Mr. Airy's Treatise *On Tides and Waves* (in the *Encyclopædia Metropolitana*) much has been done to bring the hydrodynamical theory of oceanic tides into agreement with observation. In this admirable work, Mr. Airy has, by peculiar artifices, solved problems which come so near the actual cases that they may represent them. He has, in this way, deduced the laws of the semi-diurnal and the diurnal tide, and the other features of the tides which the equilibrium theory in some degree imitates; but he has also, taking into account the effect of friction, shown that the actual tide may be represented as the tide of an earlier epoch;—that the relative mass of the moon and sun, as inferred from the tides, would depend upon the depth of the ocean (Art. 455);—with many other results remarkably explaining the observed phenomena. He has also shown that the relation of the cotidal lines to the tide waves really propagated is, in complex cases, very obscure, because different waves of different magnitudes, travelling in different directions, may coexist, and the cotidal line is the compound result of all these.

With reference to the *Maps of Cotidal Lines*, mentioned in the text, I may add, that we are as yet destitute of observations which should supply the means of drawing such lines on a large scale in the Pacific Ocean. Admiral Lütke has however supplied us with some valuable materials and remarks on this subject in his *Notice sur les Marées Périodiques dans le grand Océan Boréal et dans la Mer Glaciale;* and has drawn them, apparently on sufficient data, in the White Sea.]

CHAPTER V.

Discoveries added to the Newtonian Theory.

Sect. 1.—*Tables of Astronomical Refraction.*

WE have travelled over an immense field of astronomical and mathematical labor in the last few pages, and have yet, at the end of every step, still found ourselves under the jurisdiction of the Newtonian laws. We are reminded of the universal monarchies, where a man could not escape from the empire without quitting the world. We have now to notice some other discoveries, in which this reference to the law of universal gravitation is less immediate and obvious; I mean the astronomical discoveries respecting Light.

The general truths to which the establishment of the true laws of Atmospheric Refraction led astronomers, were the law of Deflection of the rays of light, which applies to all refractions, and the real structure and size of the Atmosphere, so far as it became known. The great discoveries of Römer and Bradley, namely, the Velocity of Light, the Aberration of Light, and the Nutation of the earth's axis, gave a new distinctness to the conceptions of the propagation of light in the minds of philosophers, and confirmed the doctrines of Copernicus, Kepler, and Newton, respecting the motions which belong to the earth.

The true laws of Atmospheric Refraction were slowly discovered. Tycho attributed the apparent displacement of the heavenly bodies to the low and gross part of the atmosphere only, and hence made it cease at a point half-way to the zenith; but Kepler rightly extended it to the zenith itself. Dominic Cassini endeavored to discover the law of this correction by observation, and gave his result in the form which, as we have said, sound science prescribes, a Table to be habitually used for all observations. But great difficulties at this time embarrassed this investigation, for the parallaxes of the sun and of the planets were unknown, and very diverse values had been assigned them by different astronomers. To remove some of these difficulties, Richer, in 1762, went to observe at the equator; and on his return, Cassini was able to confirm and amend his former estimations of parallax and refraction. But there were still difficulties. According to La Hire, though the phenomena of twilight give an altitude of 34,000 toises to the atmosphere,[1] those of refraction make it only 2000. John Cassini undertook to support and improve the calculations of his father Dominic, and took the true supposition, that the light follows a curvilinear path through the air. The Royal Society of London had already ascertained experimentally the refractive power of air.[2] Newton calculated a Table of Refractions, which was published under Halley's name in the *Philosphical Transactions* for 1721, without any indication of the method by which it was constructed. But M. Biot has recently shown,[3] by means of the published correspondence of Flamsteed, that Newton had solved the problem in a manner nearly corresponding to the most improved methods of modern analysis.

Dominic Cassini and Picard proved,[4] Le Monnier in 1738 confirmed more fully, the fact that the variations of the Thermometer affect the Refraction. Mayer, taking into account both these changes, and the changes indicated by the Barometer, formed a theory, which Lacaille, with immense labor, applied to the construction of a Table of Refractions from observation. But Bradley's Table (published in 1763 by Maskelyne) was more commonly adopted in England; and his formula, originally obtained empirically, has been shown by Young to result from the most probable suppositions we can make respecting the atmosphere. Bessel's Refraction Tables are now considered the best of those which have appeared.

Sect. 2.—*Discovery of the Velocity of Light.—Römer.*

The astronomical history of Refraction is not marked by any great discoveries, and was, for the most part, a work of labor only. The progress of the other portions of our knowledge respecting light is

---

1 Bailly, ii. 612. 2 Ibid. ii. 607.
3 Biot, *Acad. Sc. Compte Rendu*, Sept. 5, 1836. 4 Bailly, iii. 92.

more striking. In 1676, a great number of observations of eclipses of Jupiter's satellites were accumulated, and could be compared with Cassini's Tables. Römer, a Danish astronomer, whom Picard had brought to Paris, perceived that these eclipses happened constantly later than the calculated time at one season of the year, and earlier at another season;—a difference for which astronomy could offer no account. The error was the same for all the satellites; if it had depended on a defect in the Tables of Jupiter, it might have affected all, but the effect would have had a reference to the velocities of the satellites. The cause, then, was something extraneous to Jupiter. Römer had the happy thought of comparing the error with the earth's distance from Jupiter, and it was found that the eclipses happened later in proportion as Jupiter was further off.[5] Thus we see the eclipse later, as it is more remote; and thus light, the messenger which brings us intelligence of the occurrence, travels over its course in a measurable time. By this evidence, light appeared to take about eleven minutes in describing the diameter of the earth's orbit.

This discovery, like so many others, once made, appears easy and inevitable; yet Dominic Cassini had entertained the idea for a moment,[6] and had rejected it; and Fontenelle had congratulated himself publicly on having narrowly escaped this seductive error. The objections to the admission of the truth arose principally from the inaccuracy of observation, and from the persuasion that the motions of the satellites were circular and uniform. Their irregularities disguised the fact in question. As these irregularities became clearly known, Römer's discovery was finally established, and the "Equation of Light" took its place in the Tables.

Sect. 3.—Discovery of Aberration.—Bradley.

Improvements in instruments, and in the art of observing, were requisite for making the next great step in tracing the effect of the laws of light. It appears clear, on consideration, that since light and the spectator on the earth are both in motion, the apparent direction of an object will be determined by the composition of these motions. But yet the effect of this composition of motions was (as is usual in such cases) traced as a fact in observation, before it was clearly seen as a consequence of reasoning. This fact, the Aberration of Light, the greatest astronomical discovery of the eighteenth century, belongs to Bradley,

[5] Bailly, ii. 17. [6] Ib. ii. 419.

who was then Professor of Astronomy at Oxford, and afterwards Astronomer Royal at Greenwich. Molyneux and Bradley, in 1725, began a series of observations for the purpose of ascertaining, by observations near the zenith, the existence of an annual parallax of the fixed stars, which Hooke had hoped to detect, and Flamsteed thought he had discovered. Bradley[7] soon found that the star observed by him had a minute apparent motion different from that which the annual parallax would produce. He thought of a nutation of the earth's axis as a mode of accounting for this; but found, by comparison of a star on the other side of the pole, that this explanation would not apply. Bradley and Molyneux then considered for a moment an annual alteration of figure in the earth's atmosphere, such as might affect the refractions, but this hypothesis was soon rejected.[8] In 1727, Bradley resumed his observations, with a new instrument, at Wanstead, and obtained empirical rules for the changes of declination of different stars. At last, accident turned his thoughts to the direction in which he was to find the cause of the variations which he had discovered. Being in a boat on the Thames, he observed that the vane on the top of the mast gave a different apparent direction to the wind, as the boat sailed one way or the other. Here was an image of his case: the boat represented the earth moving in different directions at different seasons, and the wind represented the light of a star. He had now to trace the consequences of this idea; he found that it led to the empirical rules, which he had already discovered, and, in 1729, he gave his discovery to the Royal Society. His paper is a very happy narrative of his labors and his thoughts. His theory was so sound that no astronomer ever contested it; and his observations were so accurate, that the quantity which he assigned as the greatest amount of the change (one nineteenth of a degree) has hardly been corrected by more recent astronomers. It must be noticed, however, that he considered the effects in declination only; the effects in right ascension required a different mode of observation, and a consummate goodness in the machinery of clocks, which at that time was hardly attained.

Sect. 4.—Discovery of Nutation.

When Bradley went to Greenwich as Astronomer Royal, he continued with perseverance observations of the same kind as those by which he had detected Aberration. The result of these was another

---

[7] Rigaud's Bradley.

[8] Rigaud, p. xxiii.

discovery; namely, that very Nutation which he had formerly rejected. This may appear strange, but it is easily explained. The aberration is an annual change, and is detected by observing a star at different seasons of the year: the Nutation is a change of which the cycle is eighteen years; and which, therefore, though it does not much change the place of a star in one year, is discoverable in the alterations of several successive years. A very few years' observations showed Bradley the effect of this change;[9] and long before the half cycle of nine years had elapsed, he had connected it in his mind with the true cause, the motion of the moon's nodes. Machin was then Secretary to the Royal Society,[10] and was "employed in considering the theory of gravity, and its consequences with regard to the celestial motions:" to him Bradley communicated his conjectures; from him he soon received a Table containing the results of his calculations; and the law was found to be the same in the Table and in observation, though the quantities were somewhat different. It appeared by both, that the earth's pole, besides the motion which the precession of the equinoxes gives it, moves, in eighteen years, through a small circle;—or rather, as was afterwards found by Bradley, an ellipse, of which the axes are nineteen and fourteen seconds.[11]

For the rigorous establishment of the mechanical theory of that effect of the moon's attraction from which the phenomena of Nutation flow, Bradley rightly and prudently invited the assistance of the great mathematicians of his time. D'Alembert, Thomas Simpson, Euler, and others, answered this call, and the result was, as we have already said in the last chapter (Sect. 7), that this investigation added another to the recondite and profound evidences of the doctrine of universal gravitation.

It has been said[12] that Bradley's discoveries "assure him the most distinguished place among astronomers after Hipparchus and Kepler." If his discoveries had been made before Newton's, there could have been no hesitation as to placing him on a level with those great men. The existence of such suggestions as the Newtonian theory offered on all astronomical subjects, may perhaps dim, in our eyes, the brilliance of Bradley's achievements; but this circumstance cannot place any other person above the author of such discoveries, and therefore we may consider Delambre's adjudication of precedence as well warranted, and deserving to be permanent.

---

[9] Rigaud, lxiv. [10] Ib. 25. [11] Ib. lxvi.

[12] Delambre, *Ast. du* 18 *Siè*c. p. 420. Rigaud, xxxvii.

Sect. 5.—Discovery of the Laws of Double Stars.—The two Herschels.

No truth, then, can be more certainly established, than that the law of gravitation prevails to the very boundaries of the solar system. But does it hold good further? Do the fixed stars also obey this universal sway? The idea, the question, is an obvious one—but where are we to find the means of submitting it to the test of observation?

If the Stars were each insulated from the rest, as our Sun appears to be from them, we should have been quite unable to answer this inquiry. But among the stars, there are some which are called *Double Stars*, and which consist of two stars, so near to each other that the telescope alone can separate them. The elder Herschel diligently observed and measured the relative positions of the two stars in such pairs; and as has so often happened in astronomical history, pursuing one object he fell in with another. Supposing such pairs to be really unconnected, he wished to learn, from their phenomena, something respecting the annual parallax of the earth's orbit. But in the course of twenty years' observations he made the discovery (in 1803) that some of these couples were turning round each other with various angular velocities. These revolutions were for the most part so slow that he was obliged to leave their complete determination as an inheritance to the next generation. His son was not careless of the bequest, and after having added an enormous mass of observations to those of his father, he applied himself to determine the laws of these revolutions. A problem so obvious and so tempting was attacked also by others, as Savary and Encke, in 1830 and 1832, with the resources of analysis. But a problem in which the data are so minute and inevitably imperfect, required the mathematician to employ much judgment, as well as skill in using and combining these data; and Sir John Herschel, by employing positions only of the line joining the pair of stars (which can be observed with comparative exactness), to the exclusion of their distances (which cannot be measured with much correctness), and by inventing a method which depended upon the whole body of observations, and not upon selected ones only, for the determination of the motion, has made his investigations by far the most satisfactory of those which have appeared. The result is, that it has been rendered very probable, that in several of the double stars the two stars describe ellipses about each other; and therefore that here also, at an immeasurable distance from our system, the law of attraction according to the inverse square of the distance, prevails. And, according to the practice of astronomers when a law has been established, Tables have been calculated for the future motions; and we have Ephemerides of the revolutions of suns round each other, in a region so remote, that the whole circle of our earth's orbit, if placed there, would be imperceptible by our strongest telescopes. The permanent comparison of the observed with the predicted motions, continued for more than one revolution, is the severe and decisive test of the truth of the theory; and the result of this test astronomers are now awaiting.

[2d Ed.] [In calculating the orbits of revolving systems of double stars, there is a peculiar difficulty, arising from the plane of the orbit being in a position unknown, but probably oblique, to the visual ray. Hence it comes to pass that even if the orbit be an ellipse described about the focus by the laws of planetary motion, it will appear otherwise; and the true orbit will have to be deduced from the apparent one.

With regard to a difficulty which has been mentioned, that the two stars, if they are governed by gravity, will not revolve the one about the other, but both about their common centre of gravity;—this circumstance adds little difficulty to the problem. Newton has shown (*Princip.* lib. i. Prop. 61) in the *problem of two bodies*, the relation between the relative orbits and the orbit about the common centre of gravity.

How many of the apparently double stars have orbitual motions? Sir John Herschel in 1833 gave, in his *Astronomy* (Art. 606), a list of *nine* stars, with periods extending from 43 years (η Coronæ) to 1200 years (γ Leonis), which he presented as the chief results then obtained in this department. In his work on Double Stars, the fruit of his labors in both hemispheres, which the astronomical world are looking for with eager expectation, he will, I believe, have a few more to add to these.

Is it well established that such double stars attract each other according to the law of the inverse square of the distance? The answer to this question must be determined by ascertaining whether the above cases are regulated by the laws of elliptical motion. This is a matter which it must require a long course of careful observation to determine in such a number of cases as to prove the universality of the rule. Perhaps the minds of astronomers are still in suspense upon the subject. When Sir John Herschel's work shall appear, it will probably be found that with regard to some of these stars, and γ Virginis in particular, the conformity of the observations with the laws of elliptical motion amounts to a degree of exactness which must give astronomers a strong conviction of the truth of the law. For since Sir W. Herschel's first measures in 1781, the arc described by one star about the other is above 305 degrees; and during this period the angular annual motion has been very various, passing through all gradations from about 20 minutes to 80 degrees. Yet in the whole of this change, the two curves constructed, the one from the observations, the other from the elliptical elements, for the purpose of comparison, having a total ordinate of 305 parts, do not, in any part of their course, deviate from each other so much as *two* such parts.]

The verification of Newton's discoveries was sufficient employment for the last century; the first step in the extension of them belongs to this century. We cannot at present foresee the magnitude of this task, but every one must feel that the law of gravitation, before verified in all the particles of our own system, and now probably extended to the all but infinite distance of the fixed stars, presses upon our minds with a strong claim to be accepted as a universal law of the whole material creation.

Thus, in this and the preceding chapter, I have given a brief sketch of the history of the verification and extension of Newton's great discovery. By the mass of labor and of skill which this head of our subject includes, we may judge of the magnitude of the advance in our knowledge which that discovery made. A wonderful amount of talent and industry have been requisite for this purpose; but with these, external means have co-operated. Wealth, authority, mechanical skill, the division of labor, the power of associations and of governments, have been largely and worthily applied in bringing astronomy to its present high and flourishing condition. We must consider briefly what has thus been done.

CHAPTER VI.

The Instruments and Aids of Astronomy during the Newtonian Period.

Sect. 1.—*Instruments.*

SOME instruments or other were employed at all periods of astronomical observation. But it was only when observation had attained a considerable degree of delicacy, that the exact construction of instruments became an object of serious care. Gradually, as the possibility and the value of increased exactness became manifest, it was seen that every thing which could improve the astronomer's instruments was of high importance to him. And hence in some cases a vast increase of size and of expense was introduced; in other cases new combinations, or the result of improvements in other sciences, were brought into play. Extensive knowledge, intense thought, and great ingenuity, were requisite in the astronomical instrument maker. Instead of ranking with artisans, he became a man of science, sharing the honor and dignity of the astronomer himself.

1. *Measure of Angles.*—Tycho Brahe was the first astronomer who acted upon a due appreciation of the importance of good instruments. The collection of such at Uraniburg was by far the finest which had ever existed. He endeavored to give steadiness to the frame, and accuracy to the divisions of his instruments. His Mural Quadrant was well adapted for this purpose; its radius was five cubits: it is clear, that as we enlarge the instrument we are enabled to measure smaller arcs. On this principle many large *gnomons* were erected. Cassini's celebrated one in the church of St. Petronius at Bologna, was eighty-three feet (French) high. But this mode of obtaining accuracy was soon abandoned for better methods. Three great improvements were introduced about the same time. The application of the Micrometer to the telescope, by Huyghens, Malvasia, and Auzout; the application of the Telescope to the astronomical quadrant; and the fixation of the centre of its field by a Cross of fine wires placed in the focus by Gascoigne, and afterwards by Picard. We may judge how great was the improvement which these contrivances introduced into the art of ob-

serving, by finding that Hevelius refused to adopt them because they would make all the old observations of no value. He had spent a laborious and active life in the exercise of the old methods, and could not bear to think that all the treasures which he had accumulated had lost their worth by the discovery of a new mine of richer ore.

[2d Ed.] [Littrow, in his *Die Wunder des Himmels*, Ed. 2, pp. 684, 685, says that Gascoigne invented and used the telescope with wires in the common focus of the lenses in 1640. He refers to *Phil. Trans.* xxx. 603. Picard reinvented this arrangement in 1667. I have already spoken of Gascoigne as the inventor of the micrometer.

Römer (already mentioned, p. 464) brought into use the Transit Instrument, and the employment of complete Circles, instead of the Quadrants used till then; and by these means gave to practical astronomy a new form, of which the full value was not discovered till long afterwards.

The apparent place of the object in the instrument being so precisely determined by the new methods, the exact Division of the arc into degrees and their subdivisions became a matter of great consequence. A series of artists, principally English, have acquired distinguished places in the lists of scientific fame by their performances in this way; and from that period, particular instruments have possessed historical interest and individual reputation. Graham was one of the first of these artists. He executed a great Mural Arc for Halley at Greenwich; for Bradley he constructed the Sector which detected aberration. He also made the Sector which the French academicians carried to Lapland; and probably the goodness of this instrument, compared with the imperfection of those which were sent to Peru, was one main cause of the great difference of duration in the two series of observations. Bird, somewhat later[1] (about 1750), divided several Quadrants for public observatories. His method of dividing was considered so perfect, that the knowledge of it was purchased by the English government, and published in 1767. Ramsden was equally celebrated. The error of one of his best Quadrants (that at Padua) is said to be never greater than two seconds. But at a later period, Ramsden constructed Mural Circles only, holding this to be a kind of instrument far superior to the quadrant. He made one of five feet diameter, in 1788, for M. Piazzi at Palermo; and one of eight feet for the observatory of Dublin. Troughton, a worthy successor of the art-

[1] Mont. iv. 337.

ists we have mentioned, has invented a method of dividing the circle still superior to the former ones; indeed, one which is theoretically perfect, and practically capable of consummate accuracy. In this way, circles have been constructed for Greenwich, Armagh, Cambridge, and many other places; and probably this method, carefully applied, offers to the astronomer as much exactness as his other implements allow him to receive; but the slightest casualty happening to such an instrument, after it has been constructed, or any doubt whether the method of graduation has been rightly applied, makes it unfit for the jealous scrupulosity of modern astronomy.

The English artists sought to attain accurate measurements by continued bisection and other aliquot subdivision of the limb of their circle; but Mayer proposed to obtain this end otherwise, by *repeating* the measure on different parts of the circumference till the error of the division becomes unimportant, instead of attempting to divide an instrument without error. This invention of the Repeating Circle was zealously adopted by the French, and the relative superiority of the rival methods is still a matter of difference of opinion.

[2d Ed.] [In the series of these great astronomical mechanists, we must also reckon George Reichenbach. He was born Aug. 24, 1772, at Durlach; became Lieutenant of Artillery in the Bavarian service in 1794; (Salinenrath) Commissioner of Salt-works in 1811; and in 1820, First Commissioner of Water-works and Roads. He became, with Fraunhofer, the ornament of the mechanical and optical Institute erected in 1805 at Benedictbeuern by Utzschneider; and his astronomical instruments, meridian circles, transit instruments, equatorials, heliometers, make an epoch in Observing Astronomy. His contrivances in the Salt-works at Berchtesgaden and Reichenhall, in the Arms Manufactory at Amberg, and in the works for boring cannon at Vienna, are enduring monuments of his rare mechanical talent. He died May 21, 1826, at Munich.]

2. *Clocks.*—The improvements in the measures of space require corresponding improvements in the measure of time. The beginning of any thing which we can call accuracy, in this subject, was the application of the Pendulum to clocks, by Huyghens, in 1656. That the successive oscillations of a pendulum occupy equal times, had been noticed by Galileo; but in order to take advantage of this property, the pendulum must be connected with machinery by which its motion is kept from languishing, and by which the number of its swings is recorded. By inventing such machinery, Huyghens at once obtained

a measure of time more accurate than the sun itself. Hence astronomers were soon led to obtain the right ascension of a star, not directly, by measuring a Distance in the heavens, but indirectly, by observing the Moment of its Transit. This observation is now made with a degree of accuracy which might, at first sight, appear beyond the limits of human sense, being noted to a *tenth of a second of time:* but we may explain this, by remarking that though the number of the second at which the transit happens is given by the clock, and is reckoned according to the course of time, the subdivision of the second of time into smaller fractions is performed by the eye,—by seeing the space described by the heavenly body in a whole second, and hence estimating a smaller time, according to the space which its description occupies.

But in order to make clocks so accurate as to justify this degree of precision, their construction was improved by various persons in succession. Picard soon found that Huyghens' clocks were affected in their going by temperature, for heat caused expansion of the metallic pendulum. This cause of error was remedied by combining different metals, as iron and copper, which expand in a different degree, in such a way that their effects compensate each other. Graham afterwards used quicksilver for the same purpose. The *Escapement* too (which connects the force which impels the clock with the pendulum which regulates it), and other parts of the machinery, had the most refined mechanical skill and ingenuity of the best artists constantly bestowed upon then. The astronomer of the present day, constantly testing the going of such a clock by the motions of the fixed stars, has a scale of time as stable and as minutely exact as the scales on which he measures distance.

The construction of good Watches, that is, portable or marine clocks, was important on another account, namely, because they might be used in determining the longitude of places. Hence the improvement of this little machine became an object of national interest, and was included in the reward of 20,000*l.* which we have already noticed as offered by the English parliament for the discovery of the longitude. Harrison,[2] originally a carpenter, turned his mind to this subject with success. After thirty years of labor, in which he was encouraged by many eminent persons, he produced, in 1758, a time-keeper, which was sent on a voyage to Jamaica for trial. After 161 days, the error

[2] Mont. iv. 554.

of the watch was only one minute five seconds, and the artist received from the nation 5000*l.* At a later period,[3] at the age of seventy-five years, after a life devoted to this object, having still further satisfied the commissioners, he received, in 1765, 10,000*l.*, at the same time that Euler and the heirs of Mayer received each 3000*l.* for the lunar tables which they had constructed.

The two methods of finding the longitude, by Chronometers and by Lunar Observations, have solved the problem for all practical purposes; but the latter could not have been employed at sea without the aid of that invaluable instrument, the Sextant, in which the distance of two objects is observed, by bringing one to coincide apparently with the reflected image of the other. This instrument was invented by Hadley, in 1731. Though the problem of finding the longitude be, in fact, one of geography rather than astronomy, it is an application of astronomical science which has so materially affected the progress of our knowledge, that it deserves the notice we have bestowed upon it.

3. *Telescopes.*—We have spoken of the application of the telescope to astronomical measurements, but not of the improvement of the telescope itself. If we endeavor to augment the optical power of this instrument, we run, according to the path we take, into various inconveniences;—distortion, confusion, want of light, or colored images. Distortion and confusion are produced, if we increase the magnifying power, retaining the length and the aperture of the object-glass. If we diminish the aperture we suffer from loss of light. What remains then is to increase the focal length. This was done to an extraordinary extent, in telescopes constructed in the beginning of the last century. Huyghens, in his first attempts, made them 22 feet long;[4] afterwards, Campani, by order of Louis the Fourteenth, made them of 86, 100, and 136 feet. Huyghens, by new exertions, made a telescope 210 feet long. Auzout and Hartsoecker are said to have gone much further, and to have succeeded in making an object-glass of 600 feet focus. But even such telescopes as those of Campani are almost unmanageable: in that of Huyghens, the object-glass was placed on a pole, and the observer was placed at the focus with an eye-glass.

The most serious objection to the increase of the aperture of object-glasses, was the coloration of the image produced, in consequence of the unequal refrangibility of differently colored rays. Newton, who discovered the principle of this defect in lenses, had maintained that

---

[3] Mont. iv, 560. [4] Bailly, ii. 253.

the evil was irremediable, and that a compound lens could no more refract without producing color, than a single lens could. Euler and Klingenstierna doubted the exactness of Newton's proposition; and, in 1755, Dollond disproved it by experiment. This discovery pointed out a method of making object-glasses which should give no color;—which should be *achromatic*. For this purpose Dollond fabricated various kinds of glass (flint and crown glass); and Clairaut and D'Alembert calculated formulæ. Dollond and his son[5] succeeded in constructing telescopes of three feet long (with a triple object-glass) which produced an effect as great as those of forty-five feet on the ancient principles. At first it was conceived that these discoveries opened the way to a vast extension of the astronomer's power of vision; but it was found that the most material improvement was the compendious size of the new instruments; for, in increasing the dimensions, the optician was stopped by the impossibility of obtaining lenses of flint-glass of very large dimensions. And this branch of art remained long stationary; but, after a time, its epoch of advance again arrived. In the present century, Fraunhofer, at Munich, with the help of Guinand and the pecuniary support of Utzschneider, succeeded in forming lenses of flint-glass of a magnitude till then unheard of. Achromatic object-glasses, of a foot in diameter, and twenty feet focal length, are now no longer impossible; although in such attempts the artist cannot reckon on certain success.

[2d Ed.] [Joseph Fraunhofer was born March 6, 1787, at Straubing in Bavaria, the son of a poor glazier. He was in his earlier years employed in his father's trade, so that he was not able to attend school, and remained ignorant of writing and arithmetic till his fourteenth year. At a later period he was assisted by Utzschneider, and tried rapidly to recover his lost ground. In the year 1806 he entered the establishment of Utzschneider as an optician. In this establishment (transferred from Benedictbeuern to Munich in 1819) he soon came to be the greatest Optician of Germany. His excellent telescopes and microscopes are known throughout Europe. His greatest telescope, that in the Observatory at Dorpat, has an object-glass of 9 inches diameter, and a focal length of 13⅓ feet. His written productions are to be found in the *Memoirs* of the Bavarian Academy, in Gilbert's *Annalen der Physik*, and in Schumacher's *Astronomische Nachrichten*. He died the 7th of June, 1826.]

---

[5] Bailly, iii. 118.

Such telescopes might be expected to add something to our knowledge of the heavens, if they had not been anticipated by reflectors of an equal or greater scale. James Gregory had invented, and Newton had more efficaciously introduced, reflecting telescopes. But these were not used with any peculiar effect, till the elder Herschel made them his especial study. His skill and perseverance in grinding specula, and in contriving the best apparatus for their use, were rewarded by a number of curious and striking discoveries, among which, as we have already related, was the discovery of a new planet beyond Saturn. In 1789, Herschel surpassed all his former attempts, by bringing into action a reflecting telescope of forty feet length, with a speculum of four feet in diameter. The first application of this magnificent instrument showed a new satellite (the sixth) of Saturn. He and his son have, with reflectors of twenty feet, made a complete survey of the heavens, so far as they are visible in this country; and the latter is now in a distant region completing this survey, by adding to it the other hemisphere.

In speaking of the improvements of telescopes we ought to notice, that they have been pursued in the eye-glasses as well as in the object-glasses. Instead of the single lens, Huyghens substituted an eye-piece of two lenses, which, though introduced for another purpose, attained the object of destroying color.[6] Ramsden's eye-piece is one fit to be used with a micrometer, and others of more complex construction have been used for various purposes.

Sect. 2.—Observatories.

Astronomy, which is thus benefited by the erection of large and stable instruments, requires also the establishment of permanent Observatories, supplied with funds for their support, and for that of the observers. Such observatories have existed at all periods of the history of the science; but from the commencement of the period which we are now reviewing, they multiplied to such an extent that we cannot even enumerate them. Yet we must undoubtedly look upon such establishments, and the labors of which they have been the scene, as important and essential parts of the history of the progress of astronomy. Some of the most distinguished of the observatories of modern times we may mention. The first of these were that of Tycho Brahe

---

[6] Coddington's *Optics*, ii. 21.

at Uraniburg, and that of the Landgrave of Hesse Cassel, at Cassel, where Rothman and Byrgius observed. But by far the most important observations, at least since those of Tycho, which were the basis of the discoveries of Kepler and Newton, have been made at Paris and Greenwich. The Observatory of Paris was built in 1667. It was there that the first Cassini made many of his discoveries; three of his descendants have since labored in the same place, and two others of his family, the Maraldis;[7] besides many other eminent astronomers, as Picard, La Hire, Lefêvre, Fouchy, Legentil, Chappe, Méchain, Bouvard. Greenwich Observatory was built a few years later (1675); and ever since its erection, the observations there made have been the foundation of the greatest improvements which astronomy, for the time, received. Flamsteed, Halley, Bradley, Bliss, Maskelyne, Pond, have occupied the place in succession: on the retirement of the last-named astronomer in 1835, Professor Airy was removed thither from the Cambridge Observatory. In every state, and in almost every principality in Europe, Observatories have been established; but these have often fallen speedily into inaction, or have contributed little to the progress of astronomy, because their observations have not been published. From the same causes, the numerous private observatories which exist throughout Europe have added little to our knowledge, except where the attention of the astronomer has been directed to some definite points; as, for instance, the magnificent labors of the Herschels, or the skilful observations made by Mr. Pond with the Westbury circle, which first pointed out the error of graduation of the Greenwich quadrants. The Observations, now regularly published,[8] are those of Greenwich, begun by Maskelyne, and continued quarterly by Mr. Pond; those of Königsberg, published by Bessel since 1814; of Vienna, by Littrow since 1820; of Speier, by Schwerd since 1826; those of Cambridge, commenced by Airy in 1828; of Armagh, by Robinson in 1829. Besides these, a number of useful observations have been published in journals and occasional forms; as, for instance, those of Zach, made by Seeberg, near Gotha, since 1788; and others have been employed in forming catalogues, of which we shall speak shortly.

[2d Ed.] [I have left the statement of published Observations in the text as it stood originally. I believe that at present (1847) the twelve places contained in the following list publish their Observations quite regularly, or nearly so;—Greenwich, Oxford, Cambridge, Vienna,

---

[7] Mont. iv. 346. [8] Airy, *Rep.* p. 128.

Berlin, Dorpat, Munich, Geneva, Paris, Königsberg, Madras, the Cape of Good Hope.

Littrow, in his translation, adds to the publications noticed in the text as containing astronomical Observations, Zach's *Monatliche Correspondenz*, Lindenau and Bohnenberger's *Zeitschrift für Astronomie*, Bode's *Astronomisches Jahrbuch*, Schumacher's *Astronomische Nachrichten.*]

Nor has the establishment of observatories been confined to Europe. In 1786, M. de Beauchamp, at the expense of Louis the Sixteenth, erected an observatory at Bagdad, "built to restore the Chaldean and Arabian observations," as the inscription stated; but, probably, the restoration once effected, the main intention had been fulfilled, and little perseverance in observing was thought necessary. In 1828, the British government completed the building of an observatory at the Cape of Good Hope, which Lacaille had already made an astronomical station by his observations there at an earlier period (1750); and an observatory formed in New South Wales by Sir T. M. Brisbane in 1822, and presented by him to the government, is also in activity. The East India Company has founded observatories at Madras, Bombay, and St. Helena; and observations made at the former of these places, and at St. Helena, have been published.

The bearing of the work done at such observatories upon the past progress of astronomy, has already been seen in the preceding narrative. Their bearing upon the present condition of the science will be the subject of a few remarks hereafter.

Sect. 3.—*Scientific Societies.*

The influence of Scientific Societies, or Academical Bodies, has also been very powerful in the subject before us. In all branches of knowledge, the use of such associations of studious and inquiring men is great; the clearness and coherence of a speculator's ideas, and their agreement with facts (the two main conditions of scientific truth), are severally but beneficially tested by collision with other minds. In astronomy, moreover, the vast extent of the subject makes requisite the division of labor and the support of sympathy. The Royal Societies of London and of Paris were founded nearly at the same time as the metropolitan Observatories of the two countries. We have seen what constellations of philosophers, and what activity of research, existed at those periods; these philosophers appear in the lists, their discoveries in the publications, of the above-mentioned eminent Societies. As the progress of physical science, and principally of astronomy, attracted more and more admiration, Academies were created in other countries. That of Berlin was founded by Leibnitz in 1710; that of St. Petersburg was established by Peter the Great in 1725; and both these have produced highly valuable Memoirs. In more modern times these associations have multiplied almost beyond the power of estimation. They have been formed according to divisions, both of locality and of subject, conformable to the present extent of science, and the vast population of its cultivators. It would be useless to attempt to give a view either of their number or of the enormous mass of scientific literature which their Transactions present. But we may notice, as especially connected with our present subject, the Astronomical Society of London, founded in 1820, which gave a strong impulse to the pursuit of the science in England.

Sect. 4.—Patrons of Astronomy.

The advantages which letters and philosophy derive from the patronage of the great have sometimes been questioned; that love of knowledge, it has been thought, cannot be genuine which requires such stimulation, nor those speculations free and true which are thus forced into being. In the sciences of observation and calculation, however, in which disputed questions can be experimentally decided, and in which opinions are not disturbed by men's practical principles and interests, there is nothing necessarily operating to poison or neutralize the resources which wealth and power supply to the investigation of truth.

Astronomy has, in all ages, flourished under the favor of the rich and powerful; in the period of which we speak, this was eminently the case. Louis the Fourteenth gave to the astronomy of France a distinction which, without him, it could not have attained. No step perhaps tended more to this than his bringing the celebrated Dominic Cassini to Paris. This Italian astronomer (for he was born at Permaldo, in the county of Nice, and was professor at Bologna), was already in possession of a brilliant reputation, when the French ambassador, in the name of his sovereign, applied to Pope Clement the Ninth, and to the senate of Bologna, that he should be allowed to remove to Paris. The request was granted only so far as an absence of six years; but at the end of that time, the benefits and honors which the king had conferred upon him, fixed him in France. The impulse which his arrival (in 1669) and his residence gave to astronomy, showed the wisdom of the measure. In the same spirit, the French government drew to Paris Römer from Denmark, Huyghens from Holland, and gave a pension to Hevelius, and a large sum when his observatory at Dantzic had been destroyed by fire in 1679.

When the sovereigns of Prussia and Russia were exerting themselves to encourage the sciences in their countries, they followed the same course which had been so successful in France. Thus, as we have said, the Czar Peter took Delisle to Petersburg in 1725; the celebrated Frederick the Great drew to Berlin, Voltaire and Maupertuis, Euler and Lagrange; and the Empress Catharine obtained in the same way Euler, two of the Bernoullis, and other mathematicians. In none of these instances, however, did it happen that "the generous plant did still its stock renew," as we have seen was the case at Paris, with the Cassinis, and their kinsmen the Maraldis.

[2d Ed.] [I may notice among instances of the patronage of Astronomy, the reward at present offered by the King of Denmark for the discovery of a Comet.]

It is not necessary to mention here the more recent cases in which sovereigns or statesmen have attempted to patronize individual astronomers.

Sect. 5.—Astronomical Expeditions.

Besides the pensions thus bestowed upon resident mathematicians and astronomers, the governments of Europe have wisely and usefully employed considerable sums upon expeditions and travels undertaken by men of science for some appropriate object. Thus Picard, in 1671, was sent to Uraniburg, the scene of Tycho's observations, to determine its latitude and its longitude. He found that "the City of the Skies" had utterly disappeared from the earth; and even its foundations were retraced with difficulty. With the same object, that of accurately connecting the labors of the places which had been at different periods the metropolis of astronomy, Chazelles was sent, in 1693, to Alexandria. We have already mentioned Richer's astronomical expedition to Cayenne in 1672. Varin and Deshayes[9] were sent a few years later into the same regions for similar purposes. Halley's expedition to St.

---

[9] Bailly, ii. 374.

Helena in 1677, with the view of observing the southern stars, was at his own expense; but at a later period (in 1698), he was appointed to the command of a small vessel by King William the Third, in order that he might make his magnetical observations in all parts of the world. Lacaille was maintained by the French government four years at the Cape of Good Hope (1750–4), for the purpose of observing the stars of the southern hemisphere. The two transits of Venus in 1761 and 1769, occasioned expeditions to be sent to Kamtschatka and Tobolsk by the Russians; to the Isle of France, and to Coromandel, by the French;[10] to the isles of St. Helena and Otaheite by the English; to Lapland and to Drontheim, by the Swedes and Danes. I shall not here refer to the measures of degrees executed by various nations, still less the innumerable surveys by land and sea; but I may just notice the successive English expeditions of Captains Basil Hall, Sabine, and Foster, for the purpose of determining the length of the seconds' pendulum in different latitudes; and the voyages of M. Biot and others, sent by the French government for the same purpose. Much has been done in this way, but not more than the progress of astronomy absolutely required; and only a small portion of that which the completion of the subject calls for.

Sect. 6.—Present State of Astronomy.

Astronomy, in its present condition, is not only much the most advanced of the sciences, but is also in far more favorable circumstances than any other science for making any future advance, as soon as this is possible. The general methods and conditions by which such an advantage is to be obtained for the various sciences, we shall endeavor hereafter to throw some light upon; but in the mean time, we may notice here some of the circumstances in which this peculiar felicity of the present state of astronomy may be traced.

The science is cultivated by a number of votaries, with an assiduity and labor, and with an expenditure of private and public resources, to which no other subject approaches; and the mode of its cultivation in all public and most private observatories, has this character—that it forms, at the same time, a constant process of verification of existing discoveries, and a strict search for any new discoverable laws. The observations made are immediately referred to the best tables, and cor-

---

10 Bailly, iii. 107.

rected by the best formulæ which are known; and if the result of such a reduction leaves any thing unaccounted for, the astronomer is forthwith curious and anxious to trace this deviation from the expected numbers to its rule and its origin; and till the first, at least, of these things is performed, he is dissatisfied and unquiet. The reference of observations to the state of the heavens as known by previous researches, implies a great amount of calculation. The exact places of the stars at some standard period are recorded in *Catalogues;* their movements, according to the laws hitherto detected, are arranged in *Tables;* and if these tables are applied to predict the numbers which observation on each day ought to give, they form *Ephemerides.* Thus the catalogues of fixed stars of Flamsteed, of Piazzi, of Maskelyne, of the Astronomical Society, are the basis of all observation. To these are applied the Corrections for Refraction of Bradley or Bessel, and those for Aberration, for Nutation, for Precession, of the best modern astronomers. The observations so corrected enable the observer to satisfy himself of the delicacy and fidelity of his measures of time and space; his Clocks and his Arcs. But this being done, different stars so observed can be compared with each other, and the astronomer can then endeavor further to correct his fundamental Elements;—his Catalogue, or his Tables of Corrections. In these Tables, though previous discovery has ascertained the law, yet the exact quantity, the *constant* or *coefficient* of the formula, can be exactly fixed only by numerous observations and comparisons. This is a labor which is still going on, and in which there are differences of opinion on almost every point; but the amount of these differences is the strongest evidence of the certainty and exactness of those doctrines in which all agree. Thus Lindenau makes the coefficient of Nutation rather less than nine seconds, which other astronomers give as about nine seconds and three-tenths. The Tables of Refraction are still the subject of much discussion, and of many attempts at improvement. And after or amid these discussions, arise questions whether there be not other corrections of which the law has not yet been assigned. The most remarkable example of such questions is the controversy concerning the existence of an Annual Parallax of the fixed stars, which Brinkley asserted, and which Pond denied. Such a dispute between two of the best modern observers, only proves that the quantity in question, if it really exist, is of the same order as the hitherto unsurmounted errors of instruments and corrections.

[2d Ed.] [The belief in an appreciable parallax of some of the fixed stars appears to gain ground among astronomers. The parallax of 61 *Cygni*, as determined by Bessel, is 0″·34; about one-third of a second, or 1-10000 of a degree. That of *α Centauri*, as determined by Maclear, is 0″·9, or 1-4000 of a degree.]

But besides the fixed stars and their corrections, the astronomer has the motions of the planets for his field of action. The established theories have given us tables of these, from which their daily places are calculated and given in our Ephemerides, as the *Berliner Jahrbuch* of Encke, or the *Nautical Almanac*, published by the government of this country, the *Connaissance des Tems* which appears at Paris, or the *Effemeridi di Milano.* The comparison of the observed with the tabular place, gives us the means of correcting the coefficients of the tables; and thus of obtaining greater exactness in the constants of the solar system. But these constants depend upon the mass and form of the bodies of which the system is composed; and in this province, as well as in sidereal astronomy, different determinations, obtained by different paths, may be compared; and doubts may be raised and may be solved. In this way, the perturbations produced by Jupiter on different planets gave rise to a doubt whether his attraction be really proportional to his mass, as the law of universal gravitation asserts. The doubt has been solved by Nicolai and Encke in Germany, and by Airy in England. The mass of Jupiter, as shown by the perturbations of Juno, of Vesta, and of Encke's Comet, and by the motion of his outermost Satellite, is found to agree, though different from the mass previously received on the authority of Laplace. Thus also Burckhardt, Littrow, and Airy, have corrected the elements of the Solar Tables. In other cases, the astronomer finds that no change of the coefficients will bring the Tables and the observations to a coincidence;—that a new term in the formula is wanting. He obtains, as far as he can, the law of this unknown term; if possible, he traces it to some known or probable cause. Thus Mr. Airy, in his examination of the Solar Tables, not only found that a diminution of the received mass of Mars was necessary, but perceived discordances which led him to suspect the existence of a new inequality. Such an inequality was at length found to result theoretically from the attraction of Venus. Encke, in his examination of his comet, found a diminution of the periodic time in the successive revolutions; from which he inferred the existence of a resisting medium. Uranus still deviates from his tabular place, and the cause remains yet to be discovered. (But see the *Additions* to this volume.)

Thus it is impossible that an assertion, false to any amount which the existing state of observation can easily detect, should have any abiding prevalence in astronomy. Such errors may long keep their ground in any science which is contained mainly in didactic works, and studied in the closet, but not acted upon elsewhere;—which is reasoned upon much, but brought to the test of experiment rarely or never. Here, on the contrary, an error, if it arise, makes its way into the Tables, into the Ephemeris, into the observer's nightly List, or his sheet of Reductions; the evidence of sense flies in its face in a thousand observatories; the discrepancy is traced to its source, and soon disappears forever.

In this favored branch of knowledge, the most recondite and delicate discoveries can no more suffer doubt or contradiction, than the most palpable facts of sense which the face of nature offers to our notice. The last great discovery in astronomy—the motion of the stars arising from Aberration—is as obvious to the vast population of astronomical observers in all parts of the world, as the motion of the stars about the pole is to the casual night wanderer. And this immunity from the danger of any large error in the received doctrines, is a firm platform on which the astronomer can stand and exert himself to reach perpetually further and further into the region of the unknown.

The same scrupulous care and diligence in recording all that has hitherto been ascertained, has been extended to those departments of astronomy in which we have as yet no general principles which serve to bind together our acquired treasures. These records may be considered as constituting a *Descriptive Astronomy;* such are, for instance, Catalogues of Stars, and Maps of the Heavens, Maps of the Moon, representations of the appearance of the Sun and Planets as seen through powerful telescopes, pictures of Nebulæ, of Comets, and the like. Thus, besides the Catalogue of Fundamental Stars which may be considered as standard points of reference for all observations of the Sun, Moon, and Planets, there exist many large catalogues of smaller stars. Flamsteed's *Historia Celestis*, which much surpassed any previous catalogue, contained above 3000 stars. But in 1801, the French *Histoire Céleste* appeared, comprising observations of 50,000 stars. Catalogues or charts of other special portions of the sky have been published more recently; and in 1825, the Berlin Academy proposed to the astronomers of Europe to carry on this work by portioning out the heavens among them.

[2d Ed.] [Before Flamsteed, the best Catalogue of the Stars was

Tycho Brahe's, containing the places of about 1000 stars, determined very roughly with the naked eye. On the occasion of a project of finding the longitude, which was offered to Charles II., in 1674, Flamsteed represented that the method was quite useless, in consequence, among other things, of the inaccuracy of Tycho's places of the stars. Flamsteed's letters being shown King Charles, he was startled at the assertion of the fixed stars' places being false in the Catalogue, and said, with some vehemence, "He must have them anew observed, examined, and corrected for the use of his seamen." This was the immediate occasion of building Greenwich Observatory, and placing Flamsteed there as an observer. Flamsteed's *Historia Celestis* contained above 3000 stars, observed with telescopic sights. It has recently been republished with important improvements by Mr. Baily. See Baily's *Flamsteed*, p. 38.

The French *Histoire Céleste* was published in 1801 by Lalande, containing 50,000 stars, simply as observed by himself and other French astronomers. The reduction of the observations contained in this Catalogue to the mean places at the beginning of the year 1800 may be effected by means of Tables published by Schumacher for that purpose in 1825.

In 1807, Piazzi's Catalogue of 6748 stars, founded on Maskelyne's Catalogue of 1700, was published; afterwards extended to 7646 stars in 1814. This is considered as the greatest work undertaken by any modern astronomer; the observations being well made, reduced, and compared with those of former astronomers. Piazzi's Catalogue is the standard and accurate Catalogue, as the *Histoire Céleste* is the standard approximate Catalogue for small stars. But the new planets were discovered mostly by a comparison of the heavens with Bode's (Berlin) Catalogue.

I may mention other Catalogues of Stars which have recently been published. Pond's Catalogue contains 1112 Northern stars; Johnson's, 606; Wrottesley's, 1318 (in Right Ascension only); Airy's First Cambridge Catalogue, 726; his Greenwich Catalogue, 1439. Pearson's has 520 zodiacal stars; Groombridge's, 4243 circumpolar stars as far as 50 degrees of North Polar distance; Santini's, a zone 18 degrees North of the equator. Besides these, Mr. Taylor has published, by order of the Madras government, a Catalogue of 11,000 stars observed by him at Madras; and Rumker, who observed in the Observatory established by Sir Thomas Brisbane at Paramatta (in Australia), has commenced a Catalogue which is to contain 12,000. Mr. Baily published two Standard Catalogues; that of the Royal Astronomical Society, containing 2881 stars; and that of the British Association, containing 8377 stars. I omit other Catalogues, as those of Argelander, &c., and Catalogues of Southern Stars.

Of the Berlin Maps, fourteen hours in Right Ascension have been published; and their value may be judged of by this circumstance, that it was in a great measure by comparing the heavens with these Maps that the new planet Astræa was discovered. The Zone observations made at Königsberg, by the late illustrious astronomer Bessel, deserve to be mentioned, as embracing a vast number of stars.

The common mode of *designating the Stars* is founded upon the ancient constellations as given by Ptolemy; to which Bayer, of Augsburg, in his *Uranometria*, added the artifice of designating the brightest stars in each constellation by the Greek letters, α, β, γ, &c., applied in order of brightness, and when these were exhausted, the Latin letters. Flamsteed used numbers. As the number of observed stars increased, various methods were employed for designating them; and the confusion which has been thus introduced, both with regard to the boundaries of the constellations and the nomenclature of the stars in each, has been much complained of lately. Some attempts have been made to remedy this variety and disorder. Mr. Argelander has recently recorded stars, according to their magnitudes as seen by the naked eye, in a *Neue Uranometrie.*

Among representations of the Moon I may mention Hevelius's *Selenographia*, a work of former times, and Beer and Mädler's Map of the Moon, recently published.]

I have already said something of the observations of the two Herschels on *Double Stars*, which have led to a knowledge of the law of the revolution of such systems. But besides these, the same illustrious astronomers have accumulated enormous treasures of observations of *Nebulæ;* the materials, it may be, hereafter, of some vast new generalization with respect to the history of the system of the universe.

[2d Ed.] [A few measures of Double Stars are to be found in previous astronomical records. But the epoch of the creation of this part of the science of astronomy must be placed at the beginning of the present century, when Sir William Herschel (in 1802) published in the *Phil. Trans.* a Catalogue of 500 new Nebulæ of various classes, and in the *Phil. Trans.* 1803, a paper "On the changes in the relative situation of the Double Stars in 25 years." In succeeding papers he pursued the subject. In one in 1814 he noticed the breaking up of the

Milky Way in different places, apparently from some principle of Attraction; and in this, and in one in 1817, he published those remarkable views on the distribution of the stars in our own cluster as forming a large stratum, and on the connection of stars and nebulæ (the stars appearing sometimes to be accompanied by nebulæ, sometimes to have absorbed a part of the nebula, and sometimes to have been formed from nebulæ), which have been accepted and propounded by others as the *Nebular Theory*. Sir William Herschel's last paper was a Catalogue of 145 new Double Stars communicated to the Astronomical Society in 1822. In 1827 M. Struve, of Dorpat (in Russia), published his *Catalogus Novus*, containing the places of 3112 double stars. While this was going on, Sir John Herschel and Sir James South published (in the *Phil. Trans.* 1824) accurate measures of 380 Double and Triple Stars, to which Sir J. South afterwards added 458. Mr. Dunlop published measures of 253 Southern Double Stars. Other Observations have been published by Capt. Smyth, Mr. Dawes, &c. The great work of Struve, *Mensuræ Micrometricæ*, &c., contains 3134 such objects, including most of Sir W. Herschel's Double Stars. Sir J. Herschel in 1826, 7, and 8 presented to the Astronomical Society about 1000 measures of Double Stars; and in 1830, good measures of 1236, made with his 20-feet reflector. His paper in vol. v. of the *Ast. Soc. Mem.*, besides measures of 364 such stars, exhibits all the most striking results, as to the motion of Double Stars, which have yet been obtained. In 1835 he carried his 20-feet reflector to the Cape of Good Hope for the purpose of completing the survey of Double Stars and Nebulæ in the southern hemisphere with the same instruments which had explored the northern skies. He returned from the Cape in 1838, and is now (1846) about to give the world the results of his labors. Besides the stars just mentioned, his work will contain from 1500 to 2000 additional double stars; making a gross number of above 8000; in which of course are included a number of objects of no great scientific interest, but in which also are contained the materials of the most important discoveries which remain to be made by astronomers. The publication of Sir John Herschel's great work upon Double Stars and Nebulæ is looked for with eager interest by astronomers.

Of the observations of Nebulæ we may say what has just been said of the observations of Double Stars;—that they probably contain the materials of important future discoveries. It is impossible not to regard these phenomena with reference to the *Nebular Hypothesis*, which has been propounded by Laplace, and much more strongly insisted upon by other persons;—namely, the hypothesis that systems of revolving planets, of which the Solar System is an example, arise from the gradual contraction and separation of vast masses of nebulous matter. Yet it does not appear that any changes have been observed in nebulæ which tend to confirm this hypothesis; and the most powerful telescope in the world, recently erected by the Earl of Rosse, has given results which militate against the hypothesis; inasmuch as it has shown that what appeared a diffused nebulous mass is, by a greater power of vision, resolved, in all cases yet examined, into separate stars.

When astronomical phenomena are viewed with reference to the Nebular Hypothesis, they do not belong so properly to Astronomy, in the view here taken of it, as to Cosmogony. If such speculations should acquire any scientific value, we shall have to arrange them among those which I have called *Palætiological* Sciences; namely, those Sciences which contemplate the universe, the earth, and its inhabitants, with reference to their historical changes and the causes of those changes.]

ADDITIONS TO THE THIRD EDITION.

INTRODUCTION.

THERE is a difficulty in writing for popular readers a History of the Inductive Sciences, arising from this;—that the sympathy of such readers goes most readily and naturally along the course which leads to false science and to failure. Men, in the outset of their attempts at knowledge, are prone to rush from a few hasty observations of facts to some wide and comprehensive principles; and then, to frame a system on these principles. This is the opposite of the method by which the Sciences have really and historically been conducted; namely, the method of a gradual and cautious ascent from observation to principles of limited generality, and from them to others more general. This latter, the true Scientific Method, is *Induction*, and has led to the *Inductive Sciences.* The other, the spontaneous and delusive course, has been termed by Francis Bacon, who first clearly pointed out the distinction, and warned men of the error, *Anticipation.* The hopelessness of this course is the great lesson of his philosophy; but by this course proceeded all the earlier attempts of the Greek philosophers to obtain a knowledge of the Universe.

Laborious observation, narrow and modest inference, caution, slow and gradual advance, limited knowledge, are all unwelcome efforts and restraints to the mind of man, when his speculative spirit is once roused: yet these are the necessary conditions of all advance in the Inductive Sciences. Hence, as I have said, it is difficult to win the sympathy of popular readers to the true history of these sciences. The career of bold systems and fanciful pretences of knowledge is more entertaining and striking. Not only so, but the bold guesses and fanciful reasonings of men unchecked by doubt or fear of failure are often presented as the dictates of *Common Sense;*—as the plain, unsophisticated, unforced reason of man, acting according to no artificial rules, but following its own natural course. Such Common Sense, while it complacently plumes itself on its clear-sightedness in rejecting arbitrary systems of others, is no less arbitrary in its own arguments, and often no less fanciful in its inventions, than those whom it condemns.

We cannot take a better representative of the Common Sense of the ancient Greeks than Socrates: and we find that his Common Sense, judging with such admirable sagacity and acuteness respecting moral and practical matters, offered, when he applied it to physical questions, examples of the unconscious assumptions and fanciful reasonings which, as we have said, Common Sense on such subjects commonly involves.

Socrates, Xenophon tells us (*Memorabilia*, iv. 7), recommended his friends not to study astronomy, so as to pursue it into scientific details. This was practical advice: but he proceeded further to speak of the palpable mistakes made by those who had carried such studies farthest. Anaxagoras, for instance, he said, held that the Sun was a Fire:—he did not consider that men can look at a fire, but they cannot look at the Sun; they become dark by the Sun shining upon them, but not so by the fire. He did not consider that no plants can grow well except they have sunshine, but if they are exposed to the fire they are spoiled. Again, when he said that the Sun was a stone red-hot, he did not consider that a stone heated by the fire is not luminous, and soon cools, but the sun is always luminous and always hot.

We may easily conceive how a disciple of Anaxagoras would reply to these arguments. He would say, for example, as we should probably say at present, that if there were a mass of matter so large and so hot as Anaxagoras supposed the Sun to be, its light might be as great and its heat as permanent as the heat and light of the sun are, as yet, known to be. In this case the arguments of Socrates are at any rate no better than the doctrine of Anaxagoras.

BOOK I.

THE GREEK SCHOOL PHILOSOPHY.

CHAPTER II.

The Greek Schools.

The Platonic Doctrine of Ideas.

IN speaking of the Foundation of the Greek School Philosophy, I have referred to the dialogue entitled *Parmenides*, commonly ascribed to Plato. And the doctrines ascribed to Parmenides, in that and in other works of ancient authors, are certainly remarkable examples of the tendency which prevailed among the Greeks to rush at once to the highest generalizations of which the human mind is capable. The distinctive dogma of the Eleatic School, of which Parmenides was one of the most illustrious teachers, was that *All Things are One.* This indeed was rather a doctrine of metaphysical theology than of physical science. It tended to, or agreed with, the doctrine that All things are God:—the doctrine commonly called *Pantheism.* But the tenet of the Platonists which was commonly put in opposition to this, that we must seek *The One in the Many,* had a bearing upon physical science; at least, if we interpret it, as it is generally interpreted, that we must seek the one Law which pervades a multiplicity of Phenomena. We may however take the liberty of remarking, that to speak of a Rule which is exemplified in many cases, as being "the One in the Many" (a way of speaking by which we put out of sight the consideration what very different kinds of things *the One* and *the Many* are), is a mode of expression which makes a very simple matter look very mysterious; and is another example of the tendency which urges speculative men to aim at metaphysical generality rather than scientific truth.

The Dialogue *Parmenides* is, as I have said, commonly referred to Plato. Yet it is entirely different in substance, manner, and tendency from the most characteristic of the Platonic Dialogues. In these, Socrates is represented as finally successful in refuting or routing his adversaries, however confident their tone and however popular their assertions. They are angered or humbled; he retains his good temper and his air of superiority, and when they are exhausted, he sums up in his own way.

In the *Parmenides*, on the contrary, every thing is the reverse of this. Parmenides and Zeno exchange good-humored smiles at Socrates' criticism, when the bystanders expect them to grow angry. They listen to Socrates while he propounds Plato's doctrine of Ideas; and reply to him with solid arguments which he does not answer, and which have never yet been answered. Parmenides, in a patronizing way, lets him off; and having done this, being much entreated, he pronounces a discourse concerning the One and the Many; which, obscure as it may seem to us, was obviously intended to be irrefutable: and during the whole of this part of the Dialogue, the friend of Socrates appears only as a passive respondent, saying *Yes* or *No* as the assertions of Parmenides require him to do; just in the same way in which the opponents of Socrates are represented in other Dialogues.

These circumstances, to which other historical difficulties might be added, seem to show plainly that the *Parmenides* must be regarded as an Eleatic, not as a Platonic Dialogue;—as composed to confute, not to assert, the Platonic doctrine of Ideas.

The Platonic doctrine of Ideas has an important bearing upon the philosophy of Science, and was suggested in a great measure by the progress which the Greeks had really made in Geometry, Astronomy, and other Sciences, as I shall elsewhere endeavor to show. This doctrine has been recommended in our own time,[1] as containing "a mighty substance of imperishable truth." It cannot fail to be interesting to see in what manner the doctrine is presented by those who thus judge of it. The following is the statement of its leading features which they give us.

Man's soul is made to contain not merely a consistent scheme of its own notions, but a direct apprehension of *real and eternal laws beyond it*. These real and eternal laws are things *intelligible*, and not things sensible. The laws, impressed upon creation by its Creator, and apprehended by man, are something equally distinct from the Creator

[1] A. Butler's *Lectures*, Second Series, Lect. viii. p. 132.

and from man; and the whole mass of them may be termed the World of Things purely Intelligible.

Further; there are qualities in the Supreme and Ultimate Cause of all, which are manifested in his creation; and not merely manifested, but in a manner—after being brought out of his super-essential nature into the stage of being which is below him, but next to him—are then, by the causative act of creation, deposited in things, differencing them one from the other, so that the things participate of them (*μετέχουσι*), communicate with them (*κοινωνοῦσι*).

The Intelligence of man, excited to reflection by the impressions of these objects, thus (though themselves transitory) participant of a divine quality, may rise to higher conceptions of the perfections thus faintly exhibited; and inasmuch as the perfections are unquestionably *real* existences, and known to be such in the very act of contemplation, this may he regarded as a distinct intellectual apprehension of them; —a union of the Reason with the Ideas in that sphere of being which is common to both.

Finally, the Reason, in proportion as it learns to contemplate the Perfect and Eternal, desires the enjoyment of such contemplations in a more consummate degree, and cannot be fully satisfied except in the actual fruition of the Perfect itself.

These propositions taken together constitute the THEORY OF IDEAS. When we have to treat of the Philosophy of Science, it may be worth our while to resume the consideration of this subject.

In this part of the History, the *Timæus* of Plato is referred to as an example of the loose notions of the Greek philosophers in their physical reasonings. And undoubtedly this Dialogue does remarkably exemplify the boldness of the early Greek attempts at generalization on such subjects. Yet in this and in other parts the writings of Plato contain speculations which may be regarded as containing germs of true physical science; inasmuch as they assume that the phenomena of the world are governed by mathematical laws;—by relations of space and number;—and endeavor, too boldly, no doubt, but not vaguely or loosely, to assign those laws. The Platonic writings offer, in this way, so much that forms a Prelude to the Astronomy and other Physical Sciences of the Greeks, that they will deserve our notice, as supplying materials for the next two Books of the History, in which these subjects are treated of.

CHAPTER III.

Failure of the Greek Physical Philosophy.

Francis Bacon's Remarks.

THOUGH we do not accept, as authority, even the judgments of Francis Bacon, and shall have to estimate the strong and the weak parts of his, no less than of other philosophies, we shall find his remarks on the Greek philosophers very instructive. Thus he says of Aristotle (*Nov. Org.* 1. Aph. lxiii.):

"He is an example of the kind of philosophy in which much is made out of little; so that the basis of experience is too narrow. He corrupted Natural Philosophy by his Logic, and made the world out of his Categories. He disposed of the distinction of *dense* and *rare*, by which bodies occupy more or less dimensious or space, by the frigid distinction of *act* and *power*. He assigned to each kind of body a single proper motion, so that if they have any other motion they must receive it from some extraneous source; and imposed many other arbitrary rules upon Nature; being everywhere more careful how one may give a ready answer, and make a positive assertion, than how he may apprehend the variety of nature.

"And this appears most evidently by the comparison of his philosophy with the other philosophies which had any vogue in Greece. For the *Homoiomeria*[1] of Anaxagoras, the *Atoms* of Leucippus and Democritus, the Heaven and Earth of Parmenides, the Love and Hate of Empedocles, the Fire of Heraclitus, had some trace of the thoughts of a natural philosopher; some savor of experience, and nature, and bodily things; while the Physics of Aristotle, in general, sound only of Logical Terms.

"Nor let any one be moved by this—that in his books *Of Animals*, and in his *Problems*, and in others of his tracts, there is often a quoting of experiments. For he had made up his mind beforehand; and did not consult experience in order to make right propositions and axioms, but when he had settled his system to his will, he twisted experience

[1] For these technical forms of the Greeks, see Sec. 3 of this chapter.

round, and made her bend to his system: so that in this way he is even more wrong than his modern followers, the Schoolmen, who have deserted experience altogether."

We may note also what Bacon says of the term *Sophist.* (Aph. lxxi.) "The wisdom of the Greeks was professorial, and prone to run into disputations: which kind is very adverse to the discovery of Truth. And the name of *Sophists*, which was cast in the way of contempt, by those who wished to be reckoned philosophers, upon the old professors of rhetoric, Gorgias, Protagoras, Hippias, Polus, does, in fact, fit the whole race of them, Plato,[2] Aristotle, Zeno, Epicurus, Theophrastus; and their successors, Chrysippus, Carneades, and the rest."

That these two classes of teachers, as moralists, were not different in their kind, has been urged by Mr. Grote in a very striking and amusing manner. But Bacon speaks of them here as physical philosophers; in which character he holds that all of them were *sophists*, that is, illusory reasoners.

Aristotle's Account of the Rainbow.

To exemplify the state of physical knowledge among the Greeks, we may notice briefly Aristotle's account of the *Rainbow;* a phenomenon so striking and definite, and so completely explained by the optical science of later times. We shall see that not only the explanations there offered were of no value, but that even the observation of facts, so common and so palpable, was inexact. In his *Meteorologics* (lib. iii. c. 2) he says, "The Rainbow is never more than a semicircle. And at sunset and sunrise, the circle is least, but the arch is greatest; when the sun is high, the circle is larger, but the arch is less." This is erroneous, for the diameter of the circle of which the arch of the rainbow forms a part, is always the same, namely 82°. "After the autumnal equinox," he adds, "it appears at every hour of the day; but in the summer season, it does not appear about noon." It is curious that he did not see the reason of this. The centre of the circle of which the rainbow is part, is always opposite to the sun. And therefore if the sun be more than 41° above the horizon, the centre of the rainbow will be so much below the horizon, that the place of the rainbow will

[2] It is curious that the attempt to show that Plato's opponents were not commonly illusive and immoral reasoners, has been represented as an attempt to obliterate the distinction of "Sophist" and "Philosopher."—See A. Butler's *Lectures*, i. 357. Note.

be entirely below the horizon. In the latitude of Athens, which is 38°, the equator is 52° above the horizon, and the rainbow can be visible only when the sun is 11° lower than it is at the equinoctial noon. These remarks, however, show a certain amount of careful observation; and so do those which Aristotle makes respecting the colors. "Two rainbows at most appear: and of these, each has three colors; but those in the outer bow are duller; and their order opposite to those in the inner. For in the inner bow the first and largest arch is red; but in the outer bow the smallest arch is red, the nearest to the inner; and the others in order. The colors are red, green, and purple, such as painters cannot imitate." It is curious to observe how often modern painters disregard even the order of the colors, which they could imitate, if they attended to it.

It may serve to show the loose speculation which we oppose to science, if we give Aristotle's attempt to explain the phenomenon of the Rainbow. It is produced, he says (c. iv.), by Reflexion (*ἀνάκλασις*) from a cloud opposite to the sun, when the cloud forms into drops. And as a reason for the red color, he says that a bright object seen through darkness appears red, as the flame through the smoke of a fire of green wood. This notion hardly deserves notice; and yet it was taken up again by Göthe in our own time, in his speculations concerning colors.

BOOK II.

THE PHYSICAL SCIENCES IN ANCIENT GREECE.

Plato's Timæus and Republic.

ALTHOUGH a great portion of the physical speculations of the Greek philosophers was fanciful, and consisted of doctrines which were rejected in the subsequent progress of the Inductive Sciences; still many of these speculations must be considered as forming a Prelude to more exact knowledge afterwards attained; and thus, as really belonging to the Progress of knowledge. These speculations express, as we have already said, the conviction that the phenomena of nature are governed by laws of space and number; and commonly, the mathematical laws which are thus asserted have some foundation in the facts of nature. This is more especially the case in the speculations of Plato. It has been justly stated by Professor Thompson (A. Butler's *Lectures*, Third Series, Lect. i. Note 11), that it is Plato's merit to have discovered that the laws of the physical universe are resolvable into numerical relations, and therefore capable of being represented by mathematical formulæ. Of this truth, it is there said, Aristotle does not betray the slightest consciousness.

The *Timæus* of Plato contains a scheme of mathematical and physical doctrines concerning the universe, which make it far more analogous than any work of Aristotle to Treatises which, in modern times, have borne the titles of *Principia*, *System of the World*, and the like. And fortunately the work has recently been well and carefully studied, with attention, not only to the language, but to the doctrines and their bearing upon our real knowledge. Stallbaum has published an edition of the Dialogue, and has compared the opinions of Plato with those of Aristotle on the like subjects. Professor Archer Butler of Dublin has devoted to it several of his striking and eloquent Lectures; and these have been furnished with valuable annotations by Professor Thompson of Cambridge; and M. The. Henri Martin, then Professor at Rennes, published in 1841 two volumes of *Etudes sur le Timée de Platon*, in which the bearings of the work on Science are very fully discussed. The Dialogue treats not only concerning the numerical laws of harmonical sounds, of visual appearances, and of the motions of planets and stars, but also concerning heat, as well as light; and concerning water, ice, gold, gems, iron, rust, and other natural objects;—concerning odors, tastes, hearing, sight, light, colors, and the powers of sense in general:—concerning the parts and organs of the body, as the bones, the marrow, the brain, the flesh, muscles, tendons, ligaments, nerves; the skin, the hair, the nails; the veins and arteries; respiration; generation; and in short, every obvious point of physiology.

But the opinions delivered in the *Timæus* upon these latter subjects have little to do with the progress of real knowledge. The doctrines, on the other hand, which depend upon geometrical and arithmetical relations, are portions or preludes of the sciences which, in the fulness of time, assumed a mathematical form for the expression of truth.

Among these may be mentioned the arithmetical relations of harmonical sounds, to which I have referred in the History. These occur in various parts of Plato's writings. In the *Timæus*, in which the numbers are most fully given, the meaning of the numbers is, at first sight, least obvious. The numbers are given as representing the proportion of the parts of the Soul (*Tim.* pp. 35, 36), which does not immediately refer us to the relations of Sounds. But in a subsequent part of the Dialogue (47, D), we are told that music is a privilege of the hearing given on account of Harmony; and that Harmony has Cycles corresponding to the movements of the Soul (referring plainly to those already asserted). And the numbers which are thus given by Plato as elements of harmony, are in a great measure the same as those which express the musical relations of the tones of the musical scale at this day in use, as M. Henri Martin shows (*Et. sur le Timée*, note xxiii). The intervals C to D, C to F, C to G, C to C, are expressed by the fractions $\frac{9}{8}$, $\frac{4}{3}$, $\frac{3}{2}$, $\frac{2}{1}$, and are now called a Tone, a Fourth, a Fifth, an Octave. They were expressed by the same fractions among the Greeks, and were called *Tone*, *Diatessaron*, *Diapente*, *Diapason*. The Major and Minor Third, and the Major and Minor Sixth, were however wanting, it is conceived, in the musical scale of Plato.

The *Timæus* contains also a kind of theory of vision by reflexion from a plane, and in a concave mirror; although the theory is in this case less mathematical and less precise than that of Euclid, referred to in chap. ii. of this Book.

One of the most remarkable speculations in the *Timæus* is that in which the Regular Solids are assigned as the forms of the Elements of which the Universe is composed. This curious branch of mathematics, Solid Geometry, had been pursued with great zeal by Plato and his friends, and with remarkable success. The five Regular Solids, the Tetrahedron or regular Triangular Pyramid, the Cube, the Octahedron, the Dodecahedron, and the Icosahedron, had been discovered; and the remarkable theorem, that of regular solids there can be just so many, these and no others, was known. And in the *Timæus* it is asserted that the particles of the various elements have the forms of these solids. Fire has the Pyramid; Earth has the Cube; Water the Octahedron; Air the Icosahedron; and the Dodecahedron is the plan of the Universe itself. It was natural that when Plato had learnt that other mathematical properties had a bearing upon the constitution of the Universe, he should suppose that the singular property of space, which the existence of this limited and varied class of solids implied, should have some corresponding property in the Universe, which exists in space.

We find afterwards, in Kepler and others, a recurrence to this assumption; and we may say perhaps that Crystallography shows us that there are properties of bodies, of the most intimate kind, which involve such spatial relations as are exhibited in the Regular Solids. If the distinctions of Crystalline System in bodies were hereafter to be found to depend upon the chemical elements which predominate in their composition, the admirers of Plato might point to his doctrine, of the different form of the particles of the different elements of the Universe, as a remote Prelude to such a discovery.

But the mathematical doctrines concerning the parts and elements of the Universe are put forwards by Plato, not so much as assertions concerning physical facts, of which the truth or falsehood is to be determined by a reference to nature herself. They are rather propounded as examples of a truth of a higher kind than any reference to observation can give or can test, and as revelations of principles such as must have prevailed in the mind of the Creator of the Universe; or else as contemplations by which the mind of man is to be raised above the region of sense, and brought nearer to the Divine Mind. In the *Timæus* these doctrines appear rather in the former of the two lights; as an exposition of the necessary scheme of creation, so far as its leading features are concerned. In the seventh Book of the *Polity*, the same doctrines are regarded more as a mental discipline; as the necessary study of the true philosopher. But in both places these mathematical propositions are represented as Realities more real than the Phenomena;—as a Natural Philosophy of a higher kind than the study of Nature itself can teach. This is no doubt an erroneous assumption: yet even in this there is a germ of truth; namely, that the mathematical laws, which prevail in the universe, involve mathematical truths; which being demonstrative, are of a higher and more cogent kind than mere experimental truths.

Notions, such as these of Plato, respecting a truth at which science is to aim, which is of an exact and demonstrative kind, and is imperfectly manifested in the phenomena of nature, may help or may mislead inquirers; they may be the impulse and the occasion to great discoveries; or they may lead to the assertion of false and the loss of true doctrines. Plato considers the phenomena which nature offers to the senses as mere suggestions and rude sketches of the objects which the philosophic mind is to contemplate. The heavenly bodies and all the splendors of the sky, though the most beautiful of visible objects, being only visible objects, are far inferior to the true objects of which they are the representatives. They are merely diagrams which may assist in the study of the higher truth; as we might study geometry by the aid of diagrams constructed by some consummate artist. Even then, the true object about which we reason is the conception which we have in the mind.

We have, I conceive, an instance of the error as well as of the truth, to which such views may lead, in the speculations of Plato concerning Harmony, contained in that part of his writings (the seventh Book of the *Republic*), in which these views are especially urged. He there, by way of illustrating the superiority of philosophical truth over such exactness as the senses can attest, speaks slightingly of those who take immense pains in measuring musical notes and intervals by the ear, as the astronomers measure the heavenly motions by the eye. "They screw their pegs and pinch their strings, and dispute whether two notes are the same or not." Now, in truth, the ear *is* the final and supreme judge whether two notes are the same or not. But there is a case in which notes which are nominally the same, are different really and to the ear; and it is probably to disputes on this subject, which we know did prevail among the Greek musicians, that Plato here refers. We may ascend from a note A_1 to a note C_3 by two octaves and a third. We may also ascend from the same note A_1 to C_3 by fifths four times repeated. But the two notes C_3 thus arrived at are not the same: they differ by a small interval, which the Greeks called a Comma, of which the notes are in the ratio of 80 to 81. That the ear really detects this defect of musical coincidence of the two notes under the proper conditions, is a proof of the coincidence of our musical perceptions with the mathematical relations of the notes; and is therefore an experimental confirmation of the mathematical principles of harmony. But it seems to be represented by Plato, that to look out for such confirmation of mathematical principles, implies a disposition to lean on the senses, which he regards as very unphilosophical.

Hero of Alexandria.

The other branches of mathematical science which I have spoken of in the History as cultivated by the Greeks, namely Mechanics and Hydrostatics, are not treated expressly by Plato; though we know from Aristotle and others that some of the propositions of those sciences were known about his time. Machines moved not only by weights and springs, but by water and air, were constructed at an early period. Ctesibius, who lived probably about B. C. 250, under the Ptolemies, is said to have invented a clepsydra or water-clock, and an hydraulic organ; and to have been the first to discover the elastic power of air, and to apply it as a moving power. Of his pupil Hero, the name is to this day familiar, through the little pneumatic instrument called *Hero's Fountain.* He also described pumps and hydraulic machines of various kinds; and an instrument which has been spoken of by some modern writers as a *steam-engine*, but which was merely a toy made to whirl round by the steam emitted from holes in its arms. Concerning mechanism, besides descriptions of *Automatons*, Hero composed two works: the one entitled *Mechanics*, or *Mechanical Introductions;* the other *Barulcos*, the *Weight-lifter.* In these works the elementary contrivances by which weights may be lifted or drawn were spoken of as the *Five Mechanical Powers*, the same enumeration of such machines as prevails to this day; namely, the Lever, the Wheel and Axle, the Pulley, the Wedge, and the Screw. In his Mechanics, it appears that Hero reduced all these machines to one single machine, namely to the lever. In the *Barulcos*, Hero proposed and solved the problem which it was the glory of Archimedes to have solved: To move any object (however large) by any power (however small). This, as may easily be conceived by any one acquainted with the elements of Mechanics, is done by means of a combination of the mechanical powers, and especially by means of a train of toothed-wheels and axles.

The remaining writings of Hero of Alexandria have been the subject of a special, careful, and learned examination by M. Th. H. Martin (Paris, 1854), in which the works of this writer, Hero the Ancient, as he is sometimes called, are distinguished from those of another writer of the same name of later date.

Hero of Alexandria wrote also, as it appears, a treatise on *Pneumatics*, in which he described machines, either useful or amusing, moved by the force of air and vapor.

He also wrote a work called *Catoptrics*, which contained proofs of properties of the rays of reflected light.

And a treatise *On the Dioptra;* which subject, however, must be carefully distinguished from the subject entitled *Dioptrics* by the moderns. This latter subject treats of the properties of *refracted* light; a subject on which the ancients had little exact knowledge till a later period, as I have shown in the History. The *Dioptra*, as understood by Hero, was an instrument for taking angles so as to measure the position, and hence to determine the distance of inaccessible objects; as is done by the *Theodolite* in our times.

M. Martin is of opinion that Hero of Alexandria lived at a later period than is generally supposed; namely, after B. C. 81.

BOOK III.

THE GREEK ASTRONOMY.

INTRODUCTION.

THE mathematical opinions of Plato respecting the philosophy of nature, and especially respecting what we commonly call "the heavenly bodies," the Sun, Moon, and Planets, were founded upon the view which I have already described: namely, that it is the business of philosophy to aim at a truth higher than observation can teach; and to solve problems which the phenomena of the universe only suggest. And though the students of nature in more recent times have learnt that this is too presumptuous a notion of human knowledge, yet the very boldness and hopefulness which it involved impelled men in the pursuit of truth, with more vigor than a more timorous temper could have done; and the belief that there must be, in nature, mathematical laws more exact than experience could discover, stimulated men often to discover true laws, though often also to invent false laws. Plato's writings, supplying examples of both these processes, belong to the Prelude of true Astronomy, as well as to the errors of false philosophy. We may find specimens of both kinds in those parts of his Dialogues to which we have referred in the preceding Book of our History.

To Plato's merits in preparing the way for the Theory of Epicycles, I have already referred in Chapter ii. of this Book. I conceive that he had a great share in that which is an important step in every discovery, the proposing distinctly the problem to be solved; which was, in this case, as he states it, To account for the apparent movements of the planets by a combination of two circular motions for each:—the motion of identity, and the motion of difference. (*Tim.* 39, A.) In the tenth Book of the *Republic*, quoted in our text, the spindle which Destiny or Necessity holds between her knees, and on which are rings, by means of which the planets revolve round it as an axis, is a step towards the conception of the problem, as the construction of a machine.

It will not be thought surprising that Plato expected that Astronomy, when further advanced, would be able to render an account of many things for which she has not accounted even to this day. Thus, in the passage in the seventh Book of the *Republic*, he says that the philosopher requires a reason for the proportion of the day to the month, and the month to the year, deeper and more substantial than mere observation can give. Yet Astronomy has not yet shown us any reason why the proportion of the times of the earth's rotation on its axis, the moon's revolution round the earth, and the earth's revolution round the sun, might not have been made by the Creator quite different from what they are. But in thus asking Mathematical Astronomy for reasons which she cannot give, Plato was only doing what a great astronomical discoverer, Kepler, did at a later period. One of the questions which Kepler especially wished to have answered was, why there are five planets, and why at such particular distances from the sun? And it is still more curious that he thought he had found the reason of these things, in the relations of those Five Regular Solids which, as we have seen, Plato was desirous of introducing into the philosophy of the universe. We have Kepler's account of this, his imaginary discovery, in the *Mysterium Cosmographicum*, published in 1596, as stated in our History, Book v. Chap. iv. Sect. 2.

Kepler regards the law which thus determines the number and magnitude of the planetary orbits by means of the five regular solids as a discovery no less remarkable and certain than the Three Laws which give his name its imperishable place in the history of astronomy.

We are not on this account to think that there is no steady criterion of the difference between imaginary and real discoveries in science. As discovery becomes possible by the liberty of guessing, it becomes real by allowing observation constantly and authoritatively to determine the value of guesses. Kepler added to Plato's boldness of fancy his own patient and candid habit of testing his fancies by a rigorous and laborious comparison with the phenomena; and thus his discoveries led to those of Newton.

CHAPTER I.

EARLIEST STAGES OF ASTRONOMY.

The Globular Form of the Earth.

THERE are parts of Plato's writings which have been adduced as bearing upon the subsequent progress of science; and especially upon the globular form of the earth, and the other views which led to the discovery of America. In the *Timæus* we read of a great continent lying in the Ocean west of the Pillars of Hercules, which Plato calls *Atlantis*. He makes the personage in his Dialogue who speaks of this put it forward as an Egyptian tradition. M. H. Martin, who has discussed what has been written respecting the Atlantis of Plato, and has given therein a dissertation rich in erudition and of the most lively interest, conceives that Plato's notions on this subject arose from his combining his conviction of the spherical form of the earth, with interpretations of Homer, and perhaps with traditions which were current in Egypt (*Etudes sur le Timée*, Note xiii. § ix.). He does not consider that the belief in Plato's Atlantis had any share in the discoveries of Columbus.

It may perhaps surprise modern readers who have a difficulty in getting rid of the persuasion that there is a natural direction *upwards* and a natural direction *downwards*, to learn that both Plato and Aristotle, and of course other philosophers also, had completely overcome this difficulty. They were quite ready to allow and to conceive that *down* meant nothing but towards some centre, and *up*, the opposite direction. (Aristotle has, besides, an ingenious notion that while heavy bodies, as earth and water, tend to the centre, and light bodies, as fire, tend from the centre, the fifth element, of which the heavenly bodies are composed, tends to move *round* the centre.)

Plato explains this in the most decided manner in the *Timæus* (62, c). "It is quite erroneous to suppose that there are two opposite regions in the universe, one *above* and the other *below*; and that heavy things naturally tend to the latter place. The heavens are spherical, and every thing tends to the centre; and thus *above* and *below* have no real meaning. If there be a solid globe in the middle, and if a person walk round it, he will become the antipodes to himself, and the direction which is *up* at one time will be *down* at another."

The notion of *antipodes*, the inhabitants of the part of the globe of the earth opposite to ourselves, was very familiar. Thus in Cicero's *Academic Questions* (ii. 39) one of the speakers says, "Etiam dicitis esse e regione nobis, e contraria parte terræ, qui adversis vestigiis stant contra nostra vestigia, quos Antipodas vocatis." See also *Tusc. Disp.* i. 28 and v. 24.

The Heliocentric System among the Ancients.

As the more clear-sighted of the ancients had overcome the natural prejudice of believing that there is an absolute *up* and *down*, so had they also overcome the natural prejudice of believing that the earth is at rest. Cicero says (*Acad. Quest.* ii. 39), "Hicetas of Syracuse, as Theophrastus tells us, thinks that the heavens, the sun, the moon, the stars do not move; and that nothing does move but the earth. The earth revolves about her axis with immense velocity; and thus the same effect is produced as if the earth were at rest and the heavens moved; and this, he says, Plato teaches in the *Timæus*, though somewhat obscurely." Of course the assertion that the moon and planets do not move, was meant of the diurnal motion only. The passage referred to in the *Timæus* seems to be this (40, c)—"As to the Earth, which is our nurse, and which *clings to* the axis which stretches through the universe, God made her the producer and preserver of day and night." The word εἰλλομένην, which I have translated *clings to*, some translate *revolves;* and an extensive controversy has prevailed, both in ancient and modern times (beginning with Aristotle), whether Plato did or did not believe in the rotation of the earth on her axis. (See M. Cousin's Note on the *Timæus*, and M. Henri Martin's Dissertation, Note xxxvii., in his *Etudes sur le Timée.*) The result of this discussion seems to be that, in the *Timæus*, the Earth is supposed to be at rest. It is however related by Plutarch (*Platonic Questions*, viii. 1), that Plato in his old age repented of having given to the Earth the place in the centre of the universe which did not belong to it.

In describing the Prelude to the Epoch of Copernicus (Book v. Chap. i.), I have spoken of Philolaus, one of the followers of Pythagoras, who lived at the time of Socrates, as having held the doctrine that the earth revolves about the sun. This has been a current opinion;—so current, indeed, that the Abbé Bouillaud, or Bullialdus, as we more commonly call him, gave the title of *Philolaus* to the defence of Copernicus which he published in 1639; and Chiaramonti, an Aristotelian, published his answer under the title of *Antiphilolaus.* In 1645 Bullialdus published his *Astronomia Philolaica*, which was another exposition of the heliocentric doctrine.

Yet notwithstanding this general belief, it appears to be tolerably certain that Philolaus did not hold the doctrine of the earth's motion round the sun. (M. H. Martin, *Etudes sur le Timée*, 1841, Note xxxvii. Sect. i.; and Bœckh, *De vera Indole Astronomiœ Philolaicœ*, 1810.) In the system of Philolaus, the earth revolved about *the central fire;* but this central fire was not the sun. The Sun, along with the moon and planets, revolved in circles external to the earth. The Earth had the *Antichthon* or *Counter-Earth* between it and the centre; and revolving round this centre in one day, the Antichthon, being always between it and the centre, was, during a portion of the revolution, interposed between the Earth and the Sun, and thus made night; while the Sun, by his proper motion, produced the changes of the year.

When men were willing to suppose the earth to be in motion, in order to account for the recurrence of day and night, it is curious that they did not see that the revolution of a spherical earth about an axis passing through its centre was a scheme both simple and quite satisfactory. Yet the illumination of a globular earth by a distant sun, and the circumstances and phenomena thence resulting, appear to have been conceived in a very confused manner by many persons. Thus Tacitus (*Agric.* xii.), after stating that he has heard that in the northern part of the island of Britain, the night disappears in the height of summer, says, as his account of this phenomenon, that "the extreme parts of the earth are low and level, and do not throw their shadow upwards; so that the shade of night falls below the sky and the stars." But, as a little consideration will show, it is the globular form of the earth, and not the level character of the country, which produces this effect.

It is not in any degree probable that Pythagoras taught that the Earth revolves round the Sun, or that it rotates on its own axis. Nor did Plato hold either of these motions of the Earth. They got so far as to believe in the Spherical Form of the Earth; and this was apparently such an effort that the human mind made a pause before going any further. "It required," says M. H. Martin, "a great struggle for men to free themselves from the prejudices of the senses, and to interpret their testimony in such a manner as to conceive the sphericity of the earth. It is natural that they should have stopped at this point, before putting the earth in motion in space."

Some of the expressions which have been understood as describing a system in which the Sun is the *centre of motion*, do really imply merely the Sun is the *middle term* of the series of heavenly bodies which revolve round the earth : the series being Moon, Mercury, Venus, Sun, Mars, Jupiter, Saturn. This is the case, for instance, in a passage of Cicero's *Vision of Scipio*, which has been supposed to imply (as I have stated in the History), that Mercury and Venus revolve about the Sun.

But though the doctrine of the diurnal rotation and annual revolution of the earth is not the doctrine of Pythagoras, or of Philolaus, or of Plato, it was nevertheless held by some of the philosophers of antiquity. The testimony of Archimedes that this doctrine was held by his contemporary Aristarchus of Samos, is unquestionable ; and there is no reason to doubt Plutarch's assertion that Seleucus further enforced it.

It is curious that Copernicus appears not to have known any thing of the opinions of Aristarchus and Seleucus, which were really anticipations of his doctrine ; and to have derived his notion from passages which, as I have been showing, contain no such doctrine. He says, in his Dedication to Pope Paul III., "I found in Cicero that Nicetas [or Hicetas] held that the earth was in motion : and in Plutarch I found that some others had been of that opinion ; and his words I will transcribe, that any one may read them: 'Philosophers in general hold that the earth is at rest. But Philolaus the Pythagorean teaches that it moves round the central fire in an oblique circle, in the same direction as the Sun and the Moon. Heraclides of Pontus and Ecphantus the Pythagorean give the earth a motion, but not a motion of translation ; they make it revolve like a wheel about its own centre from west to east.' " This last opinion was a correct assertion of the diurnal motion.

The Eclipse of Thales.

"The Eclipse of Thales" is so remarkable a point in the history of astronomy, and has been the subject of so much discussion among astronomers, that it ought to be more especially noticed. The original record is in the first Book of Herodotus's History (chap. lxxiv.) He says that there was a war between the Lydians and the Medes; and after various turns of fortune, "in the sixth year a conflict took place; and on the battle being joined, it happened that the day suddenly became night. And this change, Thales of Miletus had predicted to them, definitely naming this year, in which the event really took place. The Lydians and the Medes, when they saw day turned into night, ceased from fighting; and both sides were desirous of peace." Probably this prediction was founded upon the Chaldean period of eighteen years, of which I have spoken in Section 11. It is probable, as I have already said, that this period was discovered by noticing the recurrence of eclipses. It is to be observed that Thales predicted only the year of the eclipse, not the day or the month. In fact, the exact prediction of the circumstances of an eclipse of the sun is a very difficult problem; much more difficult, it may be remarked, than the prediction of the circumstance of an eclipse of the moon.

Now that the Theory of the Moon is brought so far towards completeness, astronomers are able to calculate backwards the eclipses of the sun which have taken place in former times; and the question has been much discussed in what year this Eclipse of Thales really occurred. The Memoir of Mr. Airy, the Astronomer Royal, on this subject, in the *Phil. Trans.* for 1853, gives an account of the modern examinations of this subject. Mr. Airy starts from the assumption that the eclipse must have been one decidedly total; the difference between such a one and an eclipse only *nearly* total being very marked. A total eclipse alone was likely to produce so strong an effect on the minds of the combatants. Mr. Airy concludes from his calculations that the eclipse predicted by Thales took place B. C. 585.

Ancient eclipses of the Moon and Sun, if they can be identified, are of great value for modern astronomy; for in the long interval of between two and three thousand years which separates them from our time, those of the *inequalities*, that is, accelerations or retardations of the Moon's motion, which go on increasing constantly,[1] accumulate to a large amount; so that the actual time and circumstances of the eclipse give astronomers the means of determining what the rate of these accelerations or retardations has been. Accordingly Mr. Airy has discussed, as even more important than the eclipse of Thales, an eclipse which Diodorus relates to have happened during an expedition of

[1] Or at least for very long periods.

Agathocles, the ruler of Sicily, and which is hence known as the Eclipse of Agathocles. He determines it to have occurred B. C. 310.

M. H. Martin, in Note xxxvii. to his *Etudes sur le Timée*, discusses among other astronomical matters, the Eclipse of Thales. He does not appear to render a very cordial belief to the historical fact of Thales having delivered the prediction before the event. He says that even if Thales did make such a prediction of an eclipse of the sun, as he might do, by means of the Chaldean period of 18 years, or 223 lunations, he would have to take the chance of its being visible in Greece, about which he could only guess:—that no author asserts that Thales, or his successors Anaximander and Anaxagoras, ever tried their luck in the same way again:—that "en revanche" we are told that Anaximander predicted an earthquake, and Anaxagoras the fall of aërolites, which are plainly fabulous stories, though as well attested as the Eclipse of Thales. He adds that according to Aristotle, Thales and Anaximenes were so far from having sound notions of cosmography, that they did not even believe in the roundness of the earth.

BOOK IV.

PHYSICAL SCIENCE IN THE MIDDLE AGES.

GENERAL REMARKS.

IN the twelfth Book of the *Philosophy*, in which I have given a Review of Opinions on the Nature of Knowledge and the method of seeking it, I have given some account of several of the most important persons belonging to the ages now under consideration. I have there (vol. ii. b. xii. p. 146) spoken of the manner in which remarks made by Aristotle came to be accepted as fundamental maxims in the schools of the middle ages, and of the manner in which they were discussed by the greatest of the schoolmen, as Thomas Aquinas, Albertus Magnus, and the like. I have spoken also (p. 149) of a certain kind of recognition of the derivation of our knowledge from experience; as shown in Richard of St. Victor, in the twelfth century. I have considered (p. 152) the plea of the admirers of those ages, that religious authority was not claimed for physical science.

I have noticed that the rise of Experimental Philosophy exhibited two features (chap. vii. p. 155), the Insurrection against Authority, and the Appeal to Experience: and as exemplifying these features, I have spoken of Raymond Lully and of Roger Bacon. I have further noticed the opposition to the prevailing Aristotelian dogmatism manifested (chap. viii.) by Nicolas of Cus, Marsilius Ficinus, Francis Patricius, Picus of Mirandula, Cornelius Agrippa, Theophrastus Paracelsus, Robert Fludd. I have gone on to notice the Theoretical Reformers of Science (chap. ix.), Bernardinus Telesius, Thomas Campanella, Andreas Cæsalpinus, Peter Ramus; and the Protestant Reformers, as Melancthon. After these come the Practical Reformers of Science, who have their place in the subsequent history of Inductive Philosophy; Leo nardo da Vinci, and the Heralds of the dawning light of real science, whom Francis Bacon welcomes, as Heralds are accosted in Homer:

Χαίρετε Κήρυκες Διὸς ἄγγελοι ἠδὲ καὶ ἀνδρῶν.
Hail, Heralds, messengers of Gods and men!

I have, in the part of the *Philosophy* referred to, discussed the merits and defects of Francis Bacon's *Method*, and I shall have occasion, in the next Book, to speak of his mode of dealing with the positive science of his time. There is room for much more reflection on these subjects, but the references now made may suffice at present.

CHAPTER V.

Progress in the Middle Ages.

Thomas Aquinas.

AQUINAS wrote (besides the *Summa* mentioned in the text) a Commentary on the Physics of Aristotle: *Commentaria in Aristotelis Libros Physicorum*, Venice, 1492. This work is of course of no scientific value; and the commentary consists of empty permutations of abstract terms, similar to those which constitute the main substance of the text in Aristotle's physical speculations. There is, however, an attempt to give a more technical form to the propositions and their demonstrations. As specimens of these, I may mention that in Book vi. c. 2, we have a demonstration that when bodies move, the time and the magnitude (that is, the space described), are divided similarly; with many like propositions. And in Book viii. we have such propositions as this (c. 10): "Demonstration that a finite mover (*movens*) cannot move any thing in an infinite time." This is illustrated by a diagram in which two hands are represented as engaged in moving a whole sphere, and one hand in moving a hemisphere.

This mode of representing force, in diagrams illustrative of mechanical reasonings, by human hands pushing, pulling, and the like, is still employed in elementary books. Probably this is the first example of such a mode of representation.

Roger Bacon.

This writer, a contemporary of Thomas Aquinas, exhibits to us a kind of knowledge, speculation, and opinion, so different from that of any known person near his time, that he deserves especial notice here;

and I shall transfer to this place the account which I have given of him in the *Philosophy*. I do this the more willingly because I regard the existence of such a work as the *Opus Majus* at that period as a problem which has never yet been solved. Also I may add, that the scheme of the Contents of this work which I have given, deserves, as I conceive, more notice than it has yet received.

"Roger Bacon was born in 1214, near Ilchester, in Somersetshire, of an old family. In his youth he was a student at Oxford, and made extraordinary progress in all branches of learning. He then went to the University of Paris, as was at that time the custom of learned Englishmen, and there received the degree of Doctor of Theology. At the persuasion of Robert Grostête, bishop of Lincoln, he entered the brotherhood of Franciscans in Oxford, and gave himself up to study with extraordinary fervor. He was termed by his brother monks *Doctor Mirabilis*. We know from his own works, as well as from the traditions concerning him, that he possessed an intimate acquaintance with all the science of his time which could be acquired from books; and that he had made many remarkable advances by means of his own experimental labors. He was acquainted with Arabic, as well as with the other languages common in his time. In the title of his works, we find the whole range of science and philosophy, Mathematics and Mechanics, Optics, Astronomy, Geography, Chronology, Chemistry, Magic, Music, Medicine, Grammar, Logics, Metaphysics, Ethics, and Theology; and judging from those which are published, these works are full of sound and exact knowledge. He is, with good reason, supposed to have discovered, or to have had some knowledge of, several of the most remarkable inventions which were made generally known soon afterwards; as gunpowder, lenses, burning specula, telescopes, clocks, the correction of the calendar, and the explanation of the rainbow.

"Thus possessing, in the acquirements and habits of his own mind, abundant examples of the nature of knowledge and of the process of invention, Roger Bacon felt also a deep interest in the growth and progress of science, a spirit of inquiry respecting the causes which produced or prevented its advance, and a fervent hope and trust in its future destinies; and these feelings impelled him to speculate worthily and wisely respecting a Reform of the Method of Philosophizing. The manuscripts of his works have existed for nearly six hundred years in many of the libraries of Europe, and especially in those of England; and for a long period the very imperfect portions of them which were generally known, left the character and attainments of the author shrouded in a kind of mysterious obscurity. About a century ago, however, his *Opus Majus* was published[1] by Dr. S. Jebb, principally from a manuscript in the library of Trinity College, Dublin; and this contained most or all of the separate works which were previously known to the public, along with others still more peculiar and characteristic. We are thus able to judge of Roger Bacon's knowledge and of his views, and they are in every way well worthy our attention.

"The *Opus Majus* is addressed to Pope Clement the Fourth, whom Bacon had known when he was legate in England as Cardinal-bishop of Sabina, and who admired the talents of the monk, and pitied him for the persecutions to which he was exposed. On his elevation to the papal chair, this account of Bacon's labors and views was sent, at the earnest request of the pontiff. Besides the *Opus Majus*, he wrote two others, the *Opus Minus* and *Opus Tertium;* which were also sent to the pope, as the author says,[2] 'on account of the danger of roads, and the possible loss of the work.' These works still exist unpublished, in the Cottonian and other libraries.

"The *Opus Majus* is a work equally wonderful with regard to its general scheme, and to the special treatises with which the outlines of the plan are filled up. The professed object of the work is to urge the necessity of a reform in the mode of philosophizing, to set forth the reasons why knowledge had not made a greater progress, to draw back attention to the sources of knowledge which had been unwisely neglected, to discover other sources which were yet almost untouched, and to animate men in the undertaking, by a prospect of the vast advantages which it offered. In the development of this plan, all the leading portions of science are expounded in the most complete shape which they had at that time assumed; and improvements of a very wide and striking kind are proposed in some of the principal of these departments. Even if the work had had no leading purpose, it would have been highly valuable as a treasure of the most solid knowledge and soundest speculations of the time; even if it had contained no such details, it would have been a work most remarkable for its general views and scope. It may be considered as, at the same time, the *Encyclopedia* and the *Novum Organon* of the thirteenth century.

---

[1] *Fratris Rogeri Bacon Ordinis Minorum* Opus Majus *ad Clementem Quartum, Pontificem Romanum, ex MS. Codice Dubliniensi cum aliis quibusdam collato nunc primum edidit* S. Jebb, M.D. Londini, 1733.

[2] *Opus Majus*, Præf.

"Since this work is thus so important in the history of Inductive Philosophy, I shall give, in a Note, a view[3] of its divisions and contents. But I must now endeavor to point out more especially the way in which the various principles, which the reform of scientific method involved, are here brought into view.

"One of the first points to be noticed for this purpose, is the resistance to authority; and at the stage of philosophical history with which we here have to do, this means resistance to the authority of Aristotle, as adopted and interpreted by the Doctors of the Schools. Bacon's work[4] is divided into Six Parts; and of these Parts, the First is, Of the four universal Causes of all Human Ignorance. The causes thus enumerated[5] are:—the force of unworthy authority;—traditionary habit;—the imperfection of the undisciplined senses;—and the disposition to conceal our ignorance and to make an ostentatious show of our knowledge. These influences involve every man, occupy every condition. They prevent our obtaining the most useful and large and fair doctrines of wisdom, the secrets of all sciences and arts. He then proceeds to argue, from the testimony of philosophers themselves, that the authority of antiquity, and especially of Aristotle, is not infallible. 'We find[6] their books full of doubts, obscurities, and perplexities. They

[3] Contents of Roger Bacon's *Opus Majus:*

Part I. On the four causes of human ignorance:—Authority, Custom, Popular Opinion, and the Pride of supposed Knowledge.

Part II. On the source of perfect wisdom in the Sacred Scripture.

Part III. On the Usefulness of Grammar.

Part IV. On the Usefulness of Mathematics.

(1.) The Necessity of Mathematics in Human Things (published separately as the *Specula Mathematica*).

(2.) The Necessity of Mathematics in Divine Things.—1°. This study has occupied holy men: 2°. Geography: 3°. Chronology: 4°. Cycles; the Golden Number, &c.: 5°. Natural Phenomena, as the Rainbow: 6°. Arithmetic: 7°. Music.

(3.) The Necessity of Mathematics in Ecclesiastical Things. 1°. The Certification of Faith: 2°. The Correction of the Calendar.

(4.) The Necessity of Mathematics in the State.—1°. Of Climates: 2°. Hydrography: 3°. Geography: 4°. Astrology.

Part V. On Perspective (published separately as *Perspectiva*).

(1.) The organs of vision.

(2.) Vision in straight lines.

(3.) Vision reflected and refracted.

(4.) De multiplicatione specierum (on the propagation of the impressions of light, heat, &c.)

Part VI. On Experimental Science.

[4] *Op. Maj.* p. 1. [5] Ib. p. 2. [6] Ib. p. 10.

scarce agree with each other in one empty question or one worthless sophism, or one operation of science, as one man agrees with another in the practical operations of medicine, surgery, and the like arts of secular men. Indeed,' he adds, 'not only the philosophers, but the saints have fallen into errors which they have afterwards retracted,' and this he instances in Augustin, Jerome, and others. He gives an admirable sketch of the progress of philosophy from the Ionic School to Aristotle; of whom he speaks with great applause. 'Yet,' he adds,[7] 'those who came after him corrected him in some things, and added many things to his works, and shall go on adding to the end of the world.' Aristotle, he adds, is now called peculiarly[8] the Philosopher, 'yet there was a time when his philosophy was silent and unregarded, either on account of the rarity of copies of his works, or their difficulty, or from envy; till after the time of Mahomet, when Avicenna and Averroes, and others, recalled this philosophy into the full light of exposition. And although the Logic and some other works were translated by Boethius from the Greek, yet the philosophy of Aristotle first received a quick increase among the Latins at the time of Michael Scot; who, in the year of our Lord 1230, appeared, bringing with him portions of the books of Aristotle on Natural Philosophy and Mathematics. And yet a small part only of the works of this author is translated, and a still smaller part is in the hands of common students.' He adds further[9] (in the Third Part of the *Opus Majus*, which is a Dissertation on Language), that the translations which are current of these writings, are very bad and imperfect. With these views, he is moved to express himself somewhat impatiently[10] respecting these works: 'If I had,' he says, 'power over the works of Aristotle, I would have them all burnt; for it is only a loss of time to study in them, and a course of error, and a multiplication of ignorance beyond expression.' 'The common herd of students,' he says, 'with their heads, have no principle by which they can be excited to any worthy employment; and hence they mope and make asses of themselves over their bad translations, and lose their time, and trouble, and money.'

---

[7] *Op. Maj.* p. 36. [8] *Autonomaticè.* [9] *Op. Maj.* p. 46.

[10] See *Pref.* to Jebb's edition. The passages there quoted, however, are not extracts from the *Opus Majus*, but (apparently) from the *Opus Minus* (*MS. Cott.* Tib. c. 5). "Si haberem potestatem supra libros Aristotelis, ego facerem omnes cremari; quia non est nisi temporis amissio studere in illis, et causa erroris, et multiplicatio ignorantiæ ultra id quod valeat explicari. . . . Vulgus studentum cum capitibus suis non habet unde excitetur ad aliquid dignum, et ideo languet et *asininat* circa male translata, et tempus et studium amittit in omnibus et expensas."

"The remedies which he recommends for these evils, are, in the first place, the study of that only perfect wisdom which is to be found in the Sacred Scripture;[11] in the next place, the study of mathematics and the use of experiment.[12] By the aid of these methods, Bacon anticipates the most splendid progress for human knowledge. He takes up the strain of hope and confidence which we have noticed as so peculiar in the Roman writers; and quotes some of the passages of Seneca which we adduced in illustration of this:—that the attempts in science were at first rude and imperfect, and were afterwards improved;—that the day will come, when what is still unknown shall be brought to light by the progress of time and the labors of a longer period;—that one age does not suffice for inquiries so wide and various;—that the people of future times shall know many things unknown to us;—and that the time shall arrive when posterity will wonder that we overlooked what was so obvious. Bacon himself adds anticipations more peculiarly in the spirit of his own time. 'We have seen,' he says, at the end of the work, 'how Aristotle, by the ways which wisdom teaches, could give to Alexander the empire of the world. And this the Church ought to take into consideration against the infidels and rebels, that there may be a sparing of Christian blood, and especially on account of the troubles that shall come to pass in the days of Antichrist; which by the grace of God it would be easy to obviate, if prelates and princes would encourage study, and join in searching out the secrets of nature and art.'

"It may not be improper to observe here that this belief in the appointed progress of knowledge, is not combined with any overweening belief in the unbounded and independent power of the human intellect. On the contrary, one of the lessons which Bacon draws from the state and prospects of knowledge, is the duty of faith and humility. 'To him,' he says,[13] 'who denies the truth of the faith because he is unable to understand it, I will propose in reply the course of nature, and as we have seen it in examples.' And after giving some instances, he adds, 'These, and the like, ought to move men and to excite them to the reception of divine truths. For if, in the vilest objects of creation, truths are found, before which the inward pride of man must bow, and believe though it cannot understand, how much more should man humble his mind before the glorious truths of God!' He had before said:[14] 'Man is incapable of perfect wisdom in this life; it is hard for

[11] Part ii. [12] Parts iv. v. and vi. [13] *Op. Maj.* p. 476. [14] Ib. p. 15.

him to ascend towards perfection, easy to glide downwards to falsehoods and vanities: let him then not boast of his wisdom, or extol his knowledge. What he knows is little and worthless, in respect of that which he believes without knowing; and still less, in respect of that which he is ignorant of. He is mad who thinks highly of his wisdom; he most mad, who exhibits it as something to be wondered at.' He adds, as another reason for humility, that he has proved by trial, he could teach in one year, to a poor boy, the marrow of all that the most diligent person could acquire in forty years' laborious and expensive study.

"To proceed somewhat more in detail with regard to Roger Bacon's views of a Reform in Scientific Inquiry, we may observe that by making Mathematics and Experiment the two great points of his recommendation, he directed his improvement to the two essential parts of all knowledge, Ideas and Facts, and thus took the course which the most enlightened philosophy would have suggested. He did not urge the prosecution of experiment, to the comparative neglect of the existing mathematical sciences and conceptions; a fault which there is some ground for ascribing to his great namesake and successor, Francis Bacon: still less did he content himself with a mere protest against the authority of the schools, and a vague demand for change, which was almost all that was done by those who put themselves forward as reformers in the intermediate time. Roger Bacon holds his way steadily between the two poles of human knowledge; which, as we have seen, it is far from easy to do. 'There are two modes of knowing,' says he;[15] 'by argument, and by experiment. Argument concludes a question; but it does not make us feel certain, or acquiesce in the contemplation of truth, except the truth be also found to be so by experience.' It is not easy to express more decidedly the clearly seen union of exact conceptions with certain facts, which, as we have explained, constitutes real knowledge.

"One large division of the *Opus Majus* is 'On the Usefulness of Mathematics,' which is shown by a copious enumeration of existing branches of knowledge, as Chronology, Geography, the Calendar, and (in a separate Part) Optics. There is a chapter,[16] in which it is proved

[15] *Op. Maj.* p. 445; see also p. 448. "Scientiæ aliæ sciunt sua principia invenire per experimenta, sed conclusiones per argumenta facta ex principiis inventis. Si vero debeant habere experientiam conclusionum suarum particularem et completam, tunc oportet quod habeant per adjutorium istius scientiæ nobilis (experimentalis)."

[16] Ib. p. 60.

by reason, that all science requires mathematics. And the arguments which are used to establish this doctrine, show a most just appreciation of the office of mathematics in science. They are such as follows:—That other sciences use examples taken from mathematics as the most evident:—That mathematical knowledge is, as it were, innate to us, on which point he refers to the well-known dialogue of Plato, as quoted by Cicero:—That this science, being the easiest, offers the best introduction to the more difficult:—That in mathematics, things as known to us are identical with things as known to nature:—That we can here entirely avoid doubt and error, and obtain certainty and truth:—That mathematics is prior to other sciences in nature, because it takes cognizance of quantity, which is apprehended by intuition (*intuitu intellectus*). 'Moreover,' he adds,[17] 'there have been found famous men, as Robert, bishop of Lincoln, and Brother Adam Marshman (de Marisco), and many others, who by the power of mathematics have been able to explain the causes of things; as may be seen in the writings of these men, for instance, concerning the Rainbow and Comets, and the generation of heat, and climates, and the celestial bodies.'

"But undoubtedly the most remarkable portion of the *Opus Majus* is the Sixth and last Part, which is entitled 'De Scientia experimentali.' It is indeed an extraordinary circumstance to find a writer of the thirteenth century, not only recognizing experiment as one source of knowledge, but urging its claims as something far more important than men had yet been aware of, exemplifying its value by striking and just examples, and speaking of its authority with a dignity of diction which sounds like a foremurmur of the Baconian sentences uttered nearly four hundred years later. Yet this is the character of what we here find.[18] 'Experimental science, the sole mistress of speculative sciences, has three great Prerogatives among other parts of knowledge: First she tests by experiment the noblest conclusions of all other sciences: Next she discovers respecting the notions which other sciences deal with, magnificent truths to which these sciences of themselves can by no means attain: her Third dignity is, that she by her own power and without respect of other sciences, investigates the secrets of nature.'

---

[17] *Op. Maj.* p. 64.

[18] "Veritates magnificas in terminis aliarum scientiarum in quas per nullam viam possunt illæ scientiæ, hæc sola scientiarum domina speculativarum, potest dare."—*Op. Maj.* p. 465.

"The examples which Bacon gives of these 'Prerogatives' are very curious, exhibiting, among some error and credulity, sound and clear views. His leading example of the First Prerogative is the Rainbow, of which the cause, as given by Aristotle, is tested by reference to experiment with a skill which is, even to us now, truly admirable. The examples of the Second Prerogative are three :—*first*, the art of making an artificial sphere which shall move with the heavens by natural influences, which Bacon trusts may be done, though astronomy herself cannot do it—'et tunc,' he says, 'thesaurum unius regis valeret hoc instrumentum ;'—*secondly*, the art of prolonging life, which experiment may teach, though medicine has no means of securing it except by regimen ;[19]—*thirdly*, the art of making gold finer than fine gold, which goes beyond the power of alchemy. The Third Prerogative of experimental science, arts independent of the received sciences, is exemplified in many curious examples, many of them whimsical traditions. Thus it is said that the character of a people may be altered by altering the air.[20] Alexander, it seems, applied to Aristotle to know whether he should exterminate certain nations which he had discovered, as being irreclaimably barbarous; to which the philosopher replied, 'If you can alter their air, permit them to live; if not, put them to death.' In this part, we find the suggestion that the fire-works made by children, of saltpetre, might lead to the invention of a formidable military weapon.

"It could not be expected that Roger Bacon, at a time when experimental science hardly existed, could give any *precepts* for the discovery of truth by experiment. But nothing can be a better *example* of the method of such investigation, than his inquiry concerning the cause of the Rainbow. Neither Aristotle, nor Avicenna, nor Seneca, he says, have given us any clear knowledge of this matter, but experimental science can do so. Let the experimenter (*experimentator*) consider the cases in which he finds the same colors, as the hexagonal crystals from Ireland and India; by looking into these he will see colors like those of the rainbow. Many think that this arises from some

[19] One of the ingredients of a preparation here mentioned, is the flesh of a dragon, which, it appears, is used as food by the Ethiopians. The mode of preparing this food cannot fail to amuse the reader. "Where there are good flying dragons, by the art which they possess, they draw them out of their dens, and have bridles and saddles in readiness, and they ride upon them, and make them bound about in the air in a violent manner, that the hardness and toughness of the flesh may be reduced, as boars are hunted and bulls are baited before they are killed for eating."—*Op. Maj.* p. 470.

[20] *Op. Maj.* p. 473.

special virtue of these stones and their hexagonal figure; let therefore the experimenter go on, and he will find the same in other transparent stones, in dark ones as well as in light-colored. He will find the same effect also in other forms than the hexagon, if they be furrowed in the surface, as the Irish crystals are. Let him consider too, that he sees the same colors in the drops which are dashed from oars in the sunshine;—and in the spray thrown by a mill-wheel;—and in the dewdrops which lie on the grass in a meadow on a summer morning;—and if a man takes water in his mouth and projects it on one side into a sunbeam;—and if in an oil lamp hanging in the air, the rays fall in certain positions upon the surface of the oil;—and in many other ways, are colors produced. We have here a collection of instances, which are almost all examples of the same kind as the phenomena under consideration; and by the help of a principle collected by induction from these facts, the colors of the rainbow were afterwards really explained.

"With regard to the form and other circumstances of the bow he is still more precise. He bids us measure the height of the bow and of the sun, to show that the centre of the bow is exactly opposite to the sun. He explains the circular form of the bow,—its being independent of the form of the cloud, its moving when we move, its flying when we follow,—by its consisting of the reflections from a vast number of minute drops. He does not, indeed, trace the course of the rays through the drop, or account for the precise magnitude which the bow assumes; but he approaches to the verge of this part of the explanation; and must be considered as having given a most happy example of experimental inquiry into nature, at a time when such examples were exceedingly scanty. In this respect, he was more fortunate than Francis Bacon, as we shall hereafter see.

"We know but little of the biography of Roger Bacon, but we have every reason to believe that his influence upon his age was not great. He was suspected of magic, and is said to have been put into close confinement in consequence of this charge. In his work he speaks of Astrology, as a science well worth cultivating. 'But,' says he, 'Theologians and Decretists, not being learned in such matters, and seeing that evil as well as good may be done, neglect and abhor such things, and reckon them among Magic Arts.' We have already seen, that at the very time when Bacon was thus raising his voice against the habit of blindly following authority, and seeking for all science in Aristotle, Thomas Aquinas was employed in fashioning Aristotle's tenets into that fixed form in which they became the great impediment to the progress of knowledge. It would seem, indeed, that something of a struggle between the progressive and stationary powers of the human mind was going on at this time. Bacon himself says,[21] 'Never was there so great an appearance of wisdom, nor so much exercise of study in so many Faculties, in so many regions, as for this last forty years. Doctors are dispersed everywhere, in every castle, in every burgh, and especially by the students of two Orders (he means the Franciscans and Dominicans, who were almost the only religious orders that distinguished themselves by an application to study),[22] which has not happened except for about forty years. And yet there was never so much ignorance, so much error.' And in the part of his work which refers to Mathematics, he says of that study,[23] that it is the door and the key of the sciences; and that the neglect of it for thirty or forty years has entirely ruined the studies of the Latins. According to these statements, some change, disastrous to the fortunes of science, must have taken place about 1230, soon after the foundation of the Dominican and Franciscan Orders.[24] Nor can we doubt that the adoption of the Aristotelian philosophy by these two Orders, in the form in which the Angelical Doctor had systematized it, was one of the events which most tended to defer, for three centuries, the reform which Roger Bacon urged as a matter of crying necessity in his own time."

It is worthy of remark that in the *Opus Majus* of Roger Bacon, as afterwards in the *Novum Organon* of Francis Bacon, we have certain features of experimental research pointed out conspicuously as *Prærogativæ:* although in the former, this term is employed to designate the superiority of experimental science in general to the science of the schools; in the latter work, the term is applied to certain classes of experiments as superior to others.

---

[21] Quoted by Jebb, Pref. to *Op. Maj.*

[22] Mosheim, *Hist.* iii. 161.

[23] *Op. Maj.* p. 57.

[24] Mosheim, iii. 161.

BOOK V.

FORMAL ASTRONOMY.

CHAPTER I.

Prelude to Copernicus.

Nicolas of Cus.

I WILL quote the passage, in the writings of this author, which bears upon the subject in question. I translate it from the edition of his book *De Docta Ignorantia*, from his works published at Basil in 1565. He praises *Learned Ignorance*—that is, Acknowledged Ignorance—as the source of knowledge. His ground for asserting the motions of the earth is, that there is no such thing as perfect rest, or an exact centre, or a perfect circle, nor perfect uniformity of motion. "Neque verus circulus dabilis est, quinetiam verior dari possit, neque unquam uno tempore sicut alio æqualiter præcisè, aut movetur, aut circulum veri similem, æqualem describit, etiamsi nobis hoc non appareat. Et ubicumque quis fuerit, se in centro esse credit." (Lib. i. cap. xi. p, 39.) He adds, "The Ancients did not attain to this knowledge, because they were wanting in Learned Ignorance. Now it is manifest to us that the Earth is truly in motion, although this do not appear to us; since we do not apprehend motion except by comparison with something fixed. For if any one were in a boat in the middle of a river, ignorant that the water was flowing, and not seeing the banks, how could he apprehend that the boat was moving? And thus since every one, whether he be in the Earth, or in the Sun, or in any other star, thinks that he is in an immovable centre, and that every thing else is moving; he would assign different poles for himself, others as being in the Sun, others in the Earth, and others in the Moon, and so of the rest. Whence the machine of the world is as if it had its centre everywhere and its circumference nowhere." This train of thought might be a preparation for the reception of the Copernican system; but it is very different from the doctrine that the Sun is the centre of the Planetary Motions.

CHAPTER II.

THE COPERNICAN THEORY.

The Moon's Rotation.

I HAVE said, in page 264, that a confusion of mind produced by the double reference of motion to absolute space, and to a centre of revolution, often leads persons to dispute whether the Moon, while she revolves about the Earth, always turning to it the same face, revolves about her axis or not.

This dispute has been revived very lately, and has been conducted in a manner which shows that popular readers and writers have made little progress in the clearness of their notions during the last two or three centuries; and that they have accepted the Newtonian doctrines in words with a very dim apprehension of their real import.

If the Moon were carried round the Earth by a rigid arm revolving about the Earth as a centre, being rigidly fastened to this arm, as a mimic Moon might be, in a machine constructed to represent her motions, this contrivance, while it made her revolve round the Earth, would make her also turn the same face to the Earth: and if we were to make such a machine the standard example of rotation, the Moon might be said not to rotate on her axis.

But we are speedily led to endless confusion by taking this case as the standard of rotation. For the selection of the centre of rotation in a system which includes several bodies is arbitrary. The Moon turns all her faces successively to the Sun, and therefore with regard to the Sun, she does rotate on her axis; and yet she revolves round the Sun as truly as she revolves round the Earth. And the only really simple and consistent mode of speaking of rotation, is to refer the motion not to any relative centre, but to absolute space.

This is the argument merely on the ground of simplicity and consistency. But we find physical reasons, as well as mathematical, for referring the motion to absolute space. If a cup of water be carried round a centre so as to describe a circle, a straw floating on the surface of the water, if it point to the centre of the circle at first, does not continue to do so, but remains parallel to itself during the whole revolution. Now there is no cause to make the water (and therefore the straw) rotate on its axis; and therefore it is not a clear or convenient way of speaking, to say that the water in this case does revolve on its axis. But if the water in this case do not revolve on its axis, a body in the case of the Moon does revolve on its axis.

The difficulty, as I have said in the text, is of the same nature as that which the Copernicans at first found in the parallel motion of the Earth's axis. In order to make the axis of the Earth's rotation remain parallel to itself while the Earth revolves about the Sun, in a mechanical representation, some machinery is needed *in addition to* the machinery which produces the revolution round the centre (the Sun): but the simplest way of regarding the parallel motion is, to conceive that the axis has no motion except that which carries it round the central Sun. And it was seen, when the science of Mechanics was established, that no force was needed in nature to produce this parallelism of the Earth's axis. It was therefore the only scientific course, to conceive this parallelism as not being a rotation: and in like manner we are to conceive the parallelism of a revolving body as not being a rotation.

M. Foucault's Proofs of the Earth's Motion.

It was hardly to be expected that we should discover, in our own day, a new physical proof of the earth's motion, yet so it has been. The experiments of M. Foucault have enabled us to see the Rotation of the Earth on its axis, as taking place, we may say, before our eyes. These experiments are, in fact, a result of what has been said in speaking of the Moon's rotation: namely, That the mechanical causes of motion operate with reference to absolute, not relative, space; so that where there is no cause operating to change a motion, it will retain its direction in *absolute* space; and may on that account seem to change, if regarded relatively in a *limited* space.

In M. Foucault's first experiment, the motion employed was that of a pendulum. If a pendulum oscillate quite freely, there is no cause acting to change the vertical plane of oscillation *absolutely;* for the forces which produce the oscillation are *in* the vertical plane. But if the vertical plane remain the same *absolutely*, at a spot on the surface of the revolving Earth, it will change *relatively* to the spectator. He will see the pendulum oscillate in a vertical plane which gradually turns away from its first position. Now this is what really happens; and thus the revolution of the Earth in absolute space is experimentally proved.

In subsequent experiments, M. Foucault has used the rotation of a body to prove the same thing. For when a body rotates freely, acted upon by no power, there is nothing to change the position of the axis of rotation in absolute space. But if the position of the axis remain the same in absolute space, it will, in virtue of its relative motion, change as seen by a spectator at any spot on the rotating Earth. By taking a heavy disk or globe and making it rotate on its axis rapidly, the force of absolute permanence (as compared with the inevitable casual disturbances arising from the machinery which supports the revolving disk) becomes considerable; and hence the relative motion can, in this way also, be made visible.

Mr. De Morgan has said (*Comp. to Brit. Alm.* 1836, p. 18) that astronomy does not supply any argument for the earth's motion which is absolutely and demonstrably conclusive, till we come to the Aberration of Light. But we may now venture to say that the experiments of M. Foucault prove the *diurnal* motion of the Earth in the most conclusive manner, by palpable and broad effects, if we accept the doctrines of the science of Mechanics: while Aberration proves the *annual* motion, if we suppose that we can observe the places of the fixed stars to the accuracy of a few seconds; and if we accept, in addition to the doctrines of Mechanics, the doctrine of the motion of light with a certain great velocity.

CHAPTER III.

SEQUEL TO COPERNICUS.

English Copernicans.

PROFESSOR DE MORGAN has made numerous and interesting contributions to the history of the progress and reception of the Copernican System. These are given mainly in the *Companion to the British Almanac;* especially in his papers entitled "Old Arguments against the Motion of the Earth" (1836); "English Mathematical and Astronomical Writers" (1837); "On the Difficulty of Correct

Description of Books" (1853); "The Progress of the Doctrine of the Earth's Motion between the Times of Copernicus and Galileo" (1855). In these papers he insists very rightly upon the distinction between the *mathematical* and the *physical* aspect of the doctrines of Copernicus: a distinction corresponding very nearly with the distinction which we have drawn between Formal and Physical Astronomy; and in accordance with which we have given the history of the Heliocentric Doctrine as a Formal Theory in Book v., and as a Physical Theory in Book vii.

Another interesting part of Mr. De Morgan's researches are the notices which he has given of the early assertors of the heliocentric doctrine in England. These make their appearance as soon as it was well possible they should exist. The work of Copernicus was published, as we have said, in 1543. In September, 1556, John Field published an Ephemeris for 1557, "juxta Copernici et Reinholdi Canones," in the preface to which he avows his conviction of the truth of the Copernican hypothesis. Robert Recorde, the author of various works on Arithmetic, published among others, "The Pathway to Knowledge" in 1551. In this book, the author discusses the question of the "quietnes of the earth," and professes to leave it undecided; but Mr. De Morgan (*Comp. A.*, 1837, p. 33) conceives that it appears from what is said, that he was really a Copernican, but did not think the world ripe for any such doctrine.

Mr. Joseph Hunter also has brought to notice[1] the claims of Field, whom he designates as the *Proto-Copernican* of England. He quotes the Address to the Reader prefixed to his first *Ephemeris*, and dated May 31, 1556, in which he says that, since abler men decline the task, "I have therefore published this Ephemeris of the year 1557, following in it as my authorities, N. Copernicus and Erasmus Reinhold, whose writings are established and founded on true, certain, and authentic demonstrations." I conceive that this passage, however, only shows that Field had adopted the Copernican scheme as a basis for the calculation of Ephemerides; which, as Mr. De Morgan has remarked, is a very different thing from accepting it as a physical truth. Field, in this same address, makes mention of the errors "illius turbæ quæ Alphonsi utitur hypothesi;" but the word *hypothesis* is still indecisive.

As evidence that Field was regarded in his own day as a man who

[1] *Ast. Soc. Notices*, vol. iii. p. 3 (1833).

had rendered good service to science, Mr. Hunter notices that, in 1558, the Heralds granted to him the right of using, with his arms, the crest or additional device of a red right arm issuing from the clouds, and presenting a golden armillary sphere.

Recorde's claims depend upon a passage in a Dialogue between *Master* and *Scholar*, in which the Master expounds the doctrine of Copernicus, and the authorities against it; to which the Scholar answers, taking the common view: "Nay, sir, in good faith I desire not to hear such vaine phantasies, so far against common reason, and repugnant to all the learned multitude of wryters, and therefore let it passe forever and a day longer." The Master, more sagely, warns him against a hasty judgment, and says, "Another time I will so declare his supposition, that you shall not only wonder to hear it, but also peradventure be as earnest then to credit it, as you now are to condemne it." I conceive that this passage proves Mr. De Morgan's assertion, that Recorde was a Copernican, and very likely the first in England.

In 1555, also, Leonard Digges published his "Prognostication Everlasting;" but this is, as Mr. De Morgan says (*Comp. A.*, 1837, p. 40), a meteorological work. It was republished in 1592 by his son Thomas Digges with additions; and as these have been the occasion of some confusion among those who have written on the history of astronomy, I am glad to be able, through the kindness of Professor Walker, of Oxford, to give a distinct account of the editions of the work.

In the Bodleian Library, besides the editions of 1555 and 1592 of the "Prognostication Everlasting," there is an edition of 1564. It is still decidedly Ptolemaic, and contains a diagram representing a number of concentric circles, which are marked, in order, as—

"The Earth,
Moone,
Venus,
Mercury,
Sunne,
Mars,
Jupiter,
Saturne,
The Starrie Firmament,
The Crystalline Heavens,
The First Mover,
The Abode of God and the Elect. Here the Learned do approve."

The third edition, of 1592, contains an Addition, by the son, of twenty pages. He there speaks of having found, apparently among his father's papers, "A description or modile of the world and situation of Spheres Cœlestiall and elementare according to the doctrine of Ptolemie, whereunto all universities (led thereunto chiefly by the authoritie of Aristotle) do consent." He adds: "But in this our age, one rare witte (seeing the continuall errors that from time to time more and more continually have been discovered, besides the infinite absurdities in their Theoricks, which they have been forced to admit that would not confesse any Mobilitie in the ball of the Earth) hath by long studye, paynfull practise, and rare invention, delivered a new Theorick or Model of the world, shewing that the Earth resteth not in the Center of the whole world or globe of elements, which encircled and enclosed in the Moone's orbe, and together with the whole globe of mortalitye is carried yearely round about the Sunne, which like a king in the middest of all, raygneth and giveth lawes of motion to all the rest, sphærically dispersing his glorious beames of light through all this sacred cœlestiall Temple. And the Earth itselfe to be one of the Planets, having his peculiar and strange courses, turning every 24 hours rounde upon his owne centre, whereby the Sunne and great globe of fixed Starres seem to sway about and turne, albeit indeed they remaine fixed—So many ways is the sense of mortal man abused."

This Addition is headed:

"A Perfit Description of the Cœlestiall Orbes, according to the most ancient doctrine of the Pythagoreans: lately revived by Copernicus, and by Geometrical Demonstrations approved." Mr. De Morgan, not having seen this edition, and knowing the title-page only as far as the word "Pythagoreans," says "their *astrological* doctrines we presume, not their reputed *Copernican* ones." But it now appears that in this, as in other cases, the authority of the Pythagoreans was claimed for the Copernican system. Antony a Wood quotes the latter part of the title thus: "Cui subnectitur *orbium* Copernicarum accurata descriptio;" which is inaccurate. Weidler, still more inaccurately, cites it, "Cui subnectitur *operum* Copernici accurata descriptio." Lalande goes still further, attempting, it would seem, to recover the English title-page from the Latin: we find in the *Bibl. Astron.* the following: "1592 . . Leonard Digges, Accurate Description of the Copernican System to the Astronomical perpetual Prognostication."

Thomas Digges appears, by others also of his writings, to have been a clear and decided Copernican. In his "Alæ sive Scalæ Mathematicæ," 1573, he bestows high praise upon Copernicus and upon his system: and appears to have been a believer in the real motion of the Earth, and not merely an admirer of the system of Copernicus as an explanatory hypothesis.

Giordano Bruno.

THE complete title of the work referred to is:

"Jordani Bruni Nolani De Monade Numero et Figura liber consequens Quinque De Minimo Magno et Mensura, item De Innumerabilibus, Immenso et Infigurabili; seu De Universo et Mundis libro octo. (Francofurti, 1591.)"

That the Reader may judge of the value of Bruno's speculations, I give the following quotations:

Lib. iv. c. 11 (Index). "Tellurem totam habitabilem esse *intus* et extra, et innumerabilia animantium complecti tum nobis sensibilium tum *occultorum* genera."

C. 13. "Ut Mundorum Synodi in Universo et particulares Mundi in Synodis ordinentur," &c.

He says (Lib. v. c. 1, p. 461): "Besides the stars and the great worlds there are smaller living creatures carried through the ethereal space, in the form of a small sphere which has the aspect of a bright fire, and is by the vulgar regarded as a fiery beam. They are below the clouds, and I saw one which seemed to touch the roofs of the houses. Now this sphere, or beam as they call it, was really a living creature (*animal*), which I once saw moving in a straight path, and grazing as it were the roofs of the city of Nola, as if it were going to impinge on Mount Cicada; which however it went over."

There are two recent editions of the works of Giordano Bruno; by Adolf Wagner, Leipsick, 1830, in two volumes; and by Gfrörer, Berlin, 1833. Of the latter I do not know that more than one volume (vol. ii.) has appeared.

Did Francis Bacon reject the Copernican System?

MR. DE MORGAN has very properly remarked (*Comp. B. A.* 1855, p. 11) that the notice of the heliocentric question in the *Novum Organon* must be considered one of the most important passages in his works upon this point, as being probably the latest written and best matured. It occurs in Lib. ii. Aphorism xxxvi., in which he is speaking of *Prerogative Instances*, of which he gives twenty-seven species. In the passage now referred to, he is speaking of a kind of Prerogative Instances, better known to ordinary readers than most of the kinds by name, the *Instantia Crucis:* though probably the metaphor from which this name is derived is commonly wrongly apprehended. Bacon's meaning is *Guide-Post Instances:* and the *Crux* which he alludes to is not a Cross, but a Guide-Post at Cross-roads. And among the cases to which such Instances may be applied, he mentions the diurnal motion of the heavens from east to west, and the special motion of the particular heavenly bodies from west to east. And he suggests what he conceives may be an *Instantia Crucis* in each case. If, he says, we find any motion from east to west in the bodies which surround the earth, slow in the ocean, quicker in the air, quicker still in comets, gradually quicker in planets according to their greater distance from the earth; *then* we may suppose that there is a cosmical diurnal motion, and the motion of the earth must be denied.

With regard to the special motions of the heavenly bodies, he first remarks that each body not coming quite so far westwards as before, after one revolution of the heavens, and going to the north or the south, does not imply any special motion; since it may be accounted for by a modification of the diurnal motion in each, which produces a defect of the return, and a spiral path; and he says that if we look at the matter as common people[2] and disregard the devices of astronomers, the motion is really so to the senses; and that he has made an imitation of it by means of wires. The *instantia crucis* which he here suggests is, to see if we can find in any credible history an account of any comet which did not share in the diurnal revolution of the skies.

On his assertion that the motion of each separate planet is, to sense, a spiral, we may remark that it is certainly true; but that the business of science, here, as elsewhere, consists in *resolving* the complex phenomenon into simple phenomena; the complex spiral motion into simple circular motions.

With regard to the diurnal motion of the earth, it would seem as if Bacon himself had a leaning to believe it when he wrote this passage; for neither is he himself, nor are any of the Anticopernicans, accus-

---

[2] Et certissimum est si paulisper pro plebeiis nos geramus (missis astronomorum et scholæ commentis, quibus illud in more est, ut sensui in multis immerito vim faciant et obscuriora malint) talem esse motum istum ad sensum qualem diximus.

tomed to assert that the immensely rapid motion of the sphere of the Fixed Stars graduates by a slower and slower motion of Planets, Comets, Air, and Ocean, into the immobility of the Earth. So that the conditions are not satisfied on which he hypothetically says, "tum abnegandus est motus terræ."

With regard to the proper motions of the planets, this passage seems to me to confirm what I have already said of him; that he does not appear to have seen the full value and meaning of what had been done, up to his time, in Formal Astronomy.

We may however fully agree with Mr. De Morgan; that the whole of what he has said on this subject, when put together, does not justify Hume's assertion that he rejected the Copernican system "with the most positive disdain."

Mr. De Morgan, in order to balance the Copernican argument derived from the immense velocity of the stars in their diurnal velocity on the other supposition, has reminded us that those who reject this great velocity as improbable, accept without scruple the greater velocity of light. It is curious that Bacon also has made this comparison, though using it for a different purpose; namely, to show that the transmission of the visual impression may be instantaneous. In Aphorism xlvi. of Book ii. of the *Novum Organon* he is speaking of what he calls *Instantiæ curriculi*, or *Instantiæ ad aquam*, which we may call *Instances by the clock:* and he says that the great velocity of the diurnal sphere makes the marvellous velocity of the rays of light more credible.

"Immensa illa velocitas in ipso corpore, quæ cernitur in motu diurno (quæ etiam viros graves ita obstupefecit ut *mallent credere motum terræ*), facit motum illum ejaculationis ab ipsis [stellis] (licet celeritate ut diximus admirabilem) magis credibilem." This passage shows an inclination towards the opinion of the earth's being at rest, but not a very strong conviction.

Kepler persecuted.

We have seen (p. 280) that Kepler writes to Galileo in 1597—"Be trustful and go forwards. If Italy is not a convenient place for the publication of your views, and if you are likely to meet with any obstacles, perhaps Germany will grant us the necessary liberty." Kepler however had soon afterwards occasion to learn that in Germany also, the cultivators of science were exposed to persecution. It is true that in his case the persecution went mainly on the broad ground of his being a Protestant, and extended to great numbers of persons at that time. The circumstances of this and other portions of Kepler's life have been brought to light only recently through an examination of public documents in the Archives of Würtemberg and unpublished letters of Kepler. (Johann Keppler's Leben und Wirken, nach neuerlich aufgefundenen Manuscripten bearbeitet von J. L. C. Freiherrn v. Breitschwart, K. Würtemberg. Staats-Rath. Stuttgart, 1831.)

Schiller, in his *History of the Thirty Years' War*, says that when Ferdinand of Austria succeeded to the Archduchy of Stiria, and found a great number of Protestants among his subjects, he suppressed their public worship without cruelty and almost without noise. But it appears now that the Protestants were treated with great severity. Kepler held a professorship in Stiria, and had married, in 1597, Barbara Müller, who had landed property in that province. On the 11th of June, 1598, he writes to his friend Mæstlin that the arrival of the Prince out of Italy is looked forwards to with terror. In December he writes that the Protestants had irritated the Catholics by attacks from the pulpit and by caricatures; that hereupon the Prince, at the prayer of the Estates, had declared the Letter of License granted by his father to be forfeited, and had ordered all the Evangelical Teachers to leave the country on pain of death. They went to the frontiers of Hungary and Croatia; but after a month, Kepler was allowed to return, on condition of keeping quiet. His discoveries appear to have operated in his favor. But the next year he found his situation in Stiria intolerable, and longed to return to his native country of Würtemberg, and to find some position there. This he did not obtain. He wrote a circular letter to his Brother Protestants, to give them consolation and courage; and this was held to be a violation of the conditions on which his residence was tolerated. Fortunately, at this time he was invited to join Tycho Brahe, who had also been driven from his native country, and was living at Prague. The two astronomers worked together under the patronage of the Emperor Rudolph II.; and when Tycho died in 1601, Kepler became the Imperial *Mathematicus*.

We are not to imagine that even among Protestants, astronomical notions were out of the sphere of religious considerations. When Kepler was established in Stiria, his first official business was the calculation of the Calendar for the Evangelical Community. They protested against the new Calendar, as manifestly calculated for the furtherance of an impious papistry: and, say they, "We hold the Pope for a horrible roaring Lion. If we take his Calendar, we must needs go into the church when he rings us in." Kepler however did not fail to see, and to say, that the Papal Reformation of the Calendar was a vast improvement.

Kepler, as court-astronomer, was of course required to provide such observations of the heavens as were requisite for the calculations of the Astrologers. That he considered Astrology to be valuable only as the nurse of Astronomy, he did not hesitate to reveal. He wrote a work with a title of which the following is the best translation which I can give: "*Tertius interveniens;* or, A Warning to certain *Theologi, Medici, Philosophi,* that while they reasonably reject star-gazing superstition, they do not throw away the kernel with the shell.[3] 1610." In this he says, "You over-clever Philosophers blame this Daughter of Astronomy more than is reasonable. Do you not know that she must maintain her mother with her charms? How many men would be able to make Astronomy their business, if men did not cherish the hope to read the Future in the skies?"

Were the Papal Edicts against the Copernican System repealed?

ADMIRAL SMYTH, in his *Cycle of Celestial Objects*, vol. i. p. 65, says —"At length, in 1818, the voice of truth was so prevailing that Pius VII. repealed the edicts against the Copernican system, and thus, in the emphatic words of Cardinal Toriozzi, 'wiped off this scandal from the Church.'"

A like story is referred to by Sir Francis Palgrave, in his entertaining and instructive fiction, *The Merchant and the Friar.*

Having made inquiry of persons most likely to be well informed on this subject, I have not been able to learn that there is any further foundation for these statements than this: In 1818, on the revisal of the *Index Expurgatorius*, Galileo's writings were, after some opposition, expunged from that Catalogue.

Monsignor Marino Marini, an eminent Roman Prelate, had addressed to the *Romana Accademia di Archeologia*, certain historico-critical Memoirs, which he published in 1850, with the title *Galileo e l'Inquisizione.* In these, he confirms the conclusion which, I think, almost

[3] The German passage involves a curious image, borrowed, I suppose, from some odd story: "dass sie mit billiger Verwerfung des sternguckerischen Aberglaubens das Kind nicht mit dem Bade ausschütten." "That they do not throw away the child along with the dirty water of his bath."

all persons who have studied the facts have arrived at;[4] that Galileo trifled with authority to which he professed to submit, and was punished for obstinate contumacy, not for heresy. M. Marini renders full justice to Galileo's ability, and does not at all hesitate to regard his scientific attainments as among the glories of Italy. He quotes, what Galileo himself quoted, an expression of Cardinal Baronius, that "the intention of the Holy Spirit was to teach how to go to heaven, not how heaven goes."[5] He shows that Galileo pleaded (p. 62) that he had not held the Copernican opinion after it had been intimated to him (by Bellarmine in 1616), that he was not to hold it; and that his breach of promise in this respect was the cause of the proceedings against him.

Those who admire Galileo and regard him as a martyr because, after escaping punishment by saying "It *does not* move," he forthwith said "And yet it *does* move," will perhaps be interested to know that the former answer was suggested to him by friends anxious for his safety. Niccolini writes to Bali Cioli (April 9, 1633) that Galileo continued to be so persuaded of the truth of his opinions that "he was resolved (some moments before his sentence) to defend them stoutly; but I (continues Niccolini) exhorted him to make an end of this; not to mind defending them; and to submit himself to that which he sees that they may desire him to believe or to hold about this matter of the motion of the earth. He was extremely afflicted." But the Inquisition was satisfied with his answers, and required no more.[6]

---

4 M. Marini (p. 29) mentions Leibnitz, Guizot, Spittler, Eichhorn, Raumer, Ranke, among the "storici eterodossi" who have at last done justice to the Roman Church.

5 Come si vada al Cielo, e non come vada il Cielo.

6 Marini, p. 61.

BOOK VI.

MECHANICS.

CHAPTER III.

PRINCIPLES AND PROBLEMS.

Significance of Analytical Mechanics.

IN the text, page 372, I have stated that Lagrange, near the end of his life, expressed his sorrow that the methods of approximation employed in Physical Astronomy rested on arbitrary processes, and not on any insight into the results of mechanical action. From the recent biography of Gauss, the greatest physical mathematician of modern times, we learn that he congratulated himself on having escaped this error. He remarked[1] that many of the most celebrated mathematicians, Euler very often, Lagrange sometimes, had trusted too much to the symbolical calculation of their problems, and would not have been able to give an account of the meaning of each successive step of their investigation. He said that he himself, on the other hand, could assert that at every step which he took, he always had the aim and purpose of his operations before his eyes without ever turning aside from the way. The same, he remarked, might be said of Newton.

Engineering Mechanics.

The principles of the science of Mechanics were discovered by observations made upon bodies within the reach of men; as we have seen in speaking of the discoveries of Stevinus, Galileo, and others, up to the time of Newton. And when there arose the controversy about *vis viva* (Chap. v. Sect. 2 of this Book);—namely, whether the "living force" of a body is measured by the product of the weight into the

---

[1] Gauss, *Zum Gedächtniss, von W. Sartorius v. Waltershausen*, p. 80.

velocity, or of the weight into the square of the velocity;—still the examples taken were cases of action in machines and the like terrestrial objects. But Newton's discoveries identified celestial with terrestrial mechanics; and from that time the mechanical problems of the heavens became more important and attractive to mathematicians than the problems about earthly machines. And thus the generalizations of the problems, principles, and methods of the mathematical science of Mechanics from this period are principally those which have reference to the motions of the heavenly bodies: such as the Problem of Three Bodies, the Principles of the Conservation of Areas, and of the Immovable Plane, the Method of Variation of Parameters, and the like (Chap. vi. Sect. 7 and 14). And the same is the case in the more recent progress of that subject, in the hands of Gauss, Bessel, Hansen, and others.

But yet the science of Mechanics as applied to terrestrial machines —*Industrial Mechanics*, as it has been termed—has made some steps which it may be worth while to notice, even in a general history of science. For the most part, all the most general laws of mechanical action being already finally established, in the way which we have had to narrate, the determination of the results and conditions of any combination of materials and movements becomes really a mathematical deduction from known principles. But such deductions may be made much more easy and much more luminous by the establishment of general terms and general propositions suited to their special conditions. Among these I may mention a new abstract term, introduced because a general mechanical principle can be expressed by means of it, which has lately been much employed by the mathematical engineers of France, MM. Poncelet, Navier, Morin, &c. The abstract term is *Travail*, which has been translated *Laboring Force;* and the principle which gives it its value, and makes it useful in the solution of problems, is this;—that the *work done* (in overcoming resistance or producing any other effect) is equal to the *Laboring Force*, by whatever contrivances the force be applied. This is not a new principle, being in fact mathematically equivalent to the conservation of Vis Viva; but it has been employed by the mathematicians of whom I have spoken with a fertility and simplicity which make it the mark of a new school of *The Mechanics of Engineering.*

The Laboring Force expended and the work done have been described by various terms, as *Theoretical Effect* and *Practical Effect*, and the like. The usual term among English engineers for the work which an Engine usually does, is *Duty;* but as this word naturally signifies what the engine *ought* to do, rather than what it does, we should at least distinguish between the Theoretical and the Actual Duty.

The difference between the Theoretical and Actual Duty of a Machine arises from this: that a portion of the Laboring Force is absorbed in producing effects, that is, in doing work which is not reckoned as Duty: for instance, overcoming the resistance and waste of the machine itself. And so long as this resistance and waste are not rightly estimated, no correspondence can be established between the theoretical and the practical Duty. Though much had been written previously upon the theory of the steam-engine, the correspondence between the Force expended and the Work done was not clearly made out till Comte De Pambour published his *Treatise on Locomotive Engines* in 1835, and his *Theory of the Steam-Engine* in 1839.

Strength of Materials.

Among the subjects which have specially engaged the attention of those who have applied the science of Mechanics to practical matters, is the strength of materials: for example, the strength of a horizontal beam to resist being broken by a weight pressing upon it. This was one of the problems which Galileo took up. He was led to his study of it by a visit which he made to the arsenal and dockyards of Venice, and the conclusions which he drew were published in his *Dialogues*, in 1633. In his mode of regarding the problem, he considers the section at which the beam breaks as the short arm of a bent lever which resists fracture, and the part of the beam which is broken off as the longer arm of the lever, the lever turning about the fracture as a hinge. So far this is true; and from this principle he obtained results which are also true; as, that the strength of a rectangular beam is proportional to the breadth multiplied into the square of the depth: —that a hollow beam is stronger than a solid beam of the same mass; and the like.

But he erred in this, that he supposed the hinge about which the breaking beam turns, to be exactly at the unrent surface, that surface resisting all change, and the beam being rent all the way across. Whereas the fact is, that the unrent surface yields to compression, while the opposite surface is rent; and the hinge about which the breaking beam turns is at an intermediate point, where the extension and rupture end, and the compression and crushing begin: a point which has been called *the neutral axis*. This was pointed out by Mariotte; and the notion, once suggested, was so manifestly true that it was adopted by mathematicians in general. James Bernoulli,[2] in 1705, investigated the strength of beams on this view; and several eminent mathematicians pursued the subject; as Varignon, Parent, and Bulfinger; and at a later period, Dr. Robison in our own country.

But along with the fracture of beams, the mathematicians considered also another subject, the flexure of beams, which they undergo before they break, in virtue of their elasticity. What is the *elastic curve?*—the curve into which an elastic line forms itself under the pressure of a weight—is a problem which had been proposed by Galileo, and was fully solved, as a mathematical problem, by Euler and others.

But beams in practice are not mere lines: they are solids. And their resistance to flexure, and the amount of it, depends upon the resistance of their internal parts to extension and compression, and is different for different substances. To measure these differences, Dr. Thomas Young introduced the notion of the *Modulus of Elasticity*:[3] meaning thereby a column of the substance of the same diameter, such as would by its weight produce a compression equal to the whole length of the beam, the rate of compression being supposed to continue the same throughout. Thus if a rod of any kind, 100 inches long, were compressed 1 inch by a weight 1000 pounds, the weight of its modulus of elasticity would be 100,000 pounds. This notion assumes Hooke's law that the extension of a substance is as its tension; and extends this law to compression also.

There is this great advantage in introducing the definition of the Modulus of Elasticity,—that it applies equally to the flexure of a substance and to the minute vibrations which propagate sound, and the like. And the notion was applied so as to lead to curious and important results with regard to the power of beams to resist flexure, not only when loaded transversely, but when pressed in the direction of their length, and in any oblique direction.

But in the fracture of beams, the resistance to extension and to compression are not practically equal; and it was necessary to determine

---

[2] *Opera*, ii. p. 976.

[3] Lecture xiii. The height of the modulus is the same for the same substance, whatever its breadth and thickness may be; for atmospheric air it is about five miles, and for steel nearly 1500 miles.

the difference of these two forces by experiments. Several persons pursued researches on this subject; especially Mr. Barlow, of the Royal Military Academy,[4] who investigated the subject with great labor and skill, so far as wood is concerned. But the difference between the resistance to tension and to compression requires more special study in the case of iron; and has been especially attended to in recent times, in consequence of the vast increase in the number of iron structures, and in particular, railways. It appears that wrought iron yields to compressive somewhat more easily than to tensile force, while cast iron yields far more easily to tensile than to compressive strains. In all cases the power of a beam to resist fracture resides mainly in the upper and the under side, for there the tenacity of the material acts at the greatest leverage round the hinge of fracture. Hence the practice was introduced of making iron beams with a broad *flange* at the upper and another flange at the under side, connected by a vertical plate or *web*, of which the office was to keep the two flanges asunder. Mr. Hodgkinson made many valuable experiments on a large scale, to determine the forms and properties of such beams.

But though engineers were, by such experiments and reasonings, enabled to calculate the strength of a given iron beam, and the dimensions of a beam which should bear a given load, it would hardly have occurred to the boldest speculator, a few years ago, to predict that there might be constructed beams nearly 500 feet long, resting merely on their two extremities, of which it could be known beforehand, that they would sustain, without bending or yielding in any perceptible degree, the weight of a railroad train, and the jar of its unchecked motion. Yet of such beams, constructed beforehand with the most perfect confidence, crowned with the most complete success, is composed the great tubular bridge which that consummate engineer, Mr. Robert Stephenson, has thrown across the Menai Strait, joining Wales with the Island of Anglesey. The upper and under surfaces of this quadrangular tube are the flanges of the beam, and the two sides are the webs which connect them. In planning this wonderful structure, the point which required especial care was to make the upper surface strong enough to resist the compressive force which it has to sustain; and this was done by constructing the upper part of the beam of a series of cells, made of iron plate. The application of the arch, of the dome, and of groined vaulting, to the widest space over which they have ever been thrown,

---

[4] *An Essay on the Strength and Shape of Timber.* 3d edition, 1826.

are achievements which have, in the ages in which they occurred, been received with great admiration and applause; but in those cases the principle of the structure had been tried and verified for ages upon a smaller scale. Here not only was the space thus spanned wider than any ever spanned before, but the principle of such a beam with a cellular structure of its parts, was invented for this very purpose, experimentally verified with care, and applied with the most exact calculation of its results.

Roofs—Arches—Vaults.

The calculations of the mechanical conditions of structures consisting of several beams, as for instance, the frames of roofs, depends upon elementary principles of mechanics; and was a subject of investigation at an early period of the science. Such frames may be regarded as assemblages of levers. The parts of which they consist are rigid beams which sustain and convey force, and *Ties* which resist such force by their tension. The former parts must be made rigid in the way just spoken of with regard to iron beams; but ties may be rods merely. The wide structures of many of the roofs of railway stations, compared with the massive wooden roofs of ancient buildings, may show us how boldly and how successfully this distinction has been carried out in modern times. The investigation of the conditions and strength of structures consisting of wooden beams has been cultivated by Mathematicians and Engineers, and is often entitled *Carpentry* in our Mechanical Treatises. In our own time, Dr. Robison and Dr. Thomas Young have been two of the most eminent mathematicians who have written upon this subject.

The properties of the simple machines have been known, as we have narrated, from the time of the Ancient Greeks. But it is plain that such machines are prevented from producing their full effect by various causes. Among the rest, the rubbing of one part of the machine upon another produces an obstacle to the effectiveness of a machine: for instance, the rubbing of the axle of a wheel in the hole in which it rests, the rubbing of a screw against the sides of its hollow screw; the rubbing of a wedge against the sides of its notch; the rubbing of a cord against its pulley. In all these cases, the effect of the machine to produce motion is diminished by the friction. And this *Friction* may be measured and its effects calculated; and thus we have a new branch of mechanics, which has been much cultivated.

Among the effects of friction, we may notice the standing of a stone arch. For each of the vaulting stones of an arch is a truncated wedge; and though a collection of such stones might be so proportioned in their weights as to balance exactly, yet this balance would be a tottering equilibrium, which the slightest shock would throw down, and which would not practically subsist. But the friction of the vaulting stones against one another prevents this instability from being a practical inconvenience; and makes an equilibrated arch to be an arch strong for practical purposes. The *Theory of Arches* is a portion of Mechanics which has been much cultivated, and which has led to conclusions of practical use, as well as of theoretical beauty.

I have already spoken of the invention of the Arch, the Dome, and Groined Vaulting, as marked steps in building. In all these cases the invention was devised by practical builders; and mechanical theory, though it can afterwards justify these structures, did not originally suggest them. They are not part of the result, nor even of the application of theory, but only of its exemplification. The authors of all these inventions are unknown; and the inventions themselves may be regarded as a part of the Prelude of the science of mechanics, because they indicate that the ideas of mechanical pressure and support, in various forms, are acquiring clearness and fixity.

In this point of view, I spoke (Book iv. chap. v. sect. 5) of the Architecture of the Middle Ages as indicating a progress of thought which led men towards the formation of Statics as a science.

As particular instances of the operation of such ideas, we have the *Flying Buttresses* which support stone vaults; and especially, as already noted, the various contrivances by which stone vaults are made to intersect one another, so as to cover a complex pillared space below with *Groined Vaulting*. This invention, executed as it was by the builders of the twelfth and succeeding centuries, is the most remarkable advance in the mechanics of building, after the invention of the *Arch* itself.

It is curious that it has been the fortune of our times, among its many inventions, to have produced one in this department, of which we may say that it is the most remarkable step in the mechanics of arches which has been made since the introduction of pointed groined vaults. I speak of what are called *Skew Arches*, in which the courses of stone or brick of which the bridge is built run obliquely to the walls of the bridge. Such bridges have become very common in the works of railroads; for they save space and material, and the invention once made, the cost of the ingenuity is nothing. Of course, the mechanical principles involved in such structures are obvious to the mathematician, when the problem has been practically solved. And in this case, as in the previous cardinal inventions in structure, though the event has taken place within a few years, no single person, so far as I am aware, can be named as the inventor.[5]

---

[5] Since this was written, I have been referred to Rees's *Cyclopædia*, Article *Oblique Arches*, where this invention is correctly explained, and is claimed for an engineer named Chapman. It is there said, that the first arch of this kind was erected in 1787 at Naas, near Kildare in Ireland.

BOOK VII.

PHYSICAL ASTRONOMY.

CHAPTER I.

PRELUDE TO NEWTON.

The Ancients.

EXPRESSIONS in ancient writers which may be interpreted as indicating a notion of gravitation in the Newtonian sense, no doubt occur. But such a notion, we may be sure, must have been in the highest degree obscure, wavering, and partial. I have mentioned (Book i. Chap. 3) an author who has fancied that he traces in the works of the ancients the origin of most of the vaunted discoveries of the moderns. But to ascribe much importance to such expressions would be to give a false representation of the real progress of science. Yet some of Newton's followers put forward these passages as well deserving notice; and Newton himself appears to have had some pleasure in citing such expressions; probably with the feeling that they relieved him of some of the odium which, he seems to have apprehended, hung over new discoveries. The Preface to the *Principia* begins by quoting[1] the authority of the ancients, as well as the moderns, in favor of applying the science of Mechanics to Natural Philosophy. In the Preface to David Gregory's *Astronomiæ Physicæ et Geometricæ Elementa*, published in 1702, is a large array of names of ancient authors, and of quotations, to prove the early and wide diffusion of the doctrine of the gravity of the Heavenly Bodies. And it appears to be now made out, that this collection of ancient authorities

[1] Cum veteres *Mechanicam* (uti author est *Pappus*), in rerum Naturalium investigatione maximi fecerint, et recentiores, missis formis substantialibus et qualitatibus occultis, Phenomena Naturæ ad leges mathematicas revocare aggressa sunt; visum est in hoc Tractatu *Mathesin* excolere quatenus ea ad *Philosophiam* spectat.

was supplied to Gregory by Newton himself. The late Professor Rigaud, in his *Historical Essay on the First Publication of Sir Isaac Newton's Principia*, says (pp. 80 and 101) that having been allowed to examine Gregory's papers, he found that the quotations given by him in his Preface are copied or abridged from notes which Newton had supplied to him in his own handwriting. Some of the most noticeable of the quotations are those taken from Plutarch's Dialogue *on the Face which appears in the Moon's Disk:* it is there said, for example, by one of the speakers, that the Moon is perhaps prevented from falling to the earth by the rapidity of her revolution round it; as a stone whirled in a sling keeps it stretched. Lucretius also is quoted, as teaching that all bodies would descend with an equal celerity in a vacuum:

> Omnia quapropter debent per inane quietum
> Æque ponderibus non æquis concita ferri.
> Lib. ii. v. 238.

It is asserted in Gregory's Preface that Pythagoras was not unacquainted with the important law of gravity, the inverse squares of the distances from the centre. For, it is argued, the seven strings of Apollo's lyre mean the seven planets; and the proportions of the notes of strings are reciprocally as the inverse squares of the weights which stretch them.

I have attempted, throughout this work, to trace the progress of the discovery of the great truths which constitute real science, in a more precise manner than that which these interpretations of ancient authors exemplify.

Jeremiah Horrox.

In describing the Prelude to the Epoch of Newton, I have spoken (p. 395) of a group of philosophers in England who began, in the first half of the seventeenth century, to knock at the door where Truth was to be found, although it was left for Newton to force it open; and I have there noticed the influence of the civil wars on the progress of philosophical studies. To the persons thus tending towards the true physical theory of the solar system, I ought to have added Jeremy Horrox, whom I have mentioned in a former part (Book v. chap. 5) as one of the earliest admirers of Kepler's discoveries. He died at the early age of twenty-two, having been the first person who ever saw Venus pass across the disk of the Sun according to astronomical prediction, which took place in 1639. His *Venus in sole visa,* in which this is described, did not appear till 1661, when it was published by Hevelius of Dantzic. Some of his papers were destroyed by the soldiers in the English civil wars; and his remaining works were finally published by Wallis, in 1673. The passage to which I here specially wish to refer is contained in a letter to his astronomical ally, William Crabtree, dated 1638. He appears to have been asked by his friend to suggest some cause for the motion of the aphelion of a planet; and in reply, he uses an experimental illustration which was afterwards employed by Hooke in 1666. A ball at the end of a string is made to swing so that it describes an oval. This contrivance Hooke employed to show the way in which an orbit results from the combination of a projectile motion with a central force. But the oval does not keep its axis constantly in the same position. The apsides, as Horrox remarked, move in the same direction as the pendulum, though much slower. And it is true, that this experiment does illustrate, in a general way, the cause of the motion of the aphelia of the Planetary Orbits; although the form of the orbit is different in the experiment and in the solar system; being an ellipse with the centre of force in the centre of the ellipse, in the former case, and an ellipse with the centre of force in the focus, in the latter case. These two forms of orbits correspond to a central force varying directly as the distance, and a central force varying inversely as the square of the distance; as Newton proved in the *Principia*. But the illustration appears to show that Horrox pretty clearly saw how an orbit arose from a central force. So far, and no farther, Newton's contemporaries could get; and then he had to help them onwards by showing what was the law of the force, and what larger truths were now attainable.

Newton's Discovery of Gravitation.

[Page 402.] As I have already remarked, men have a willingness to believe that great discoveries are governed by casual coincidences, and accompanied by sudden revolutions of feeling. Newton had entertained the thought of the moon being retained in her orbit by gravitation as early as 1665 or 1666. He resumed the subject and worked the thought out into a system in 1684 and 5. What induced him to return to the question? What led to his success on this last occasion? With what feelings was the success attended? It is easy to make an imaginary connection of facts. "His optical discoveries had recommended him to the Royal Society, and he was now a member. He *there* learned the accurate measurement of the Earth by Picard, differing very much from the estimation by which he had made his calculation in 1666; and he thought his conjecture now more likely to be just."[2] M. Biot gives his assent to this guess.[3] The English translation of M. Biot's biography[4] converts the guess into an assertion. But, says Professor Rigaud,[5] Picard's measurement of the Earth was well known to the Fellows of the Royal Society as early as 1675, there being an account of the results of it given in the *Philosophical Transactions* for that year. Moreover, Norwood, in his *Seaman's Practice*, dated 1636, had given a much more exact measure than Newton employed in 1666. But Norwood, says Voltaire, had been buried in oblivion by the civil wars. No, again says the exact and truth-loving Professor Rigaud, Norwood was in communication with the Royal Society in 1667 and 1668. So these guesses at the accident which made the apple of 1665 germinate in 1684, are to be carefully distinguished from history.

But with what feelings did Newton attain to his success? Here again we have, I fear, nothing better than conjecture. "He went home, took out his old papers, and resumed his calculations. As they drew near to a close, he was so much agitated that he was obliged to desire a friend to finish them. His former conjecture was now found to agree with the phænomena with the utmost precision."[6] This conjectural story has been called "a tradition;" but he who relates it does not call it so. Every one must decide, says Professor Rigaud, from his view of Newton's character, how far he thinks it consistent with this statement. Is it likely that Newton, so calm and so indifferent to fame as he generally showed himself, should be thus agitated on such an occasion? "No," says Sir David Brewster; "it is not supported by what we know of Newton's character."[7] To this we may assent; and this conjectural incident we must therefore, I conceive, separate from history. I had incautiously admitted it into the text of the first Edition.

Newton appears to have discovered the method of demonstrating that a body might describe an ellipse when acted upon by a force residing in the focus, and varying inversely as the square of the distance, in 1669, upon occasion of his correspondence with Hooke. In 1684,

---

[2] Robison's *Mechanical Philosophy*, vol. iii. p. 94. (Art. 195.)

[3] *Biographie Universelle.*

[4] *Library of Useful Knowledge.*

[5] *Historical Essay on the First Publication of the Principia* (1838).

[6] Robison, ibid.

[7] *Life of Newton*, vol. i. p. 292.

at Halley's request, he returned to the subject; and in February, 1685, there was inserted in the Register of the Royal Society a paper of Newton's (*Isaaci Newtoni Propositiones de Motu*), which contained some of the principal propositions of the first two Books of the *Principia*. This paper, however, does not contain the proposition "Lunam gravitare in Terram," nor any of the propositions of the Third Book.

CHAPTER III.

The Principia.

Sect. 2.—Reception of the Principia.

LORD BROUGHAM has very recently (*Analytical View of Sir Isaac Newton's Principia*, 1855) shown a strong disposition still to maintain, what he says has frequently been alleged, that the reception of the work was not, even in this country, "such as might have been expected." He says, in explanation of the facts which I have adduced, showing the high estimation in which Newton was held immediately after the publication of the *Principia*, that Newton's previous fame was great by former discoveries. This is true; but the effect of this was precisely what was most honorable to Newton's countrymen, that they received with immediate acclamations this new and greater discovery. Lord Brougham adds, "after its appearance the *Principia* was more admired than studied;" which is probably true of the *Principia* still, and of all great works of like novelty and difficulty at all times. But, says Lord Brougham, "there is no getting over the inference on this head which arises from the dates of the two first editions. There elapsed an interval of no less than twenty-seven years between them; and although Cotes [in his Preface] speaks of the copies having become scarce and in very great demand when the second edition appeared in 1713, yet had this urgent demand been of many years' continuance, the reprinting could never have been so long delayed." But Lord Brougham might have learnt from Sir David Brewster's *Life of Newton* (vol. i. p. 312), which he extols so emphatically, that already in 1691 (only four years after the publication), a copy of the *Principia* could hardly be procured, and that even at that time an improved edition was in contemplation; that Newton had been pressed by his friends to undertake it, and had refused.

When Bentley had induced Newton to consent that a new edition should be printed, he announces his success with obvious exultation to Cotes, who was to superintend the work. And in the mean time the *Astronomy* of David Gregory, published in 1702, showed in every page how familiar the Newtonian doctrines were to English philosophers, and tended to make them more so, as the sermons of Bentley himself had done in 1692.

Newton's Cambridge contemporaries were among those who took a part in bringing the *Principia* before the world. The manuscript draft of it was conveyed to the Royal Society (April 28, 1686) by Dr. Vincent, Fellow of Clare Hall, who was the tutor of Whiston, Newton's deputy in his professorship; and he, in presenting the work, spoke of the novelty and dignity of the subject. There exists in the library of the University of Cambridge a manuscript containing the early Propositions of the *Principia* as far as Prop. xxxiii. (which is a part of Section vii., about Falling Bodies). This appears to have been a transcript of Newton's Lectures, delivered as Lucasian Professor: it is dated October, 1684.

Is Gravitation proportional to Quantity of Matter?

It was a portion of Newton's assertion in his great discovery, that all the bodies of the universe attract each other with forces which are *as the quantity of matter* in each: that is, for instance, the sun attracts the satellites of any planet just as much as he attracts the planet itself, in proportion to the quantity of matter in each; and the planets attract one another just as much as they attract the sun, according to the quantity of matter.

To prove this part of the law *exactly* is a matter which requires careful experiments; and though proved experimentally by Newton, has been considered in our time worthy of re-examination by the great astronomer Bessel. There was some ground for doubt; for the mass of Jupiter, as deduced from the perturbations of Saturn, was only $\frac{1}{1070}$ of the mass of the sun; the mass of the same planet as deduced from the perturbations of Juno and Pallas was $\frac{1}{1045}$ of that of the Sun. If this difference were to be confirmed by accurate observations and calculations, it would follow that the attractive power exercised by Jupiter upon the minor planets was greater than that exercised upon

Saturn. And in the same way, if the attraction of the Earth had any *specific* relation to different kinds of matter, the time of oscillation of a pendulum of equal length composed wholly or in part of the two substances would be different. If, for instance, it were more intense for magnetized iron than for stone, the iron pendulum would oscillate more quickly. Bessel showed[1] that it was possible to assume hypothetically a constitution of the sun, planets, and their appendages, such that the attraction of the Sun on the Planets and Satellites should be proportional to the quantity of matter in each; but that the attraction of the Planets on one another would not be on the same scale.

Newton had made experiments (described in the *Principia*, Book iii., Prop. vi.) by which it was shown that there could be no considerable or palpable amount of such specific difference among terrestrial bodies, but his experiments could not be regarded as exact enough for the requirements of modern science. Bessel instituted a laborious series of experiments (presented to the Berlin Academy in 1832) which completely disproved the conjecture of such a difference; every substance examined having given exactly the same coefficient of gravitating intensity as compared with inertia. Among the substances examined were metallic and stony masses of meteoric origin, which might be supposed, if any bodies could, to come from other parts of the solar system.

CHAPTER IV.

Verification and Completion of the Newtonian Theory.

Tables of the Moon and Planets.

THE Newtonian discovery of Universal Gravitation, so remarkable in other respects, is also remarkable as exemplifying the immense extent to which the verification of a great truth may be carried, the amount of human labor which may be requisite to do it justice, and the striking extension of human knowledge to which it may lead. I have said that it is remarked as a beauty in the first fixation of a theory that its measures or elements are established by means of a few

---

[1] *Berlin Mem.* 1824.

data; but that its excellence when established is in the number of observations which it explains. The multiplicity of observations which are explained by astronomy, and which are made because astronomy explains them, is immense, as I have noted in the text. And the multitude of observations thus made is employed for the purpose of correcting the first adopted elements of the theory. I have mentioned some of the examples of this process: I might mention many others in order to continue the history of this part of Astronomy up to the present time. But I will notice only those which seem to me the most remarkable.

In 1812, Burckhardt's *Tables de la Lune* were published by the French Bureau des Longitudes. A comparison of these and Burg's with a considerable number of observations, gave 9-100ths of a second as the mean error of the former in the Moon's longitude, while the mean error of Burg's was 18-100ths. The preference was therefore accorded to Burckhardt's.

Yet the Lunar Tables were still as much as thirty seconds wrong in single observations. This circumstance, and Laplace's expressed wish, induced the French Academy to offer a prize for a complete and purely theoretical determination of the Lunar path, instead of determinations resting, as hitherto, partly upon theory and partly upon observations. In 1820, two prize essays appeared, the one by Damoiseau, the other by Plana and Carlini. And some years afterwards (in 1824, and again in 1828), Damoiseau published *Tables de la Lune formeés sur la seule Théorie d'Attraction.* These agree very closely with observation. That we may form some notion of the complexity of the problem, I may state that the longitude of the Moon is in these Tables affected by no fewer than forty-seven *equations;* and the other quantities which determine her place are subject to inequalities not much less in number.

Still I had to state in the second Edition, published in 1847, that there remained an unexplained discordance between theory and observation in the motions of the Moon; an inequality of long period as it seemed, which the theory did not give.

A careful examination of a long series of the best observations of the Moon, compared throughout with the theory in its most perfect form, would afford the means both of correcting the numerical elements of the theory, and of detecting the nature, and perhaps the law, of any still remaining discrepancies. Such a work, however, required vast labor, as well as great skill and profound mathematical knowledge.

Mr. Airy undertook the task; employing for that purpose, the Observations of the Moon made at Greenwich from 1750 to 1830. Above 8000 observed places of the Moon were compared with theory by the computation of the same number of places, each separately and independently calculated from Plana's Formulæ. A body of calculators (sometimes sixteen), at the expense of the British Government, was employed for about eight years in this work. When we take this in conjunction with the labor which the observations themselves imply, it may serve to show on what a scale the verification of the Newtonian theory has been conducted. The first results of this labor were published in two quarto volumes; the final deductions as to correction of elements, &c., were given in the Memoirs of the Astronomical Society in 1848.[1]

Even while the calculations were going on, it became apparent that there were some differences between the observed places of the Moon, and the theory so far as it had then been developed. M. Hansen, an eminent German mathematician who had devised new and powerful methods for the mathematical determination of the results of the law of gravitation, was thus led to explore still further the motions of the Moon in pursuance of this law. The result was that he found there must exist two lunar inequalities, hitherto not known; the one of 273, and the other of 239 years, the coefficients of which are respectively 27 and 23 seconds. Both these originate in the attraction of Venus; one of them being connected with the long inequality in the Solar Tables, of which Mr. Airy had already proved the existence, as stated in Chap. vi. Sect. 6 of this Book.

These inequalities fell in with the discrepancies between the actual observations and the previously calculated Tables, which Mr. Airy had discovered. And again, shortly afterwards, M. Hansen found that there resulted from the theory two other new equations of the Moon; one in latitude and one in longitude, agreeing with two which were found by Mr. Airy in deducing from the observations the correction of the elements of the Lunar Tables. And again, a little later, there was detected by these mathematicians a theoretical correction for the mo-

---

[1] The total expense of computers, to the end of reading the proof-sheets, was 4300*l*.

Mr. Airy's estimate of days' works [made before beginning], for the heavy part of calculations only, was thirty-six years of one computer. This was somewhat exceeded, but not very greatly, in that part.

tion of the Node of the Moon's orbit, coinciding exactly with one which had been found to appear in the observations.

Nothing can more strikingly exhibit the confirmation which increased scrutiny brings to light between the Newtonian theory on the one hand, and the celestial motions on the other. We have here a very large mass of the best observations which have ever been made, systematically examined, with immense labor, and with the set purpose of correcting at once all the elements of the Lunar Tables. The corrections of the elements thus deduced imply of course some error in the theory as previously developed. But at the same time, and with the like determination thoroughly to explore the subject, the theory is again pressed to yield its most complete results, by the invention of new and powerful mathematical methods; and the event is, that residual errors of the old Tables, several in number, following the most diverse laws, occurring in several detached parts, agree with the residual results of the Theory thus newly extracted from it. And thus every additional exactness of scrutiny into the celestial motions on the one hand and the Newtonian theory on the other, has ended, sooner or later, in showing the exactness of their coincidence.

The comparison of the theory with observation in the case of the motions of the Planets, the motion of each being disturbed by the attraction of all the others, is a subject in some respects still more complicated and laborious. This work also was undertaken by the same indefatigable astronomer; and here also his materials belonged to the same period as before; being the admirable observations made at Greenwich from 1750 to 1830, during the time that Bradley, Maskelyne, and Pond were the Astronomers Royal.[2] These Planetary observations were deduced, and the observed places were compared with the tabular places: with Lindenau's Tables of Mercury, Venus, and Mars; and with Bouvard's Tables of Jupiter, Saturn, and Uranus; and thus, while the received theory and its elements were confirmed, the means of testing any improvement which may hereafter be proposed, either in the form of the theoretical results or in the constant elements which they involved, was placed within the reach of the astron-

[2] The observations of stars made by Bradley, who preceded Maskelyne at Greenwich, had already been discussed by Bessel, a great German astronomer; and the results published in 1818, with a title that well showed the estimation in which he held those materials: *Fundamenta Astronomiæ pro anno* 1775, *deducta ex Observationibus viri incomparabilis James Bradley in specula Astronomica Grenovicensi per annos* 1750–1762 *institutis.*

omers of all future time. The work appeared in 1845 ; the expense of the compilations and the publication being defrayed by the British Government.

The Discovery of Neptune.

The theory of gravitation was destined to receive a confirmation more striking than any which could arise from any explanation, however perfect, given by the motions of a known planet; namely, in revealing the existence of an unknown planet, disclosed to astronomers by the attraction which it exerted upon a known one. The story of the discovery of Neptune by the calculations of Mr. Adams and M. Le Verrier was partly told in the former edition of this History. I had there stated (vol. ii. p. 306) that "a deviation of observation from the theory occurs at the very extremity of the solar system, and that its existence appears to be beyond doubt. Uranus does not conform to the Tables calculated for him on the theory of gravitation. In 1821, Bouvard said in the Preface to the Tables of this Planet, "the formation of these Tables offers to us this alternative, that we cannot satisfy modern observations to the requisite degree of precision without making our Tables deviate from the ancient observations." But when we have done this, there is still a discordance between the Tables and the more modern observations, and this discordance goes on increasing. At present the Tables make the Planet come upon the meridian about eight seconds later than he really does. This discrepancy has turned the thoughts of astronomers to the effects which would result from a planet external to Uranus. It appears that the observed motion would be explained by applying a planet at twice the distance of Uranus from the Sun to exercise a disturbing force, and it is found that the present longitude of this disturbing body must be about 325 degrees.

I added, "M. Le Verrier (*Comptes Remdus*, Jan. 1, 1846) and, as I am informed by the Astronomer Royal, Mr. Adams, of St. John's College, Cambridge, have both arrived independently at this result."

To this Edition I added a Postscript, dated Nov. 7, 1846, in which I said :

"The planet exterior to Uranus, of which the existence was inferred by M. Le Verrier and Mr. Adams from the motions of Uranus (vol. ii. Note (L.)), has since been discovered. This confirmation of calculations founded upon the doctrine of universal gravitation, may be looked upon as the most remarkable event of the kind since the return of Halley's comet in 1757 ; and in some respects, as a more striking event even than that; inasmuch as the new planet had never been seen at all, and was discovered by mathematicians entirely by their feeling of its influence, which they perceived through the organ of mathematical calculation.

"There can be no doubt that to M. Le Verrier belongs the glory of having first published a prediction of the place and appearance of the new planet, and of having thus occasioned its discovery by astronomical observers. M. Le Verrier's first prediction was published in the *Comptes Rendus de l'Acad. des Sciences*, for *June* 1, 1846 (not *Jan.* 1, as erroneously printed in my Note). A subsequent paper on the subject was read Aug. 31. The planet was seen by M. Galle, at the Observatory of Berlin, on September 23, on which day he had received an express application from M. Le Verrier, recommending him to endeavor to recognize the stranger by its having a visible disk. Professor Challis, at the Observatory of Cambridge, was looking out for the new planet from July 29, and saw it on August 4, and again on August 12, but without recognizing it, in consequence of his plan of not comparing his observations till he had accumulated a greater number of them. On Sept. 29, having read for the first time M. Le Verrier's second paper, he altered his plan, and paid attention to the physical appearance rather than the position of the star. On that very evening, not having then heard of M. Galle's discovery, he singled out the star by its seeming to have a disk.

"M. Le Verrier's mode of discussing the circumstances of Uranus's motion, and inferring the new planet from these circumstances, is in the highest degree sagacious and masterly. Justice to him cannot require that the contemporaneous, though unpublished, labors of Mr. Adams, of St. John's College, Cambridge, should not also be recorded. Mr. Adams made his first calculations to account for the anomalies in the motion of Uranus, on the hypothesis of a more distant planet, in 1843. At first he had not taken into account the earlier Greenwich observations; but these were supplied to him by the Astronomer Royal, in 1844. In September, 1845, Mr. Adams communicated to Professor Challis values of the elements of the supposed disturbing body; namely, its mean distance, mean longitude at a given epoch, longitude of perihelion, eccentricity of orbit, and mass. In the next month, he communicated to the Astronomer Royal values of the same elements, somewhat corrected. The note (L.), vol. ii., of the present work (2d Ed.), in which the names of MM. Le Verrier and Adams are mentioned in conjunction, was in the press in August, 1846, a month before the planet was seen. As I have stated in the text, Mr. Adams and M. Le Verrier assigned to the unseen planet nearly the same position; they also assigned to it nearly the same mass; namely, 2½ times the mass of Uranus. And hence, supposing the density to be not greater than that of Uranus, it followed that the visible diameter would be about 3″, an apparent magnitude not much smaller than Uranus himself.

"M. Le Verrier has mentioned for the new planet the name *Neptunus;* and probably, deference to his authority as its discoverer, will obtain general currency for this name."

Mr. Airy has given a very complete history of the circumstances attending the discovery of Neptune, in the Memoirs of the Astronomical Society (read November 13, 1846). In this he shows that the probability of some disturbing body beyond Uranus had suggested itself to M. A. Bouvard and Mr. Hussey as early as 1834. Mr. Airy himself then thought that the time was not ripe for making out the nature of any external action on the planets. But Mr. Adams soon afterwards proceeded to work at the problem. As early as 1841 (as he himself informs me) he conjectured the existence of a planet exterior to Uranus, and recorded in a memorandum his design of examining its effect; but deferred the calculations till he had completed his preparations for the University examination which he was to undergo in January, 1843, in order to receive the Degree of Bachelor of Arts. He was the Senior Wrangler of that occasion, and soon afterwards proceeded to carry his design into effect; applying to the Astronomer Royal for recorded observations which might aid him in his task. On one of the last days of October, 1845, Mr. Adams went to the Observatory at Greenwich; and finding the Astronomer Royal abroad, he left there a paper containing the elements of the extra-Uranian Planet: the longitude was in this paper stated as 323½ degrees. It was, as we have seen, in June, 1846, that M. Le Verrier's Memoir appeared, in which he assigned to the disturbing body a longitude of 325 degrees. The coincidence was striking. "I cannot sufficiently express," says Mr. Airy, "the feeling of delight and satisfaction which I received from the Memoir of M. Le Verrier." This feeling communicated itself to others. Sir John Herschel said in September, 1846, at a meeting of the British Association at Southampton, "We see it (the probable new planet) as Columbus saw America from the shores of Spain. Its movements have been felt, trembling along the far-reaching line of our analysis, with a certainty hardly inferior to that of ocular demonstration."

In truth, at the moment when this was uttered, the new Planet had already been seen by Professor Challis; for, as we have said, he had seen it in the early part of August. He had included it in the net which he had cast among the stars for this very purpose; but employing a slow and cautious process, he had deferred for a time that examination of his capture which would have enabled him to detect the object sought. As soon as he received M. Le Verrier's paper of August 31 on September 29, he was so much impressed with the sagacity and clearness of the limitations of the field of observation there laid down, that he instantly changed his plan of observation, and noted the planet, as an object having a visible disk, on the evening of the same day.

In this manner the theory of gravitation predicted and produced the discovery. Thus to predict unknown facts found afterwards to be true, is, as I have said, a confirmation of a theory which in impressiveness and value goes beyond any explanation of known facts. It is a confirmation which has only occurred a few times in the history of science; and in the case only of the most refined and complete theories, such as those of Astronomy and Optics. The mathematical skill which was requisite in order to arrive at such a discovery, may in some measure be judged of by the account which we have had to give of the previous mathematical progress of the theory of gravitation. It there appeared that the lives of many of the most acute, clear-sighted, and laborious of mankind, had been employed for generations in solving the problem, Given the planetary bodies, to find their mutual perturbations: but here we have the inverse problem—Given the perturbations, to find the planets.[3]

The Minor Planets.

The discovery of the Minor Planets which revolve between the orbits of Mars and Jupiter was not a consequence or confirmation of the Newtonian theory. That theory gives no reason for the distance of

---

[3] This may be called the *inverse* problem with reference to the older and more familiar problem; but we may remark that the usual phraseology of the Problem of Central Forces differs from this analogy. In Newton's *Principia*, the earlier Sections, in which the motion is given to find the force, are spoken of as containing the *Direct* Problem of Central Forces: the Eighth Section of the First Book, where the Force is given to find the orbit, is spoken of as containing the *Inverse* Problem of Central Forces.

the Planets from the Sun; nor does any theory yet devised give such reason. But an empirical formula proposed by the Astronomer Bode of Berlin, gives a law of these distances (*Bode's Law*), which, to make it coherent, requires a planet between Mars and Jupiter. With such an addition, the distance of Mercury, Venus, Earth, Mars, the Missing Planet, Jupiter, Saturn, and Uranus, are nearly as the numbers

4, 7, 10, 16, 28, 52, 100, 196,

in which the excesses of each number above the preceding are the series

3, 3, 6, 12, 24, 48, 96.

On the strength of this law the Germans wrote *on the long-expected Planet*, and formed themselves into associations for the discovery of it.

Not only did this law stimulate the inquiries for the Missing Planet, and thus lead to the discovery of the Minor Planets, but it had also a share in the discovery of Neptune. According to the law, a planet beyond Uranus may be expected to be at the distance represented by 388. Mr. Adams and M. Le Verrier both of them began by assuming a distance of nearly this magnitude for the Planet which they sought; that is, a distance more than 38 times the earth's distance. It was found afterwards that the distance of Neptune is only 30 times that of the earth; yet the assumption was of essential use in obtaining the result: and Mr. Airy remarks that the history of the discovery shows the importance of using any received theory as far as it will go, even if the theory can claim no higher merit than that of being plausible.[4]

The discovery of Minor Planets in a certain region of the interval between Mars and Jupiter has gone on to such an extent, that their number makes them assume in a peculiar manner the character of representatives of a Missing Planet. At first, as I have said in the text, it was supposed that all these portions must pass through or near a common node; this opinion being founded on the very bold doctrine, that the portions must at one time have been united in one Planet, and must then have separated. At this node, as I have stated, Olbers lay in wait for them, as for a hostile army at a defile. Ceres, Pallas, and Juno had been discovered in this way in the period from 1801 to 1804; and Vesta was caught in 1807. For a time the chase for new planets in this region seemed to have exhausted the stock. But after thirty-eight years, to the astonishment of astronomers, they began to be again detected in extraordinary numbers. In 1845, M. Hencke of

[4] Account of the Discovery of Neptune, &c., *Mem. Ast. Soc.*, vol. xvi. p. 414.

Driessen discovered a fifth of these planets, which was termed Astræa. In various quarters the chase was resumed with great ardor. In 1847 were found Hebe, Iris, and Flora; in 1848, Metis; in 1849, Hygæa; in 1850, Parthenope, Victoria, and Egeria; in 1851, Irene and Eunomia; in 1852, Psyche, Thetis, Melpomene, Fortuna, Massilia, Lutetia, Calliope. To these we have now (at the close of 1856) to add *nineteen* others; making up the whole number of these Minor Planets at present known to *forty-two.*

As their enumeration will show, the ancient practice has been continued of giving to the Planets mythological names. And for a time, till the numbers became too great, each of the Minor Planets was designated in astronomical books by some symbol appropriate to the character of the mythological person; as from ancient times Mars has been denoted by a mark indicating a spear, and Venus by one representing a looking-glass. Thus, when a Minor Planet was discovered at London in 1851, the year in which the peace of the world was, in a manner, celebrated by the Great Exhibition of the Products of All Nations, held at that metropolis, the name *Irene* was given to the new star, as a memorial of the auspicious time of its discovery. And it was agreed, for awhile, that its symbol should be a dove with an olive-branch. But the vast multitude of the Minor Planets, as discovery went on, made any mode of designation, except a numerical one, practically inconvenient. They are now denoted by a small circle inclosing a figure in the order of their discovery. Thus, *Ceres* is ⓵, *Irene* is ⑭, and *Isis* is ㊷.

The rapidity with which these discoveries were made was owing in part to the formation of star-maps, in which all known fixed stars being represented, the existence of a new and movable star might be recognized by comparison of the sky with the map. These maps were first constructed by astronomers of different countries at the suggestion of the Academy of Berlin; but they have since been greatly extended, and now include much smaller stars than were originally laid down.

I will mention the number of planets discovered in each year. After the start was once made, by Hencke's discovery of Astræa in 1845, the same astronomer discovered Hebe in 1847; and in the same year Mr. Hind, of London, discovered two others, Iris and Flora. The years 1848 and 1849 each supplied one; the year 1850, three; 1851, two; 1852 was marked by the extraordinary discovery of *eight* new members of the planetary system. The year 1853 supplied four; 1854, six; 1855, four; and 1856 has already given us five.

These discoveries have been distributed among the observatories of Europe. The bright sky of Naples has revealed seven new planets to the telescope of Signor Gasparis. Marseilles has given us one; Germany, four, discovered by M. Luther at Bilk; Paris has furnished seven; and Mr. Hind, in Mr. Bishop's private observatory in London, notwithstanding our turbid skies, has discovered no less than ten planets; and there also Mr. Marth discovered ㉙ Amphitrite. Mr. Graham, at the private observatory of Mr. Cooper, in Ireland, discovered ⑨ Metis.

America has supplied its planet, namely ㉛ Euphrosyne, discovered by Mr. Ferguson at Washington; and the most recent of these discoveries is that by Mr. Pogson, of Oxford, who has found the forty-second of these Minor Planets, which has been named Isis.[5]

I may add that it appears to follow from the best calculations that the total mass of all these bodies is very small. Herschel reckoned the diameters of Ceres at 35, and of Pallas at 26 miles. It has since been calculated[6] that some of them are smaller still; Victoria having a diameter of 9 miles, Lutetia of 8, and Atalanta of little more than 4. It follows from this that the whole mass would probably be less than the sixth part of our moon. Hence their perturbing effects on each other or on other planets are null; but they are not the less disturbed by the action of the other planets, and especially of Jupiter.

Anomalies in the Action of Gravitation.

The complete and exact manner in which the doctrine of gravitation explains the motions of the Comets as well as of the Planets, has made astronomers very bold in proposing hypotheses to account for any deviations from the motion which the theory requires. Thus Encke's Comet is found to have its motion accelerated by about one-eighth of a day in every revolution. This result was conceived to be established by former observations, and is confirmed by the facts of the appearance of 1852.[7] The hypothesis which is proposed in order to explain this result is, that the Comet moves in a resisting medium, which makes it fall inwards from its path, towards the Sun, and thus, by narrowing its orbit, diminishes its periodic time. On the other hand, M. Le Verrier has found that Mercury's mean motion has gone on diminishing;

---

5 I take this list from a Memoir of M. Bruhns, Berlin, 1856.

6 Bruhns, as above.

7 *Berlin Memoirs*, 1854.

as if the planet were, in the progress of his revolutions, receding further from the Sun. This is explained, if we suppose that there is, in the region of Mercury, a resisting medium which moves round the Sun in the same direction as the Planets move. Evidence of a kind of nebulous disk surrounding the Sun, and extending beyond the orbits of Mercury and Venus, appears to be afforded us by the phenomenon called the *Zodiacal Light;* and as the Sun itself rotates on its axis, it is most probable that this kind of atmosphere rotates also.[8] On the other hand, M. Le Verrier conceives that the Comets which now revolve within the ordinary planetary limits have not always done so, but have been caught and detained by the Planets among which they move. In this way the action of Jupiter has brought the Comets of Faye and Vico into their present limited orbits, as it drew the Comet of Lexell out of its known orbit, when the Comet passed over the Planet in 1779, since which time it has not been seen.

Among the examples of the boldness with which astronomers assume the doctrine of gravitation even beyond the limits of the solar system to be so entirely established, that hypotheses may and must be assumed to explain any apparent irregularity of motion, we may reckon the mode of accounting for certain supposed irregularities in the proper motion of Sirius, which has been proposed by Bessel, and which M. Peters thinks is proved to be true by his recent researches (*Astr. Nach.* xxxi. p. 219, and xxxii. p. 1). The hypothesis is, that Sirius has a companion star, dark, and therefore invisible to us; and that the two, revolving round their common centre as the system moves on, the motion of Sirius is seen to be sometimes quicker and sometimes slower.

The Earth's Density.

"Cavendish's experiment," as it is commonly called—the measure of the attractions of manageable masses by the torsion balance, in order to determine the density of the Earth—has been repeated recently by Professor Reich at Freiberg, and by Mr. Baily in England, with great attention to the means of attaining accuracy. Professor Reich's result for the density of the Earth is 5·44; Mr. Baily's is 5·92. Cavendish's result was 5·48; according to recent revisions[9] it is 5·52.

---

[8] M. Le Verrier, *Annales de l'Obs. de Paris*, vol. i. p. 89.

[9] The calculation has been revised by M. Edward Schmidt. Humboldt's *Kosmos*, ii. p. 425.

But the statical effect of the attraction of manageable masses, or even of mountains, is very small. The effect of a small change in gravity may be accumulated by being constantly repeated in the oscillations of a pendulum, and thus may become perceptible. Mr. Airy attempted to determine the density of the Earth by a method depending on this view. A pendulum oscillating at the surface was to be compared with an equal pendulum at a great depth below the surface. The difference of their rates would disclose the different force of gravity at the two positions; and hence, the density of the Earth. In 1826 and 1828, Mr. Airy attempted this experiment at the copper mine of Dolcoath in Cornwall, but failed from various causes. But in 1854, he resumed it at the Harton coal mine in Durham, the depth of which is 1260 feet; having in this new trial, the advantage of transmitting the time from one station to the other by the instantaneous effect of galvanism, instead of by portable watches. The result was a density of 6·56; which is much larger than the preceding results, but, as Mr. Airy holds, is entitled to compete with the others on at least equal terms.

Tides.

I should be wanting in the expression of gratitude to those who have practically assisted me in Researches on the Tides, if I did not mention the grand series of Tide Observations made on the coast of Europe and America in June, 1835, through the authority of the Board of Admiralty, and the interposition of the late Duke of Wellington, at that time Foreign Secretary. Tide observations were made for a fortnight at all the Coast-guard stations of Great Britain and Ireland in June, 1834; and these were repeated in June, 1835, with corresponding observations on all the coasts of Europe, from the North Cape of Norway to the Straits of Gibraltar; and from the mouth of the St. Lawrence to the mouth of the Mississippi. The results of these observations, which were very complete so far as the coast tides were concerned, were given in the *Philosophical Transactions* for 1836.

Additional accuracy respecting the Tides of the North American coast may be expected from the survey now going on under the direction of Superintendent A. Bache. The Tides of the English Channel have been further investigated, and the phenomena presented under a new point of view by Admiral Beechey.

The Tides of the Coast of Ireland have been examined with great care by Mr. Airy. Numerous and careful observations were made with a view, in the first instance, of determining what was to be regarded as "the Level of the Sea;" but the results were discussed so as to bring into view the laws and progress, on the Irish coast, of the various inequalities of the Tides mentioned in Chap. iv. Sect. 9 of this Book.

I may notice as one of the curious results of the Tide Observations of 1836, that it appeared to me, from a comparison of the Observations, that there must be a point in the German Ocean, about midway between Lowestoft on the English coast, and the Brill on the Dutch coast, where the tide would vanish: and this was ascertained to be the case by observation; the observations being made by Captain Hewett, then employed in a survey of that sea.

Cotidal Lines supply, as I conceive, a good and simple method of representing the progress and connection of *littoral* tides. But to draw cotidal lines across oceans, is a very precarious mode of representing the facts, except we had much more knowledge on the subject than we at present possess. In the *Phil. Trans.* for 1848, I have resumed the subject of the Tides of the Pacific; and I have there expressed my opinion, that while the littoral tides are produced by progressive waves, the oceanic tides are more of the nature of stationary undulations.

But many points of this kind might be decided, and our knowledge on this subject might be brought to a condition of completeness, if a ship or ships were sent expressly to follow the phenomena of the Tides from point to point, as the observations themselves might suggest a course. Till this is done, our knowledge cannot be completed. Detached and casual observations, made *aliud agendo*, can never carry us much beyond the point where we at present are.

Double Stars.

Sir John Herschel's work, referred to in the History (2d Ed.) as then about to appear, was published in 1847.[10] In this work, besides a vast amount of valuable observations and reasonings on other subjects

[10] *Results of Astronomical Observations made during the years* 1834, 5, 6, 7, 8, *at the Cape of Good Hope, being the completion of a Telescopic Survey of the whole Surface of the visible Heavens commenced in* 1825.

(as Nebulæ, the Magnitude of Stars, and the like), the orbits of several double stars are computed by the aid of the new observations. But Sir John Herschel's conviction on the point in question, the operation of the Newtonian law of gravitation in the region of the stars, is expressed perhaps more clearly in another work which he published in 1849.[11] He there speaks of Double Stars, and especially of *gamma Virginis*, the one which has been most assiduously watched, and has offered phenomena of the greatest interest.[12] He then finds that the two components of this star revolve round each other in a period of 182 years; and says that the elements of the calculated orbit represent the whole series of recorded observations, comprising an angular movement of nearly nine-tenths of a complete circuit, both in angle and distance, with a degree of exactness fully equal to that of observation itself. "No doubt can therefore," he adds, "remain as to the prevalance in this remote system of the Newtonian Law of Gravitation."

Yet M. Yvon de Villarceau has endeavored to show[13] that this conclusion, however probable, is not yet proved. He holds, even for the Double Stars, which have been most observed, the observations are only equivalent to seven or eight really distinct data, and that seven data are not sufficient to determine that an ellipse is described according to the Newtonian law. Without going into the details of this reasoning, I may remark, that the more rapid relative angular motion of the components of a Double Star when they are more near each other, proves, as is allowed on all hands, that they revolve under the influence of a mutual attractive force, obeying the Keplerian Law of Areas. But that, whether this force follows the law of the inverse square or some other law, can hardly have been rigorously proved as yet, we may easily conceive, when we recollect the manner in which that law was proved for the Solar System. It was by means of an error of *eight minutes*, observed by Tycho, that Kepler was enabled, as he justly boasted, to reform the scheme of the Solar System,—to show, that is, that the planetary orbits are ellipses with the sun in the focus. Now, the observations of Double Stars cannot pretend to such accuracy as this; and therefore the Keplerian theorem cannot, as yet, have been fully demonstrated from those observations. But when we know

---

[11] *Outlines of Astronomy.* [12] *Out.* 844.

[13] *Connaissance des Temps*, for 1852; published in 1849.

that Double Stars are held together by *a* central force, to prove that this force follows a different law from the only law which has hitherto been found to obtain in the universe, and which obtains between all the known masses of the universe, would require very clear and distinct evidence, of which astronomers have as yet seen no trace.

CHAPTER VI.

Sect. 1. *Instruments.*—2. *Clocks.*

IN page 473, I have described the manner in which astronomers are able to observe the transit of a star, and other astronomical phenomena, to the exactness of a tenth of a second of time. The mode of observation there described implies that the observer at the moment of observation compares the impressions of the eye and of the ear. Now it is found that the habit which the observer must form of doing this operates differently in different observers, so that one observer notes the same fact as happening a fraction of a second earlier or later than another observer does; and this in every case. Thus, using the term *equation*, as we use it in Astronomy, to express a correction by which we get regularity from irregularity, there is a *personal equation* belonging to this mode of observation, showing that it is liable to error. Can this error be got rid of?

It is at any rate much diminished by a method of observation recently introduced into observatories, and first practised in America. The essential feature of this mode of observation consists in combining the impression of sight with that of touch, instead of with that of hearing. The observer at the moment of observation presses with his finger so as to make a mark on a machine which by its motion measures time with great accuracy and on a large scale; and thus small intervals of time are made visible.

A universal, though not a necessary, part of this machinery, as hitherto adopted, is, that a galvanic circuit has been employed in conveying the impression from the finger to the part where time is measured and marked. The facility with which galvanic wires can thus lead the impression by any path to any distance, and increase its force in any degree, has led to this combination, and almost identification, of observation by touch with its record by galvanism.

The method having been first used by Mr. Bond at Cambridge, in North America, has been adopted elsewhere, and especially at Greenwich, where it is used for all the instruments; and consequently a collection of galvanic batteries is thus as necessary a part of the apparatus of the establishment as its graduated circles and arcs.

END OF VOL. I.

www.ingramcontent.com/pod-product-compliance
Lightning Source LLC
LaVergne TN
LVHW021234110826
845150LV00002B/331